Annual Student Conference: Competitions & Events

Held at the 2024 AIChE Annual Meeting

San Diego, California, USA
27-31 October 2024

ISBN: 979-8-3313-1690-7

Printed from e-media with permission by:

Curran Associates, Inc.
57 Morehouse Lane
Red Hook, NY 12571

AIChE
120 Wall Street, FL 23
New York, NY 10005-4020

Phone: (800) 242-4363
Fax: (203) 775-5177

www.aiche.org

Additional copies of this publication are available from:

Curran Associates, Inc.
57 Morehouse Lane
Red Hook, NY 12571 USA
Phone: 845-758-0400
Fax: 845-758-2633
Email: curran@proceedings.com
Web: www.proceedings.com

TABLE OF CONTENTS

UNDERGRADUATE STUDENT POSTER SESSION: CATALYSIS AND REACTION ENGINEERING

UNDERGRADUATE STUDENT POSTER SESSION: COMPUTING AND PROCESS CONTROL

UNDERGRADUATE STUDENT POSTER SESSION: EDUCATION AND GENERAL PAPERS

UNDERGRADUATE STUDENT POSTER SESSION: ENVIRONMENTAL

UNDERGRADUATE STUDENT POSTER SESSION: FOOD, PHARMACEUTICAL, AND BIOTECHNOLOGY

UNDERGRADUATE STUDENT POSTER SESSION: FUELS, PETROCHEMICALS, AND ENERGY

UNDERGRADUATE STUDENT POSTER SESSION: MATERIALS ENGINEERING AND SCIENCES

UNDERGRADUATE STUDENT POSTER SESSION: SEPARATIONS

Accelerating Green Hydrogen: IrO_2 Nanoparticle Models for Electrocatalytic Water Splitting

Max Chen Huang

The design of active and stable electrocatalysts for the oxygen evolution reaction (OER) remains a critical bottleneck in water electrolysis for hydrogen fuel generation, primarily due to the sluggish kinetics and high overpotentials required at the anode. This study focuses on constructing IrO_2 nanoparticles, recognized as state-of-the-art OER catalysts due to their robust catalytic activity and stability. Through density functional theory (DFT), we calculate the surface energies of various IrO_2 facets and terminations to inform Wulff constructions, which model the most thermodynamically favorable geometries of these nanoparticles. Additionally, we explore how their structural properties change under different electrochemical potentials, shedding light on the relationship between nanoparticle morphology and catalytic stability.

Our ab initio thermodynamics analysis reveals that IrO_2 nanoparticles undergo a transformation in shape and stability under operational conditions. Specifically, we find a thermodynamic instability in the rutile crystal structure caused by the stabilization of highly oxidized oxygen species at the surface at OER onset potentials. As the potential increases, this instability leads to a transformation in the equilibrium shape, shifting from the traditionally studied IrO_2 (110) facets to the predominant IrO_2 (111) facets. Additionally, our findings align with experimentally synthesized nanoparticles in both shape and facet ratios, further validating our computational models.

Future steps will focus on enhancing performance through doping with 3d and 4d transition metals, which has been shown to help mitigate the trade-off between activity and stability. Additionally, extending detailed characterization studies to higher potential ranges for RuO_2, a more reactive but less stable metal oxide, is essential due to its higher natural abundance compared to iridium, making it advantageous for large-scale applications. We are also training machine learning models on our data to circumvent the computational cost of DFT calculations in predicting Pourbaix and surface phase diagrams. Our approach contributes to the ongoing effort to develop sustainable and efficient hydrogen production technologies.

Impact of Pt Encapsulation and Macroporosity in MFI Zeolite on Catalytic Plastic Upcycling

Matthew Oh, Cole Hullfish, Jun Zhi Tan, Michele Sarazen

Conventional methods of plastic recycling (e.g. mechanical recycling) typically require high-purity plastic waste and yield recycled polymers of lower quality. Chemical recycling is a promising supplement to the existing waste-processing repertoire, allowing for the conversion of polyolefin chains into value-added hydrocarbon products (e.g. fuels). Within this area, catalytic cracking has been emerging as a lower-temperature alternative to methods like pyrolysis. Previous work has shown that zeolites, with their high thermal stability and well-defined micropore structure, are particularly suited for such cracking reactions. Ongoing efforts seek to probe the impacts of structural modifications (e.g. hierarchical porosity, loading of noble metal nanoparticles) to the catalyst on relevant catalytic mechanisms. As bulkier feed hydrocarbons (i.e. polyethylene) result in highly mass transfer-limited reaction kinetics, hierarchical porosity is particularly interesting as a strategy to improve transport of hydrocarbon chains into the catalyst framework. Beyond porosity, the incorporation of metal nanoparticles has been shown to significantly impact product selectivity and coke formation, but the nature of this influence is not yet completely understood. Here, we show that introducing a macroporous (> 50 nm) structure to a metal-free, microporous MFI catalyst improves its solid conversion rate. Additionally, we provide insights into the synthesis of metal-encapsulated MFI catalysts, which allows for better understanding of the respective roles of acid and metal sites in bifunctional hydrocracking reactions.

Effects of Sulfuric Acid Molarity on Iridium Oxides

Alexandria Lam, Qingying Jia

Iridium oxide is by far the only viable catalyst for the oxygen evolution reaction (OER) in industrial PEM water electrolysis. Gaining a fundamental understanding of how Ir catalyzes OER under industrial conditions is crucial to advancing catalysts and electrodes for the green energy transition. However, there are discrepancies in results between fundamental benchtop experiments and electrolyzer stacks deployed in the field. Benchtop experiments in literature are widely performed at low molarities (e.g., 0.1 M) for convenience, whereas data taken from stack tests display trends reminiscent of higher molarities (e.g., 5 M). To examine how these discrepancies may affect catalyst performance, four different species of iridium oxide (Ir black, hydrous IrO_x, rutile IrO_2, and amorphous IrO_x) were evaluated in different molarities of sulfuric acid using a benchtop rotating disk electrode (RDE) setup. Cyclic voltammetry (*CV*) cycles taken at each of these conditions generally revealed narrower cycles in 5 M than in 0.1 M, reflecting the fact that iridium dissolved into the solution at the higher molarity. Additionally, OER performance for each of the catalysts yielded lower OER activity in 5 M. This can be attributed to the greater potential requirement at lower pH's to generate oxide species capable of OER, as illustrated in the Pourbaix diagram for iridium. Shifts in *CV* peaks at different conditions support the theory that different oxide species form in different molarities, which has great implications for studying iridium oxide performances in industrial PEM water electrolyzers. Performing benchtop studies at high molarities may better simulate real-world conditions in a membrane electrode assembly, helping researchers understand the nature of iridium oxides in devices and how they may differ from prior fundamental study results.

Electrochemical Oxidation of Ethylene Glycol on Palladium-Based Catalysts: Influences of Substrate and Selectivity for Glycolic Acid Production

Michael Galvin, Mohammad Albloushi, Wenzhen Li Sr.

Ethylene glycol (EG) is a crucial platform molecule derived from biomass, fossil fuels, and waste plastics like polyethylene terephthalate (PET). Nearly a hundred million metric tons of PET are produced globally, with roughly only 30% being recycled. Unrecycled PET can be depolymerized into EG, which can then be converted into glycolic acid (GA), a compound seven times more valuable than EG. However, current GA production suffers from poor selectivity and high energy demand. Electrooxidation of EG to GA presents a promising solution but requires energy-efficient, selective catalysts with a clear mechanistic understanding.

Through the testing of various Pd-M bimetallics for EG oxidation, it was found that PdBi @ Nickel foam (NF) showed high activity and selectivity towards GA. However, two questions arose from its high activity: what effects does NF have on the reaction, since it is an active substrate, and why does Bi boost the performance while being inactive on its own? To test the effects of the NF substrate, by electrodepositing PdBi @ Carbon Cloth (CC), an inert substrate, a direct comparison can be made. The PdBi @ CC maintained better Faradaic efficiencies, demonstrating that the NF is likely active and causing worse efficiencies. Using in situ Raman, it was found that at higher potentials, the PdBi was forming complex Ni-Pd-Bi oxides that were shifting the selectivity of the reaction from GA to formic acid (FA), which is supported by experimental data. As for Bi's role in the catalyst, a series of electrochemical control tests were completed that showed Pd @ CC did not perform well and Bi @ CC had no activity. Currently, in situ Raman and density functional theory (DFT) are being used to explain Bi's effects in the catalyst.

In addition to analyzing the catalyst's activity, the effects of temperature, potential, and other reaction parameters on selectivity and Faradaic efficiency for GA production were studied. Understanding the catalyst mechanism and reaction conditions will support future work to convert crude EG from recycled PET into GA, advancing a more sustainable plastic life cycle.

Limiting Ethanol Crossover in the Electrochemical Reduction of Carbon Dioxide By Integrating a Dual Membrane Electrolyzer

Eryn Kennedy

As the world faces the need to limit global warming significantly by 2050, the transition from fossil fuels to renewable energy is necessary. Carbon dioxide can be converted into renewable energy and carbon products via the electrochemical reduction of CO2. Through the use of a fuel cell and Cu electrocatalysts, multi-carbon products are produced. Sn dopants in the catalyst increase product selectivity to ethanol, which can be used as a fuel. This system's product yield is limited by crossover of ethanol from the cathode to the anode where it is oxidized to acetic acid. In order to limit crossover and increase product yield, a dual layer membrane consisting of an anion exchange membrane and a cation exchange membrane was implemented. An inner layer between the dual membranes was flushed with DI water. This captured ethanol before it crossed over to the anode side and facilitated conductivity across the cell. The system was optimized using 1M KOH (pH 14) through the inner layer, 1M KOH+H_3PO_4 (pH 6) anolyte, Nafion CEM, and Sustainion AEM. Performance was assessed by measuring ethanol concentration and faradaic efficiency at the anode, inner layer, and cathode via nuclear magnetic resonance and gas chromatography. Results demonstrated significant improvements in ethanol production and retention. The cathode and inner layer together produced ethanol at 51.92% Faradaic efficiency (FE), which corresponds to more than 10 wt% ethanol directly from the cathode. Importantly, crossover to the anode was minimal, with less than 1.07% FE detected. This novel approach effectively limited crossover, with the majority of the product being captured at the inner layer, thereby substantially improving the efficiency of CO_2-to-ethanol conversion in this electrochemical system

Oxygen Reduction to Hydrogen Peroxide on Hydrophilic Carbon Fiber Paper: Dependence of Mechanism and Active Site Stability on Electrolyte pH

Lydia Schultz, Connor P. Cox, Madeleine K. Wilsey, Kendra R. Watson, Teona Taseska, Yiwen Sun, Samira Siahrostami, Astrid Müller

Hydrogen peroxide (H 2 O 2) is a vital chemical for a variety of applications including water purification and chemical synthesis. Current thermochemical methods require intense energy and infrastructure needs to be manufactured industrially. Electrocatalytic H 2 O 2 synthesis eliminates extensive energy, transportation, and infrastructure needs. Electrocatalysis also offers benefits over current thermochemical processes as it can be powered by renewable energy and does not require chemical manufacturing plants which emit extensive greenhouse gases into the atmosphere. This research aimed to find a new method to synthesize hydrogen peroxide through electrocatalysis. Using hydrophilic carbon fiber paper, 96.6% selectivity for H 2 O 2 was achieved. The catalyst is non-particulate, so it does not require supporting electrodes, which allows for control over the surface chemistry. Tests in pH conditions ranging from 0.6 to 14 show the interplay of mechanism and active site stability and their effect on H 2 O 2 electrosynthesis.

The Electrochemical Conversion of CO2 into Value-Added Chemicals and Fuels to Defossilize Society

Isaac Okewole, Yuval Fishler, Wilson A. Smith, Kyra Yap, Melanie Rodriguez, Jesse Matthews

The electrochemical conversion of CO^2 into value-added chemicals and fuels has gained significant interest in the scientific community over the past decades as a potential technology to help defossilize society. Several catalysts have been found to perform the selective conversion of CO^2 into different products, where Ag and Au can make CO, Sn and In can make formic acid/formate, and Cu can uniquely make up to 18 different products, many of which are C2+.

While Cu can make many products, one main challenge is understanding the mechanisms to promote the formation of specific products while also being able to suppress the formation of unwanted products, most notably hydrogen. In situ, spectroelectrochemical techniques such as surface-enhanced infrared absorption spectroscopy (SEIRAS) can probe catalyst surfaces during electrolysis and reveal vibrational modes of reaction intermediates and near-surface electrolyte compositions.

Here, we use in situ SEIRAS to probe reaction intermediates on Cu electrocatalysts during CO^2 reduction as a function of applied potential and electrolyte composition and concentration. Specifically, we observe how reaction intermediates change in speciation and concentration while operating in electrolytes that contain divalent cations (Ca^{2+}, Mg^{2+}), which are found both in tap water and seawater. We show that the divalent cations change the surface speciation and thus reaction pathways of CO2 reduction on copper, which highlights the complexity of using non-purified electrolyte streams that may be easier to use, but more challenging to optimize, thus our goal of optimization. And eventual defossilization of the energy industry and removal of CO^2 to stop climate change.

Organic Films Amplify Cation Effects in Heterogeneous Electrochemical CO_2 Reduction

Sebastian Castro, Madeline Hicks, Nicholas Watkins, Theodor Agapie, Jonas C. Peters

Electrochemical CO_2 reduction (CO_2R) is a method to recycle waste gas from point sources or the atmosphere to renewably make industrially relevant products. Copper is a unique catalyst for CO_2R because of its ability to produce valuable C_{2+} products, albeit with poor selectivity. One strategy to improve performance is to use certain organic additive films that are electrochemically deposited on the electrode, which can dramatically boost the selectivity towards carbon-coupled products. Additionally, the electrolyte cations have a significant role in the performance of CO_2R, and mixing these cations yields a range of results. We investigate how N-tolyl pyridinium additive impacts the cation effects in Cu-catalyzed CO_2R. Electrochemical techniques, including chronoamperometry (CA) and modified pulsed voltammetry (mPV), along with attenuated total reflection surface-enhanced infrared absorption spectroscopy (ATR-SEIRAS) can provide insight into how cation size affects the electrochemical environment. Electrolysis data show that doping small amounts of Cs^+ into Li^+ electrolyte makes a noticeable impact on the performance of CO_2R towards C_{2+} products, suggesting a cation competition for accumulation on the cathode surface. To further probe the extent of cation effects, we performed experiments using Cs^+ with both NH_4^+ and NEt_4^+, which reveal that reactivity, along with hydration shell, of the cation is significant. With this mixing strategy, we find that selectivity and activity plateau at different Cs^+ compositions for bare Cu and additive-modified Cu, emphasizing how the additive creates a unique electrochemical environment close to the electrode surface. Analysis of mPV and ATR-SEIRAS data suggests that the local electric field at the Cu cathode does not change with the presence of the additive, which narrows down the difference in performance between bare and additive-modified Cu to physical, or nonelectric field, effects. These results emphasize how transport effects in electrochemical processes are equally important to consider along with electric field effects when explaining reactivity trends.

Investigating Anodic Catalyst-Ionomer Interactions in Anion Exchange Membrane Water Electrolysis (AEMWE)

Zackary Mitzel

Molecular hydrogen has an energy density per mass three times higher than gasoline and produces no carbon dioxide when burned, making it a useful alternative to polluting fossil fuels. However, over 99% of hydrogen currently comes from fossil fuels and therefore negates the carbon benefits of powering vehicles and power systems with hydrogen. Water electrolysis is a clean, alternative hydrogen production method that could be coupled with renewable energy to produce carbon-neutral hydrogen. While proton exchange membrane water electrolysis and alkaline water electrolysis are well-established technologies, the former relies on expensive titanium parts and scarce catalyst materials, and the latter suffers from poor efficiency and membrane stability, limiting their widespread adoption. Anion exchange membrane water electrolysis (AEMWE) is a promising candidate to reduce the cost of producing hydrogen, as it doesn't require expensive platinum-group metal catalysts and acid-resistant metals such as titanium for parts, yet it has reasonably good efficiency. AEMWE must further improve in efficiency and increase its lifetime for large-scale use. The anode contains higher overpotentials than the cathode; thus, much research focuses on reducing anodic overpotentials to increase efficiency. Ionomer-catalyst interactions govern the kinetics and transport through the catalyst layer, yet little research has studied these interactions. In this poster, PiperION ionomer particles were ground into various sizes before being combined with Co_3O_4 catalyst particles and solvents. The different inks were then sprayed onto nickel foam substrates to synthesize anodes with different microstructures. Cells were assembled using a Pt/C cathode and a PiperION membrane, and electrochemical tests, including electrochemical impedance spectroscopy and potentiostatic polarization curves, were used to evaluate electrolyzer performance. Overpotentials were broken down using high frequency resistance and Tafel slope analysis, while scanning electron microscopy was used to study catalyst layer morphology. Depending on KOH electrolyte concentration, ionomer particle size had different levels of influence on electrolyzer performance: with 1M KOH, ionomer particle size had little effect on performance. However, with 0.1M KOH, larger ionomer particle size in the anodic catalyst layer decreased kinetic, catalyst layer resistance, and residual—often attributed to mass transport—overpotentials, therefore increasing cell efficiency. The reductions in overpotentials were attributed to increased electrochemically active surface area, increased electronic conductivity, and increased porosity. The increase in efficiency, though, came with a tradeoff of increased degradation in the form of voltage gain after conducting a degradation test. This increased degradation was caused by significant delamination of catalyst and therefore decreased electrochemically active surface area, seen via electron microscope imaging. The results indicate that understanding ionomer-catalyst interactions is important to improving AEMWE performance, especially at low electrolyte concentrations. Future research should go into optimizing catalyst and ionomer loading simultaneously with ionomer particle size to minimize overpotentials attributed to kinetics, catalyst layer resistance, and mass transport.

Electrocatalytic Oxidative Dehydrogenation of Furfural in Alkaline Electrolyte

Isabella Hughes, Rupali Mittal, Adam Holewinski

Conventional water electrolysis to generate green hydrogen gas produces H2 at the cathode of an electrochemical cell and O2 at the anode. This process is still not economically competitive with fossil-derived hydrogen due to a high thermodynamic cell potential and a high overpotential to overcome slow anodic oxygen evolution reaction (OER) kinetics. Replacing OER with electrocatalytic oxidative dehydrogenation (EOD) reactions such as the conversion of furfural (FF) to furoic acid (FA) may lower the cost of renewably-produced H2 by simultaneously generating valuable organic products in addition to anode-derived H2.

Copper has been shown to have good catalytic activity for the EOD of FF to FA due in part to affinity for C-H bond cleavage and affinity for recombination of adsorbed hydrogen (H*) rather than for proton discharge. This study hypothesizes that the conversion and selectivity of EOD can be improved upon by strategically creating Cu-alloy catalysts, such as CuNi, as the addition of Ni would slightly compress the lattice of Cu, weakening H* interactions and promoting recombination to H2.

Herein we have examined how best to increase the conversion and selectivity of EOD electrolysis of FF to FA. We examined different electrolysis conditions, namely electrolyte pH, oxidation potential, and the type of membrane used. To enhance the catalytic activity of the pure Cu catalyst, we increased the surface area of the copper foam working electrode via electrochemical surface roughening and by electrodepositing varying amounts of additional Cu onto the Cu foam. Additionally, we investigated the effect of CuNi alloys deposited onto the Cu foam by different electrodeposition methods in varying Cu:Ni ratios.

We found that among the tested conditions, a Nafion membrane, 0.25 V vs RHE electrolysis potential, and pH 13 electrolyte were the best reaction conditions in terms of limiting side reactions. We further found that increasing the surface area of Cu-foam by both methods increased conversion of FF. Adding Ni to Cu also increased conversion of FF in some ratios of Cu:Ni deposited.

Investigating the Effect of Carbon Dioxide to Propane Ratio on Incubation Time and Gas Hydrate Conversion Yield

Ava Gross, Anish Pisipati, Devinder Mahajan

This study investigated the influence of varying carbon dioxide (CO_2) to propane (C_3H_8) ratio on the induction time and gas hydrate conversion yield measured in a 300 mL batch reactor. Gas hydrates are naturally occurring crystalline compounds that form guest-host structures from water and gas under low temperature and high-pressure conditions. These kinetically unstable compounds are attracting significant attention due to their potential in wastewater and other applications. Experiments were conducted under identical reaction parameters, including a fixed amount of deionized (DI) water, pressure, cooling rate, time, and stirring speed (rpm). By adjusting the CO_2/C_3H_8 ratio, the study aimed to optimize two critical parameters: the time required for gas hydrate nucleation (induction time) and the overall conversion yield to gas hydrates. Through five trials using CO_2/C_3H_8 ratios of 12:1, 7.2:1, 5:1, 3.2:1, and 2:1, it was found that the ratio with the highest conversion yield was 3.2:1, achieving 68.8% conversion. The shortest induction time was observed at the 7.2:1 ratio, with nucleation occurring in 383 seconds. These findings offer valuable insights in to optimizing gas hydrate formation process, which could enhance their potential applications in thermal energy storage and wastewater management, along with other applications.

Discovery and Design of Thermostable PET-Degrading Enzymes

Dawood Virk

Poly(ethylene terephthalate), or PET, is an aliphatic-aromatic copolymer of ethylene glycol and terephthalic acid. More than 80 million tons of PET-based plastic products are produced on an annual basis, 72% of which are not recycled and instead are destined for landfills or incineration. Furthermore, most PET recycling facilities require relatively pure feedstreams that may not be feasible. PETases—enzymes that can readily degrade the polymer into its oligomers and monomers—remediate challenges in PET waste management and offer a sustainable way to biodegrade the polymer without releasing carbon dioxide. Targeted PET degradation past the polymer's glass transition temperature (T_g = 75°C), where amorphous regions are readily accessible to catalysts, is achievable with thermostable enzymes.

Bioinformatics was used to discover naturally occurring PETase candidates, while computational and structural insights guided the engineering of thermostable PET-degrading enzymes. Homologs from the three classical PETases—hydrolase from *P. sakaiensis,* cutinase from *T. fusca*, and leaf compost cutinase from an unknown organism—were scanned for their species of origin. Mass sequence alignment was used to confirm the presence of a hydrolytic catalytic triad. Consequently, several PETase candidates from organisms living at up to 65°C were found. Structural analysis with both AlphaFold and crystal structures guided the installation of disulfide bonds to increase enzyme thermostability. The resulting enzymes retained activity beyond 90°C. Subsequently, MutCompute, a convolutional neural network that was trained on crystal structures from the Protein DataBank, guided single-point enzyme mutagenesis. Variants of Leaf Compost Cutinase (LCC) were created from MutCompute recommendations. When compared with the wild-type, it was found that the top MutCompute result, LCC^{L66A}, resulted in an increase in PET oligomer degradation at 70°C and 80°C by 21% and 55%, respectively; LCC^{L66A} also released 120% more PET monomers when incubated with PET at 37°C. The second most favorable variant predicted by MutCompute, LCC^{N246T}, had little activity at most temperatures but outcompeted wild-type LCC at and above the glass transition temperature. Structural insights validated the MutCompute recommendations; L66A reduces the hydrophobicity of a solvent-accessible microenvironment, while N246T improves substrate docking in the active site. The results emphasize the importance of using both quantitative models, such as neural networks and bioinformatics, as well as structural insights for the generation and validation of beneficial enzyme mutations.

Synthesis and Analysis of Molybdenum and Iron Catalysts for Non-Oxidative Methane Coupling

Bryce Clutter, Madelyn R. Ball, Mihir Kulkarni, Ethan Robey

Methane coupling is a method for producing light hydrocarbons such as ethylene, offering an alternative to oil-based processes, which can contribute to a global reduction in fossil fuel usage. Most methane is burned to create energy because it is difficult to convert methane to other molecules efficiently. Methane is a greenhouse gas that once it reaches the atmosphere contributes greatly to global warming. This study focuses on using molybdenum (Mo) and iron (Fe) catalysts to facilitate non-oxidative methane coupling. Non-oxidative methane coupling is a desirable process because it does not produce carbon monoxide as an unwanted byproduct. The monometallic Mo and bimetallic MoFe catalysts with varied Fe loading were prepared by incipient wetness impregnation. After calcination, the catalysts were evaluated for catalytic performance in a fixed-bed reactor. Prior to the reactions, the catalysts were treated with methane and hydrogen while heating to the desired temperature, to obtain the desired active phase for the methane coupling. The reactions were run under various methane partial pressures and temperatures to study the reaction kinetics. The reactor was connected to a gas chromatography to quantify the output stream. Additional techniques such as chemisorption and X-ray diffraction were also employed to study the properties of these catalysts. From these experiments, we determine how the activation energy of the reaction influences the reaction, how different Fe loadings affect production, and how different methane concentrations impact the products. Future research involving the use of platinum bimetallic catalysts and different metal loadings could be done to further enhance the understanding of this reaction system. With a better understanding of the reaction and more efficient catalysts, methane could be converted into useful products in a more cost-effective manner. This will help encourage methane to be converted to useful chemicals which will decrease methane emissions and help decrease the issue of global warming.

Synthesis of Renewably Sourced Monomers: Hydrogenation of 2,5-Furandicarboxylic Acid Using Pt and Pd Supported Catalysts

Ben Auer

Recent efforts to reduce industrial reliance on fossil fuels have focused on developing renewably sourced production routes for key monomers. This study investigates the selective hydrogenation of 2,5-furandicarboxylic acid (FDCA) to produce tetrahydrofuran-2,5-dicarboxylic acid (THFDCA), a polyester monomer and an intermediate to adipic acid (AA). Platinum (Pt) and palladium (Pd) catalysts supported on silica (SiO_2) and titania (TiO_2) were used at 5% wt. loading of the noble metal. All reactions were carried out at 34.5 bar H_2, for 4 hrs (100% conversion). Pd catalysts achieved yields of 98% (5% wt. Pd/SiO_2) and 90% (5% wt. Pd/TiO_2) towards THFDCA at 60°C, while Pt catalysts achieved yields of only 13% (5% wt. Pt/SiO_2) and 17% (5% wt. Pt/TiO_2) THFDCA. Interestingly, Pt catalysts showed preference for reductive ring opening, with 5% wt. Pt/SiO_2 achieving an 85% yield of 2-hydroxyadipic acid (2-HAA), an intermediate in the HDO of THFDCA and valuable research compound. Similarly, 5% wt. Pt/TiO_2 formed 2-HAA in 74% yield. At 24 hrs of reaction time, the product distribution of 5% wt. Pd/SiO_2 remained unchanged from 4 hrs, implying distinct reaction pathways for Pd and Pt. Increasing the temperature of reaction from 60°C to 100°C led to decreased yield of the more selective product, and decreasing the temperature to 40°C did not have a significant impact on yields. Further studies will test the hypotheses that differences in selectivity arise from differences in adsorption modes of FDCA onto the surface of the catalyst, the presence of different hydrogen species (H^* vs H^+), or a difference in electron transfer between the noble metal and support.

Elucidation of Ce/Zr Ratio Effects on the Physical Properties and Catalytic Performance of $CuO_x/Ce_yZr_{1-y}O_2$ Catalysts

Michal Luchowski, Mohammed Sifat, Amol Pophali, Wenhui Jiang, Yunfan Lu, Byeongseok Kim, Gihan Kwon, Kwangsuk Yoon, Jihun Kim, Kwangjin An, Sang-eun Shim, Hocheol Song, Tae Jin Kim

CeO_2 is widely used as a catalyst support because of its oxygen storage capacity (OSC) and Ce^{4+} and Ce^{3+} redox cycle, which help avoid catalyst deactivation by coke formation and improve catalytic activity. However, bulk CeO_2 has shown some limitations, such as low thermal stability, low number of defect sites on its surface, and deactivation of the redox cycle at high temperatures. It has been reported that zirconium (Zr) incorporation in the cerium (Ce) lattice can increase the number of oxygen vacancies, improve catalytic activity and thermal stability. Using a fixed surface density (SD) of copper (# of Cu atoms/nm^2) as a surface species, the role of the support ($Ce_yZr_{1-y}O_2$ (y = 1.0, 0.9, 0.6, 0.5, and 0.0)) and defect site effects for CO oxidation reaction were investigated. The thermal stability of $CuO_x/Ce_yZr_{1-y}O_2$ catalysts at 700 °C and 900 °C calcination temperature as well as the reduction of the surface species was also conducted in order to better understand the role of CuO_x and calcination temperature. The spectroscopic (e.g., Raman, XRD) and microscopic (e.g., SEM-EDX, TEM) characterization techniques were applied to evaluate the defect sites, crystallite size, lattice parameters, and chemical composition of catalysts. CO oxidation with varied $CO:O_2$ ratios (1:5, 1:1, and 1:0.5 (stoichiometric)) was used as a model reaction to describe the relationship between the structure and catalytic performance of each catalyst. Based on the characterization results of $Ce_yZr_{1-y}O_2$ materials, the addition of Zr causes physical and chemical changes to the overall material. The inclusion of Zr into the structure of CeO_2 decreased the overall lattice parameter of the catalyst and increased the number of defect sites, while $CuO_x/Ce_yZr_{1-y}O_2$ displayed lower catalytic ability compared to a CuO_x/CeO_2 catalyst. Based on collected Visible and UV Raman spectroscopy data there is a reduction in oxygen vacancies with higher calcination temperature. The results showed a significant change in activity and structural change with varied calcination temperature. It was concluded that defect sites of support material were not directly linked to the catalytic performance over the tested CuO_x/CeO_2 and $CuO_x/Ce_yZr_{1-y}O_2$ catalysts.

Period Four Transition Metal Effect on Catalytic Activity of CeO_2 supported Metal Oxide Catalysts Synthesized Via One Pot Chemical Vapor Deposition (OP-CVD)

Mary Calandra, William Chen, Nicole Spencer, Hannah Karkout, Kwangsuk Yoon, Hocheol Song, Amol Pophali, Tae Jin Kim

Ceria (CeO_2) is a common support material used in catalysis due to its redox properties. Thus, in this study ceria supported MO_x (M= Period four transition metals Cr, Mn, Fe, Co, Ni, Cu, and Zn) catalysts were synthesized by one-pot chemical vapor deposition (OP-CVD) method. The catalysts underwent various characterization tests and the catalytic activity of each species was analyzed using CO oxidation as a model reaction. The characterization techniques that were used include thermogravimetric analysis (TGA), X-ray diffraction (XRD) spectroscopy, Raman spectroscopy, inductively coupled plasma optical emission spectrometry (ICP-OES), and scanning electron microscopy with energy dispersive X-ray spectroscopy (SEM-EDX). Characterization testing results conclude that the MO_x species was well dispersed onto CeO_2. Activity data shows CuO/CeO_2 has the highest catalytic activity (~98% CO conversion at ~155 °C) which can be explained by the creation of oxygen vacancies and electron configuration of copper.

The obtained result provided the feasibility of OP-CVD method to synthesize a different type of supported catalysts effectively: no solvent, and relatively simple synthesis procedures (solid-mixing, evaporation, and calcination). The synthesized supported catalysts can be applied to broad ranges of catalysis areas.

Acknowledgement: The authors acknowledge funding support from the National Science Foundation (NSF-CBET-2050824). The authors would
also like to thank the Advanced Energy Research and Technology Center (AERTC) for the facilities at Stony Brook University.

High Entropy Alloys (HEAs) As Catalysts for NOx Reduction: Catalytic Activity and Stability in Pollution Control.

Harry Carrion, Matt Craps

The abatement of NOx, NH3, and volatile organic compounds (VOCs) from exhaust gases is crucial for minimizing air pollution, particularly in denitration offgas systems. This research explores the use of High Entropy Alloys (HEAs) as catalysts for the reduction of NOx to nitrogen (N_2) and water in the presence of ammonia (NH_3). HEAs, known for their enhanced catalytic activity, thermal stability, and customizable compositions, were synthesized using stock solutions of Ce^{3+}, Fe^{3+}, Pd^{2+}, Pt^{4+}, and V^{3+}. Two preparation methods—drop casting and direct mixing with graphene oxide films—were employed, followed by microwave heating. Scanning Electron Microscopy (SEM) and Energy Dispersive X-ray Spectroscopy (EDX) analyses revealed the composition and structure of the HEA films. Future work aims to test new alloy compositions and optimize catalytic performance for environmental applications.

Synthesis and Analysis of Nanostructured SiO_2 / TiO_2 Photocatalysts for Hydrogen Release from Liquid Organic Hydrogen Carriers

Simon Hedin, Justin Notestein

Hydrogen is a promising fossil fuel alternative, providing versatility, zero direct carbon emissions, and renewable production potential. However, transporting pure H_2 is energy-intensive, technically challenging, and potential dangerous. Liquid Organic Hydrogen Carriers (LOHCs) provide a solution, allowing safe H_2 storage and transportation within common industrial chemicals. Developing catalysts that efficiently and selectively release H_2 from LOHCs using only sunlight (photocatalysts) would allow H_2 to be used in underserved and off-grid communities. TiO_2 is a promising dehydrogenation photocatalyst, however its high activity makes controlling specific reactions (selectivity) difficult. Depositing layered nanostructures on TiO_2 surfaces and using organic templates to protect certain active sites has been shown to effect selectivity, reaction rates, and turnover frequency. This study investigates two methods for synthesizing layered and templated SiO_2/TiO_2 photocatalysts, exploring their impact on material properties and photocatalytic dehydrogenation behavior.

The first method uses Atomic Layer Deposition (ALD) to deposit thin SiO_2 layers from Si and O-containing vapors. The second method places two organic templates, p-tertbutyl calix[4]arene (CAL) and 3,5 di-tert-butyl catechol (TBC), and uses Chemical Liquid Deposition (CLD) to deposit SiO_2 layers on the exposed surface, creating artificial nanopores. Nine ALD catalysts (3 temperatures x 3 cycle quantities) and ten CLD catalysts (2 template sizes x 5 cycle quantities) were studied. Catalysts were analyzed for light absorption ability, Si to Ti surface ratio, and total effective surface area. Benzyl-alcohol (a model LOHC) dehydrogenation was used to measure reaction rates, catalyst activation energies, and selectivity. Initial results suggest thicker SiO_2 layers improve dehydrogenation rates for ALD catalysts and enhance active site stability for CLD catalysts. For CLD catalysts, there are indications that an intermediate SiO_2 layer thickness leads to an optimized turnover frequency. By examining synthesis-structure-performance relationships for layered SiO_2/TiO_2 materials, this study expands the field of TiO_2-based LOHC dehydrogenation photocatalysts, contributing to a future where clean H_2 can reduce society's reliance on carbon-emitting fossil fuels.

Catalyst Synthesis Via the Strong Electrostatic Adsorption Method

Sophia Pollak-Hurst

The strong electrostatic adsorption (SEA) method of catalyst synthesis leverages electrostatic attraction to achieve catalysts with a small particle size and high surface area. SEA requires no specialized materials or equipment and is scalable so it is promising for extending lab-scale studies to industry. SEA leverages the pH difference between a metal solution and the point of zero charge (PZC) of a support. While the overall procedure for SEA is gleaned from literature, we address several challenges to adapt this method for use in the Porosoff group. These challenges are determining the PZC of a support, selecting the optimal pH of the metal solution depending on the PZC, and calculating the mass of support needed for synthesis. We find that PZC can be determined by adding support to solutions of different pH, shaking, measuring the final pH of each solution, and finding the intersection between the best fit line of initial vs final pH and $\Delta pH = 0$. Optimal relative pH is calculated using the revised physical adsorption (RPA) method. After addressing these procedural challenges, synthesis is performed on activated carbon as a surrogate for a carbon-based Joule heater monolith. CO chemisorption and temperature programmed desorption are used to characterize the catalyst. Some challenges that occur during synthesis trials are the formation of a precipitate due to a reaction between the metal ions and the acid and base used to adjust pH, high pH causing silica-based supports to dissolve, and small particle sizes making filtration difficult for some syntheses. These studies on activated carbon are important proof of concept for extending our work to Joule heaters, which are promising for industrial decarbonization due to their low energy requirement for heating compared to traditional steam-based heating. SEA is a promising method of catalysis for Joule heaters due to its scalability and applicability to any surface shape.

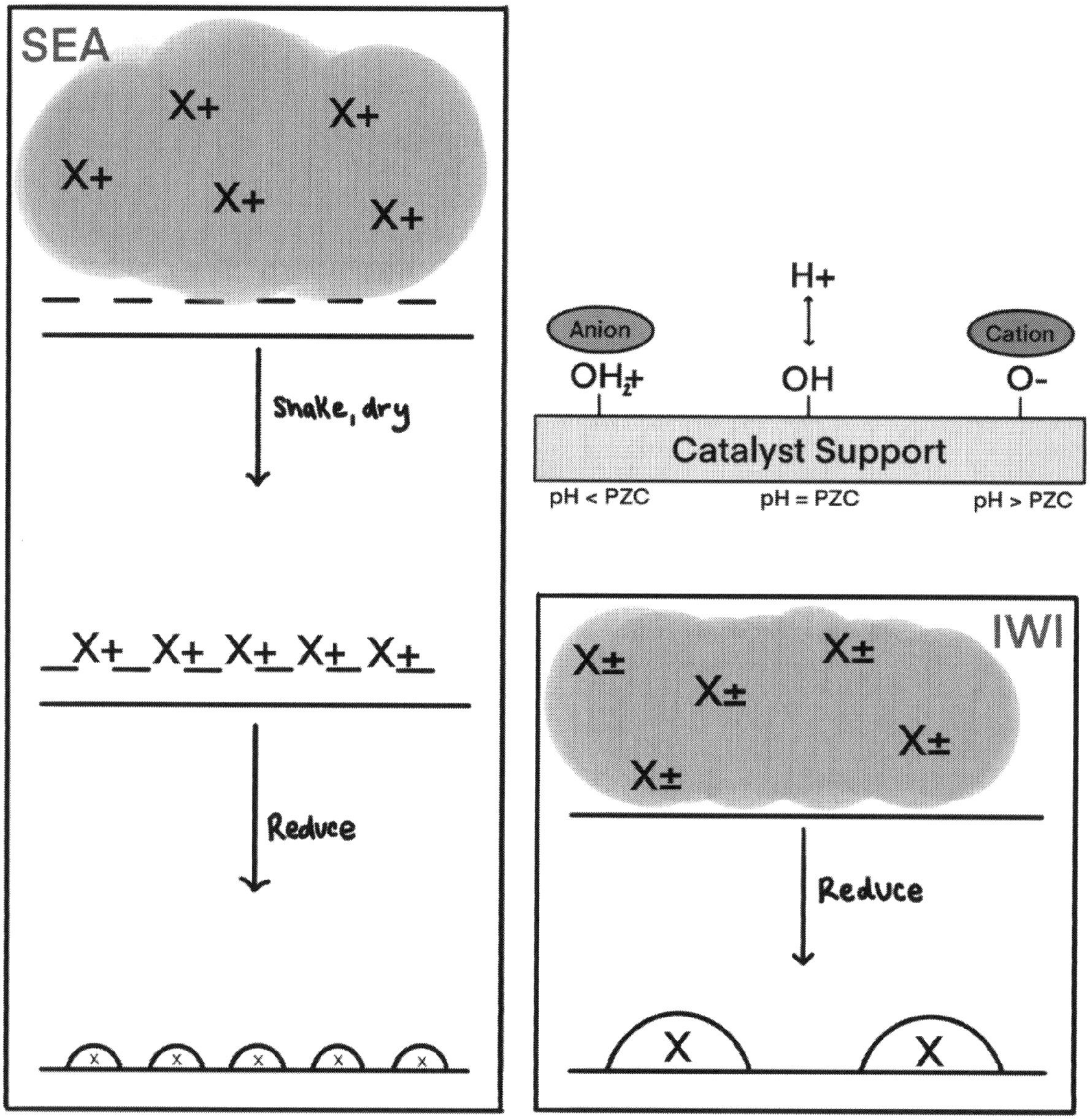
SEA
X+
Shake, dry
Reduce
H+
Anion
Cation
OH2+
OH
O-
Catalyst Support
pH < PZC
pH = PZC
pH > PZC
IWI
X±
Reduce
X

Design of New Catalysts for Generation of Clean Hydrogen from Liquid Organic Hydrogen Carriers: Dehydrogenation of Methylcyclohexane on Bimetallic Catalysts

Ahava Salomon

Hydrogen is an alternative to fossil-fuels for energy generation, a major contributor to climate change, and it is also essential in major commercial processes such as production of synthetic fuels, fertilizers, and metals. Safer and more efficient alternatives are being explored to transport and store hydrogen and one method is by hydrogenating toluene molecules to produce methylcyclohexane (MCH) which is a stable liquid at ambient conditions. MCH can then be catalytically dehydrogenated to recover the H2. Bimetallic Cu-Ni catalysts, with Cu deposited on Ni surface resulting in controlled smaller Ni surface ensembles, were shown to be highly efficient for this reaction. Further reduction of Ni ensembles can be achieved by dispersing Ni on Cu through Electroless Deposition (ED) and the development of a Ni ED bath was studied. Three tests were employed to design a working bath: no strong electrostatic adsorption (SEA) of Ni onto the silica support, significant deposition of Ni onto the surface of Cu only, and no formation of Ni particles in the bath itself. These tests required manipulation of pH, temperature, and the amount of reducing agent and complexing agent to find the correct conditions to produce the catalyst. While the ED of Ni on Cu was achieved, SEA of Ni on the silica proved problematic. The necessary conditions for an effective Ni ED on Cu are still being investigated.

Controlled Synthesis of Non-Precious Mixed Metal Oxides Using Reverse Microemulsions for Energy Storage Systems.

Klaertje Hesselink, Kunal Velinkar, Eranda Nikolla

The increasing global demand for energy necessitates the development of advanced and efficient energy storage systems. Alkali metal-oxygen (M-O_2) batteries, such as lithium-oxygen (Li-O_2), have emerged as a promising alternative to Li-ion batteries due to their exceptionally high theoretical specific energies, which stem from their unique redox chemistries. These batteries operate through the formation and breaking of chemical bonds between lithium and oxygen during discharge and charge cycles, respectively, offering the potential for significantly higher energy storage per unit mass than Li-ion batteries. However, their widespread use is hindered due to several challenges including high overpotentials, passivation of the cathode electrode, and undesirable side reactions leading to poor overall efficiency.[1,2] To address these issues, our research focuses on the strategic incorporation of electrocatalysts, which can significantly enhance the efficiency of the oxygen reduction reaction (ORR) and oxygen evolution reaction (OER) – representing the electrochemical processes during discharge and charge cycles, respectively.[2]

We specifically investigate non-precious, mixed metal oxides with tunable crystal structures, focusing on Ruddlesden-Popper ($A_2BO_{4(\pm\delta)}$) and perovskite ($ABO_{3(\pm\delta)}$) oxides, where the A site metal represents La, Sr, Ca, and the B site metal represents Ni or Co. Previous work has shown that incorporating La_2NiO_4 (LNO) nanorods as cathode electrocatalysts synthesized using a reverse microemulsion technique leads to lower overpotentials and enhanced cyclability of Li-O_2 batteries.[2] In this study, we optimize the reverse microemulsion synthesis method by incorporating a dual microemulsion system, facilitating extrapolation to different crystal structures and compositions. This approach allows tuning the oxide catalyst surface properties through various parameters such as the pH of the reducing agent, controlling the intermicellar exchange rate, and adjusting the oxophilicity and doping concentrations of the metals, to sustain rod-like morphology which can potentially reduce overpotentials further and enhance the overall efficiency and stability of Li-O_2 batteries. We hypothesize that utilizing a weaker reducing agent enables the formation of a dual microemulsion system, which enables a vast operational space and control over parameters that govern rod-like growth of the catalysts, compared to a single microemulsion system. Our findings highlight that the dual microemulsion technique yields distinct differences in particle aspect ratios compared to the single microemulsion method. This approach introduces greater control over nanoparticle growth, potentially enabling more precise investigations into the dynamic processes governing catalyst morphology.

References:

[1] Carneiro, J.S.A, et. al., *ACS Catalysis,* 2020, 10, 1, 516-527

[2] Samji, S., et. al., *Chem. Mat.,* 2019, 10, 1021

Advancing Rare-Earth Elements Separation: Continuous Flow Synthesis of Diglycolamides

Hailey Bates

Diglycolamides (DGAs) are crucial ligands for the separation of rare-earth elements (REEs). Traditional batch methods for synthesizing DGAs are labor-intensive, requiring multiple steps with manual intervention, often limited by heat and mass transfer inefficiencies. In contrast, continuous flow synthesis leverages controlled, automated operations that allow for efficient, scalable, and sustainable production.

The setup of a continuous flow reaction set up includes components such as piston pumps for precise reactant delivery, a quad mixer for uniform mixing, a coil reactor for sustained reaction, and inline analytical tools, such as infrared (IR) spectrometer for real time reaction monitoring. The reaction system is automated via Rxn Rover software, facilitating real-time adjustments and reaction monitoring. This setup enables the adjustment of various reaction parameters, including temperature, pressure, and reactant equivalents, supporting high yield optimization. Reaction mixture solutions of dimethyl 2,2'-oxydiacetate and xylene with the addition of a dialkyl amine were tested. Several dialkyl amine reactions at 8 equivalents were observed at temperatures, ranging from 180°C to 280°C for 90 minutes at each temperature.

The continuous flow reaction setup enabled expanded substrate testing, proving effective across different dialkyl amines. This approach demonstrates significant potential for enhancing the production efficiency of DGAs, offering a path towards greener and more efficient chemical manufacturing solutions for REE separation applications.

Thermocatalytic Conversion of Ethanol to Acetone: Effect of Heating Method on Selectivity and Conversion

Daichi Kobayashi, Ryan Thompson, Ben Ko, Han Wang, Erdem Sasmaz

The widespread usage of fossil fuels to supply heat to industrial processes contributes to 8% of the United States' energy-related carbon dioxide emissions. Induction heating (IH) has emerged as an alternative method to supply heating for industrial processes. By using alternating electric currents to produce oscillating magnetic fields, IH heats conductive, ferromagnetic, and ferrimagnetic materials with high efficiency. If powered by renewable energy, this electrified heating is virtually emissions-free. Additionally, IH has been demonstrated to improve catalyst activity in chemical reactions compared to conventional heating (CH), and its application in the synthesis of acetone from ethanol is of interest to the chemical industry. Acetone is a vital industrial solvent, and its versatile chemistry renders it an essential precursor for chemicals, especially plastics. Since acetone is typically produced from petroleum-derived chemicals through the cumene process, it is desirable to use more sustainable feedstocks to minimize the environmental impacts of its production. To this end, acetone can be synthesized from water and biomass-derived ethanol above 350°C with a Fe_2O_3-ZnO catalyst. In this study, the effect of induction heating on the selectivity and conversion of the thermocatalytic ethanol-to-acetone reaction was investigated. The Fe_2O_3-ZnO catalyst was synthesized, and it was characterized via XRD. The catalyst was placed in a packed-bed reactor, into which a gaseous mixture of ethanol, water, and nitrogen was fed. Time-on-stream and temperature-sweep experiments were performed using a copper solenoid for IH and a furnace for CH. The reaction products were analyzed via gas chromatography. Although both heating methods exhibited similar selectivity and deactivation trends in time-on-stream experiments, temperature-sweep experiments revealed high ethanol conversion and acetone selectivity with IH at lower temperatures than CH. As a result of the alternating magnetic field, IH causes both the stainless steel susceptor beads and ferrimagnetic catalyst to generate heat within the catalyst bed. This internal heat generation enhances heat transfer compared to CH, which heats the reactor externally. The benefits of improved heat transfer, reduced process temperature, and electrified heating demonstrate the capability of IH to decarbonize the chemical industry. Future works will probe the effects of magnetic fields, such as those produced in induction heating, on the activity of ferrimagnetic catalyst sites.

CO_2 Reduction in Acidic Medium Using Bimetallic Cu-Cd Catalysts

Evan Smith

Electrocatalytic CO_2 reduction (CO2R) offers a promising approach for recycling CO_2 into high-value multi-carbon products. The CO2R performed in basic/neutral conditions is known to result in a carbon efficiency decrease through the formation of (bi)carbonate species. Performing CO2R in acidic conditions can overcome this challenge, but most copper catalysts exhibit low single product selectivity, and CO2R must compete with the hydrogen evolution reaction (HER) in acidic media. Reducing HER in acidic conditions and improving the catalyst selectivity towards multi-carbon products are crucial for the practical deployment of CO2R.

To address these challenges, the experimental results of CO2R using several bimetallic catalysts are presented. The catalysts were synthesized using electrodeposition methods *in situ*. Notably, a copper-cadmium catalyst was found to suppress hydrogen evolution while increasing selectivity towards ethanol. Additionally, copper-tin and copper-zinc catalysts were screened, and the difference in catalyst selectivity underscores the influence of catalyst compositions (i.e., different metallic species) on the performance of copper catalysts. Continued research into optimizing bimetallic catalysts through manipulating surface morphology and deposition methods could yield higher selectivities, resulting in an effective catalyst for chemically recycling CO_2.

Dehydrogenation of Ethane Using Concentrating Solar Power

John Fawcett, Dresean Abeyta, Andrew T. DeLaRiva, Christopher Riley, H. Evan Bush, Erik Spoerke, Abhaya K. Datye

Ethylene, the world's highest volume petrochemical, is produced using steam cracking of alkanes. This reaction is carried out at high temperatures (>900°C) by the burning of natural gas which produces around a metric ton of CO_2 per ton of ethylene produced and requires energy inputs of up to 16 GJ per ton of produced ethylene [1-2]. In industry this heat and energy transfer occurs by the combustion of natural gas or other fossil fuels. The purpose of this research is to develop an alternative process that uses solar heating of a catalytic reactor to drive the dehydrogenation of ethane to ethylene. As has been previously demonstrated, ethane dehydrogenation can be performed at temperatures reachable by concentrated solar technologies (CST) [2]. The challenge then becomes what to do about the inherent intermittence of solar power. Thermal storage media already used in CST applications provide an intriguing solution. Quartz tubes were filled with solid thermal storage media, then heated to temperatures required for the dehydrogenation reaction while an ethane-carrier gas mixture was flowed through the system. Quartz tubes were chosen as they have been previously demonstrated to achieve ethylene yields of up to 68% using the open tube setup [2-4] This experimental setup demonstrated that thermal storage media can transfer sufficient energy to drive the dehydrogenation of ethane. This study also demonstrates that the packed tube is more efficient than an empty quartz tube in transferring heat to the reactant gases.

- "Net-zero carbon ethylene production via recovery of CO2 from cracking furnace flue gas." n.d. S&P Global. https://www.spglobal.com/commodityinsights/en/ci/products/net-zero-carbon-ethylene-production.html.
- R. Riley, A. De, I. L. Ibarra, A. K. Datye, and S. S. Chou, "Achieving high ethylene yield in non-oxidative ethane dehydrogenation," Applied Catalysis A: General, vol. 624, pp. 118309–118309, Aug. 2021, doi: https://doi.org/10.1016/j.apcata.2021.118309.
- Scott A. Steinmetz, Andrew T. DeLaRiva, Christopher Riley, Paul Schrader, Abhaya Datye, Erik D. Spoerke, and Christopher J. Kliewer *The Journal of Physical Chemistry C* **2022** *126* (6), 3054-3059, DOI: 10.1021/acs.jpcc.1c09955
- Efficient Low-temperature, Catalyst free Dehydrogenation Of Alkanes, by C. R. Riley. (2022, Jul. 14). *US20230037379A1*. [Online]. Available: https://ppubs.uspto.gov/pubwebapp/

Revolutionizing Waste Gas Valorization: Engineered CeO_2-Based Catalysts for Blue Hydrogen Production Via Reverse Flow Chemical Looping

Tochukwu Aniekwensi, Fanxing Li, William Martin

This research investigates the application of a reverse flow chemical looping process for repurposing waste gas into concentrated hydrogen and carbon dioxide streams, contributing to the development of sustainable technologies for mitigating industrial emissions. The study focuses on the synthesis and characterization of catalysts for blue hydrogen generation, with an emphasis on cerium oxide (CeO_2) as the base material.

CeO_2 was initially selected due to its excellent oxygen storage and release capabilities, attributed to its flexible Ce^{4+}/Ce^{3+} oxidation states. However, pure CeO_2 exhibits limitations in catalytic activity and thermal stability, necessitating further modifications to enhance its performance in the valorization process.

To address these challenges, the research explores two main approaches: surface promotion and cation doping. Surface promotion involves introducing transition metal oxides onto the CeO_2 surface, creating active sites for oxidation, reduction, and reforming reactions crucial to the process. Cation doping, on the other hand, focuses on partially substituting the cerium cation with suitable metal cations to improve overall catalyst performance.

By optimizing catalyst composition and structure, this study aims to develop highly active and stable catalysts for efficient waste gas conversion into valuable hydrogen streams. The findings of this research have the potential to revolutionize various industries, promoting a transition towards a more sustainable and circular economy through the effective utilization of waste gas resources.

Selective Decarboxylation of L-Lysine to Produce Biobased 1,5-Pentanediamine

Lucas Lisondo, Javier Chavarrio, George Huber

The ever-growing consumer preference for environmentally friendly plastic products is fueling the expansion of the biopolymer market. This trend has created a high demand for innovative processes that generate monomers from biomass feedstocks, providing a sustainable alternative to traditional fossil-based resources. This report details the synthesis of amino acid-derived 1,5-pentadiamine as a surrogate for hexamethylene diamine for Nylon synthesis. Current methods for diamines rely on toxic and environmentally damaging substances, such as cyanide and ammonia. This study explores the decarboxylation of amine-rich L-lysine for diamine synthesis using carbon supported ruthenium catalysts, aiming to improve reaction selectivity and efficiency. Various analytical techniques, including GC-MS, GC-FID, and HPLC-PDA, were employed to characterize reaction products. Preliminary results indicate a lysine conversion of up to 99.8% with a 36.0% amine selectivity attained at 200 °C and within 2.5 h. Two alternate systems were separately tested: (1) Protection of the substrate's amines with fluorenylmethyloxycarbonyl protecting groups and (2) Substitution of ruthenium catalysts with pyridoxine, a precursor to the decarboxylase enzyme cofactor pyridoxal phosphate. Future work includes improving current maximum 1,5-pentanediamine yield of 4.7% starting with further rigorous experimentation with ruthenium catalysts.

High Capacity Iron-Poor Ferrites for Syngas Generation from Carbon Dioxide and Methane

Michael Tomechko, Shang Zhai

CO_2 and methane are the two major greenhouse gasses that contribute to global warming. Dry reforming of methane converts CO_2 and methane into syngas ($CO + H_2$), an important building block for synthesizing organic compounds such as liquid fuels, utilizing the established energy industry technology. However, there is no commercial process for dry reforming of methane mainly because of catalyst deactivation by solid carbon deposition and the high cost of noble metal based catalysts.

Chemical looping technology decouples the dry reforming of methane into two steps: methane reduction of a metal oxide, and CO_2 splitting by the reduced metal oxide. Ferrites are a common type of oxide material in chemical looping for their high oxygen exchange capacity and low material cost. Iron is usually the (major) active redox element in ferrite materials such as $NiFe_2O_4$, but recent thermodynamic modeling results have predicted that iron-poor ferrite materials have greater oxygen exchange capacities than traditional iron-rich ferrites for dry reforming of methane.

Here we present the first experimental demonstration of this counterintuitive oxygen exchange capacity dependence on ferrite iron ratio in chemical looping dry reforming of methane. Thermogravimetric and fixed-bed tests were conducted on supported Co-, Ni-, and Mg-ferrites of various Fe ratios for isothermal dry reforming of methane at 500~700 °C. As predicted by the theoretical models, iron-poor ferrites exhibited greater oxygen exchange capacities than iron-rich ferrites. Additionally, a specific type of ZrO_2 as the support material was found to maximize the chemical reaction kinetics without inducing carbon deposition. The supported iron-poor ferrite has also been able to maintain less than 10% redox capacity decay in 20 cycles. The mechanistic understanding of the support material has been illustrated by x-ray diffraction and x-ray photoemission spectroscopy. To become a technology for syngas generation in remote areas with limited energy resources, the chemical looping dry reforming, which is endothermic, can also be adjusted to be autothermal by introducing a controlled amount of additional oxidant like O_2. These findings contribute to future materials design and scale-up for high capacity and energy efficient chemical looping dry reforming processes.

Conversion of Atmospheric Concentration of Methane to CO2 Via a Palladium-Zeolite Catalyst

Iain Colquhoun

Converting methane in the atmosphere into CO2 is an emerging method of combatting methane's high contribution to global warming. Our objective is to develop and test a novel catalyst to lower the energetic barriers to methane combustion at atmospheric concentrations. Using a high temperature gas reactor, we were able to convert 50% of 20 parts per million (PPM) methane in dry air at temperatures ranging from 339 to 381°C, and at flow rates from 20 to 60 standard cubic centimeters per minute (SCCM). We also experimentally determined the activation energy of methane for this lab-scale system to be 151.9 kJ/mol at 40 SCCM.

Separations in Polyethylene Terephthalate (PET) Depolymerization: Utilizing Carbon Dioxide for Terephthalic Acid Production

Maryom Rahman, Diego Trevisan Melfi, Aaron M. Scurto

Polyethylene Terephthalate (PET) is one of the most commercially available polyesters; primarily used in bottles for drinking water, shampoo bottles, packaging, and textiles. Despite its versatility, PET is not biodegradable and is often landfilled after being physically recycled a few times. Methods to chemically depolymerize PET have been proposed such as glycolysis, solvolysis, alcoholysis, and hydrolysis.

Alkaline hydrolysis achieves about 100% conversion by reacting PET with NaOH to produce disodium terephthalate (Na_2TPA) under optimized conditions. However, to precipitate terephthalic acid (TPA), the original PET precursor, large volumes of high-concentration acids are required.

Here, carbon dioxide (CO_2) will be used as a sustainable alternative to strong acids to precipitate TPA from a Na_2TPA solution. This approach can potentially help solve two major environmental issues: the accumulation of greenhouse gases in the atmosphere and the sustainable recycling of PET.

CO_2 is introduced into an aqueous Na_2TPA solution a high-pressure membrane filtration cell. A TPA suspension is formed and filtered at constant pressure. The filtrate is then analyzed using ^1H NMR to determine the residual Na_2TPA composition and solid TPA is weighed gravimetrically. Depolymerization co-products like ethylene glycol and excess reactants such as sodium hydroxide are possible contaminants, therefore, separations with varying concentrations of ethylene glycol and NaOH have been performed to test TPA yield. The ethylene glycol results produce a Na_2TPA removal of 82% at 5 weight-percent ethylene glycol. Also, there is little deviation from the removals obtained from strictly aqueous Na_2TPA for tests up to 17 weight-percent ethylene glycol. The sodium hydroxide exhibits 74% removal with only the small presence of 0.001 mols NaOH.

Further, a depolymerization reaction derived from previous literature is performed in a microwave reactor with 0.2 M sodium hydroxide (NaOH) and 10-weight percent PET to produce disodium terephthalate. Latest results include testing the monomer yield of alkali decomposition produced Na_2TPA in the precipitation reaction. The precipitation reaction produced a terephthalic acid yield of 72.3%, similar to the 75.1% TPA yield that was achieved in the absence of NaOH, and ethylene glycol with commercial disodium terephthalate. The overall terephthalic acid yield was 61%.

Future work includes optimizing the experimental procedure and repolymerizing TPA into biodegradable polymers like polybutylene adipate terephthalate (PBAT).

Temperature Effects on the Local pH of the CO_2 Reduction Reaction over Copper

Oliver Long, Victor Brandão, Carsten Sievers

The electrochemical reduction of CO_2 (CO_2RR) presents a pathway to counteract carbon emissions, supply feedstocks to the chemical industry, and thus facilitate a circular CO_2 economy to meet net-zero emissions goals by 2050. Among all CO_2RR electrocatalysts, copper is the only pure metal that produces valuable multi-carbon products, such as ethanol and ethylene, with reasonable Faradaic efficiencies. Although most copper-based CO_2RR studies are performed at room temperature, heat transfer limitations in industrial CO_2 electrolyzers will lead to temperature buildup near the copper cathode surface. Higher temperatures impact the distribution of species in the reaction microenvironment, which in turn affects product formation and reaction selectivity over the competing hydrogen evolution reaction (HER). The objective of this study, therefore, was to probe how temperature affects the concentration of local species and the reaction microenvironment pH at various potentials.

To achieve this objective, surface-enhanced Raman spectroscopy (SERS) experiments were performed in a custom-built temperature-controlled electrolytic cell charged with a CO_2-sparged $KHCO_3$ electrolyte. This cell modulated temperature between 25 and 75°C, and potential between 0.1 and -0.6 V (vs. RHE). We show that the local pH varies dramatically from its bulk value as a function of both potential and temperature. The local H^+ concentration dependence on temperature suggests that local pH is a key parameter controlling reaction selectivity at higher temperatures. These results have significant implications for the types of controllers and heat exchangers necessary for thermal management of the CO_2RR, which is a crucial consideration when designing large-scale industrial CO_2 electrolyzers.

Mechanistic Insights into the Thermal Degradation of Polytetrafluoroethylene

Charles M. Aronov, Daniel E. McGinnis, Kai D. Palam, Nathaniel F. Nelson, Aaron Ajeti, Benjamin R. Johnson, Justin Hunter, Justin R. Toole, Victor A. Jaffett, Pamela L. Sheehan, Shubham Vyas, Enoch A. Nagelli, Simuck F. Yuk

Synthetic fluoropolymers, notably poly- and perfluorinated alkyl substances (PFAS), are widely utilized in various industrial applications, such as kitchenware, textiles, and flame-resistant materials. The presence of C-F bonds endows many fluoropolymers with significant wettability, along with thermal and chemical stabilities, and distinctive friction characteristics. However, there are growing concerns regarding the toxicity of fluoropolymers in living organisms, attributed to their strong affinity for proteins and their persistence in the environment. A combustion-based methodology has been proposed to mitigate the concentrated presence of fluoropolymers in open-air conditions. As expected, identifying the key decomposed compounds produced from combustion process is not a trivial task. Therefore, theoretical approaches have often been employed to forecast the potential products formed during experimental thermal decomposition conditions.

In this study, the degradation mechanism of polytetrafluoroethylene (PTFE) has been investigated via the combination of density theory (DFT) calculations and molecular dynamics (MD) simulations. Additional analysis of the enthalpic and entropic contributions to the production of key reaction intermediates, including perfluorocyclobutane (C_4F_8), was considered. Coupling with DFT-based kinetic parameters, the microkinetic modelling was also designed to understand the concentration profiles of major products from combustion reactions. Overall, such combined theoretical approaches may serve as a reference to identify key products of the thermal degradation processes of various fluoropolymers.

Homocoupling of Carboxylic Acids through Kolbe Electrolysis

Ryan Rasmussen, MD Mosaddek Hossen, Jean-Philippe Tessonnier

Kolbe Electrolysis is a well-established electrochemical process, yet its adoption for chemical synthesis is slow compared to other methods, primarily due to its reliance on platinum electrodes. To explore more accessible alternatives, this study investigates carbon-based electrodes as substitutes for platinum in Kolbe Electrolysis. Reticulated Vitreous Carbon (RVC) is a promising candidate in preliminary research. Although Kolbe Electrolysis typically exhibits slow kinetics, it demonstrates favorable selectivity and yield during bulk electrolysis. This research also aims to explore sustainable alternatives for chemical manufacturing, mainly through biomass conversion into carbon-based molecules with carboxylic acids. The Kolbe Electrolysis mechanism is known to eliminate carboxylic acid groups, forming radicals that couple to produce longer-chain molecules; however, this mechanism has primarily been validated for linear alkane reactants. This study will investigate the feasibility of using various carboxylic acid-containing molecules with RVC electrodes in Kolbe Electrolysis to expand the understanding of its applications.

The present work investigated the Kolbe electrolysis of many carboxylic acid-containing molecules, including levulinic acid in an electrolyte comprised of acetonitrile, tetramethylammonium hydroxide (TMAH), and water. Two RVC electrodes were immersed in this solution and bulk electrolysis was performed by applying a constant current for a specified duration. Following the reaction, the reaction products were extracted and analyzed using gas chromatography-mass spectrometry (GC-MS). The resulting data was then examined to identify and quantify the products. Starting with levulinic acid, we observed the formation of 2,7-octanedione. By comparing the expected products to the measured products, we can also propose a reaction mechanism for Kolbe electrolysis on RVC electrodes that will guide future research.

Graphene and Noble Metal Hydride Clusters for Hydrogen Storage Applications

Katherine G. LaReau, Tyler Komorowski, Walter J. Cesarski, Rosemary L. Calabro, F. John Burpo, Simuck Yuk, Enoch A. Nagelli

The synthesis of graphene and noble metal composite clusters from solution-based approaches involve the interaction of charged surface chemical functional groups and dissociation of the noble metal salt structures. Our study focuses on the synthesis of the Au, Pd, and Pt nanoparticles on 2D graphene oxide sheets from chemical reduction using sodium borohydride in solution. Graphene oxide suspensions were mixed with 100 mM aqueous solutions of potassium tetrachloroplatinate (II), potassium tetrachloropalladate (II), and gold (III) chloride, respectively. The characterization of resulting graphene/noble metal nanocomposites was conducted through Fourier Transform-Infrared Spectroscopy (FT-IR), Raman Spectroscopy, and Scanning Electron Microscopy (SEM). The SEM images reveal that the metal clusters with different sizes were successfully deposited on the graphene oxide sheets. FT-IR spectra show the presence of various surface functional groups, including -OH and C=O, on the surface of graphene oxide. However, after chemical reduction of the composite graphene oxide/noble metal salts, the -OH peak in the FT-IR spectra was not present. This finding was consistent in all the composites of graphene with the three noble metals. Raman spectroscopy confirmed the characteristic D, G, and 2D bands for graphene and higher D/G ratios with the presence of noble metal nanostructures. Moreover, we conducted X-Ray Diffraction studies on the resulting graphene/noble metal cluster nanocomposites and found that there is a presence of Pt-H and Pd-H in their respective spectra with no hydride peak present in the graphene/Au diffraction. Therefore, understanding the interaction between the surface functional groups on graphene with metal salt structures is key in the chemical reduction process. Overall, this study will advance the design of 3D composite electrodes and nanostructures materials for hydrogen storage and conversion applications.

Polarity Effects on Organic Contaminant Chemical Degradation in a Continuous Gas-Liquid Flowing Reactor

Dylan Barton, Tyler Chapman, Radha Krishna Murthy Bulusu, Bruce Locke

Contaminants of emerging concern, such as per- and polyfluoroalkyl substances, pose threats to the environment and human health, and there is a need to effectively degrade these compounds. Plasma reactors can be used to enhance degradation through advanced oxidation and reduction pathways. This study uses a continuous gas-liquid flowing reactor, where the formed plasma propagates along the gas-liquid interface. This study provides insight into the effect of electrode polarity on chemical pathways for methylene blue (MB) decoloration and perfluorooctanoic acid (PFOA) defluorination while varying pulse frequency. For MB, both the co- and counter-current studies increased decolorationrates with increasing frequencies due to the increased discharge power. Decolorationwas not enhanced for methylene blue based on the polarity due to the oxidation occurring directly in the liquid phase. However, PFOA defluorination was enhanced 1.5 times for the counter current electric field. It is hypothesized that this is due to increases in the gas-phase electron density during counter-current flow because the PFOA is an interfacial species, but further studies on peak electron density are required.

Measuring Adsorption Enthalpies of PFOS and Organics on Carbon Fiber Paper

Elena Perez, Leah Coffey, Astrid Müller

In the PFAS remediation process developed by the Müller group, PFAS are electrocatalytically defluorinated using electrodes made from hydrophilic carbon fiber paper [1]. This process relies on the PFAS first adsorbing to the carbon fiber paper electrode. This study aims to investigate whether the enthalpy of this adsorption has any correlation to the efficiency of the defluorination across various PFASs. We started with perfluorooctane sulfate (PFOS) to develop a protocol. Thermogravimetric analysis (TGA) and differential scanning calorimetry (DSC) were used to identify the temperature region of desorption and subsequently measure the heat capacity change over this range of desorption. The heat capacity was then integrated according to the formula $\Delta H = \int C_p dT$ to derive the change in enthalpy of the process. It was found that PFOS desorbs from the carbon fiber paper at 400°C. However, the DSC is not able to differentiate between heat capacity changes due to desorption and those due to degradation of PFOS or the carbon fiber paper, so further research is needed to separate these events and determine the desorption enthalpy.

[1] Z. Meng, M.K. Wilsey, C.P. Cox, A.M. Müller, Complete electrocatalytic defluorination of perfluorooctane sulfonate in aqueous solution with nonprecious materials. J. Catal. 2024, 431, 115403.

Tetrabutylammonium-Based Organosulfurs: A New Family of Catalysts for Lithium-Sulfur Batteries

Victor Ribeiro Sanctis

Lithium-ion batteries hold a prominent position in the market for personal electronic devices, electric vehicles and are promising options for grid storage systems. The need for increased energy density underscores the significance of optimizing cathode materials. Among these materials, sulfur stands out for its exceptional promise. It boasts a remarkably high theoretical specific capacity and energy density, at 1,675 mAh/g and approximately 2,600 Wh/kg, respectively, over five times the energy density of current Lithium-ion batteries. Moreover, a sulfur cathode offers potentially increased safety due to lower battery voltage (around 2.4 V vs. Li^0/Li^+), reduced cost, and lower toxicity than some existing cathode materials.

However, there are challenges to be overcome. Typically, Li-S batteries employ a lithium metal anode, liquid organic electrolyte, and composite sulfur cathode. During charge/discharge, sulfur and the polysulfide intermediates exhibit high resistance, and some species are insoluble in the electrolyte. They also exhibit slow reaction rates due to their reaction mechanism, leading to limited long-term cycling. Various approaches are being pursued to address these, one of which involves catalysts. Previous research has focused on utilizing catalysts to enhance the reaction rate of polysulfide intermediates. These catalysts encompass metal oxides, sulfides, phosphides, and nitrides, which either absorb or convert the polysulfide intermediates.

For the first time, we identified a family of compounds with and without organosulfur anions as catalysts for Li-S batteries. The concept of 'families' establishes a structural baseline, enabling the observation of the effects of minor differences in molecules within these families, such as variations in elements or their quantities. Tetrabutylammonium-based compounds are soluble in traditional Li-S battery electrolytes, and their addition to the electrolyte enhances the cell's performance. These catalysts have been shown to significantly improve cycling performance, with enhancements of up to 52% observed after 50 cycles at C/5 (5 hours of charge and 5 hours of discharge) and a high coulombic efficiency of 98%. Insights into the mechanism have been proposed using operando UV-VIS and Raman spectroscopy, in addition to electrochemical methods. This novel family of catalysts paves the way for further optimization of Li-S batteries, bringing them closer to practical applications.

Trending Acid Molecular Properties with Pka

Yifei Liu, Mad Lindsey, Jean-Patrick Selo, Dean Sweeney, Bolton Tran, Bryan Goldsmith

Acid pKa encodes the proton dissociation equilibrium, an important property in various chemical processes. For example, acids can participate in reactions by donating protons to the catalyst surface in electrocatalytic reactors. Predicting the pKa of acids in different microenvironments helps advance understanding of the roles of acid proton dissociation in various chemical applications. In principle, acid proton dissociation physically relates to other molecular properties of the acid. Previous work found descriptors for predicting acid pKa such as the O-H bond length and vibrational frequency. However, the relationship between other molecular parameters and pKa has not been established.

Herein, we systematically explore the correlation between various acid molecular properties with pKa in bulk solution. We use density functional theory (DFT) modeling to determine molecular properties of 25 acids under three levels of solvation: 1) acid in the vacuum, 2) acid with explicit H_2O in vacuum, 3) acid with explicit H_2O and implicit solvation. In agreement with previous work, we found level 3 of solvation to be necessary to observe a trend between O-H bond length and vibrational frequency with acid pKa. Longer O-H bond length tends to have lower frequency, and these acids require less energy to dissociate the proton. We found no correlation between the partial charges and dipole moments of O-H with acid pKa. Future work will extend the established correlation between O-H bond length/frequency with pKa to study acid pKa at electrocatalytic interfaces.

Silver Loading Onto Rice Husk Biochar Using Avocado Seed Extract Assisted By LED Light: Catalyst and Antibacterial Capabilities.

Emilia Moreno, Arleth Gualle Brito, Karla Vizuete, Alexis Debut, Lourdes Orejuela Escobar, Sebastián Ponce

In this study, we present a novel approach for the green synthesis of biochar impregnated with silver nanoparticles (AgNPs) using agricultural waste, specifically avocado seed extract, and blue LED light. The goal of this research was to produce an environmentally friendly material that exhibits both catalytic and antibacterial properties, with potential applications in wastewater treatment, environmental remediation, and healthcare. Biochar, a carbon-rich material derived from biomass, was selected for its high porosity and ability to support metal nanoparticles, which in turn can enhance its functionality. Silver, known for its antibacterial and catalytic efficiency, was chosen as the metal for nanoparticle formation.

The synthesis method developed in this study offers a sustainable alternative to conventional chemical reduction techniques, which often rely on toxic chemicals and require high energy inputs. By utilizing an avocado seed extract as a natural reducing agent and activating the reaction with blue LED light, we successfully impregnated biochar with well-dispersed AgNPs. This eco-friendly approach aligns with the increasing demand for green chemistry solutions in material synthesis and provides a scalable method for future industrial applications.

The process begins with the production of biochar from rice husk, a lignocellulosic biomass waste commonly available in agricultural settings. The rice husk was subjected to pyrolysis, a thermochemical process that transforms the biomass into porous biochar. This biochar serves as the support matrix for silver nanoparticles. By varying the concentration of the avocado seed extract, we were able to control the size and distribution of the silver nanoparticles on the biochar surface. This aspect is critical, as the size of the nanoparticles directly influences both the catalytic activity and the antibacterial effectiveness of the material.

Once the biochar was prepared, it was impregnated with silver nitrate solution, and the reduction of silver ions to form nanoparticles was achieved using the avocado seed extract. This method leverages the natural phytochemicals in the extract, which act as reducing and capping agents, to stabilize the nanoparticles and prevent agglomeration. Blue LED light was employed to assist the reduction process, ensuring a homogeneous distribution of nanoparticles across the biochar surface. This light-assisted synthesis method is not only energy-efficient but also promotes the formation of smaller, well-dispersed nanoparticles, which are advantageous for both catalysis and antibacterial applications.

The catalytic potential of the silver-loaded biochar was assessed by testing its ability to reduce methyl orange, a common azo dye used in industrial applications. The reduction of methyl orange was chosen as a model reaction to evaluate the effectiveness of the biochar-AgNP composite as a catalyst. The results demonstrate that the biochar impregnated with smaller nanoparticles, specifically those synthesized with the lowest concentration of avocado seed extract, exhibited the highest catalytic activity. Nearly 95% of the methyl orange was reduced within five minutes, indicating that the material is highly effective in promoting the breakdown of azo dyes. This finding highlights the potential of the synthesized biochar as a sustainable catalyst for wastewater treatment, where efficient dye removal is crucial.

In addition to its catalytic properties, the biochar-AgNP composite was tested for antibacterial activity against three bacterial strains: *Staphylococcus aureus*, *Escherichia coli*, and *Pseudomonas aeruginosa*. These strains were selected to represent both Gram-positive and Gram-negative bacteria, providing a comprehensive assessment of the material's antibacterial capabilities. The results showed a significant antibacterial effect against *Staphylococcus aureus*, a Gram-positive bacterium, with an inhibition zone measuring 19.64 ± 0.247 mm for the sample synthesized with the lowest concentration of avocado seed extract. This is particularly noteworthy, as *Staphylococcus aureus* is a common pathogen responsible for various infections, and the ability to inhibit its growth suggests potential medical applications for the biochar-AgNP composite.

However, the material did not exhibit antibacterial activity against the Gram-negative bacteria *Escherichia coli* and *Pseudomonas aeruginosa*. This outcome can be attributed to the structural differences between Gram-positive and Gram-negative bacteria. Gram-negative bacteria possess an additional outer membrane that acts as a barrier, making them more resistant to silver nanoparticles. The selective antibacterial activity observed in this study suggests that the biochar-AgNP composite may be more suitable for targeting Gram-positive pathogens, though further research is needed to enhance its effectiveness against Gram-negative bacteria.

The ability to control the size of the nanoparticles by adjusting the concentration of the avocado seed extract was a key aspect of this study. Smaller nanoparticles offer a larger surface area for interaction with bacterial cells and catalytic reactions, making them more effective in both applications. The sample with the smallest AgNPs, produced using 0.15 g of avocado seed extract in 50 mL of solution, consistently outperformed the other samples in terms of both catalytic and antibacterial activity. This underscores the importance of nanoparticle size and dispersion in determining the functionality of the material.

In addition to the size and dispersion of the nanoparticles, the structural integrity of the biochar itself was preserved throughout the synthesis process. The porosity and surface area of the biochar remained largely unaffected by the impregnation of silver nanoparticles, which is essential for maintaining its adsorptive properties in catalytic and antibacterial applications. The textural analysis confirmed that the biochar retained its mesoporous structure, with a specific surface area of 5.97 m^2/g for the sample with the smallest nanoparticles. This ensures that the material can effectively interact with both contaminants and bacteria in practical applications.

The silver nanoparticles were thoroughly characterized using a range of analytical techniques, including scanning electron microscopy (SEM), energy dispersive X-ray spectroscopy (EDS), and X-ray diffraction (XRD). SEM images revealed the porous structure of the biochar and confirmed the uniform dispersion of silver nanoparticles on its surface. EDS analysis further validated the presence of silver, along with other elements naturally occurring in rice husk biochar, such as silicon, calcium, and potassium. XRD spectra indicated the successful reduction of silver ions to zero-valent silver, with characteristic diffraction peaks corresponding to crystalline silver nanoparticles. These results confirm the effectiveness of the green synthesis method in producing well-defined AgNPs on biochar.

One of the key advantages of this synthesis approach is its scalability and environmental sustainability. By utilizing agricultural waste as both the source of biomass for biochar and the reducing agent for silver nanoparticles, we have developed a process that minimizes waste and reduces the need for harmful chemicals. The use of LED light as an energy source for nanoparticle formation further enhances the eco-friendliness of the method, as it reduces energy consumption compared to traditional thermal methods. This green synthesis technique could be applied to other metals, such as gold, copper, and zinc, broadening the potential applications of biochar in catalysis, environmental remediation, and medicine.

The findings of this study suggest several promising avenues for future research. While the biochar-AgNP composite demonstrated significant antibacterial activity against Gram-positive bacteria, further work is needed to improve its efficacy against Gram-negative strains. This could involve modifying the synthesis process to produce smaller nanoparticles or incorporating other metals with known antibacterial properties. Additionally, the long-term stability and reusability of the biochar-AgNP composite should be investigated, particularly in real-world applications such as wastewater treatment and medical devices.

In conclusion, we have developed a novel, green synthesis method for producing biochar impregnated with silver nanoparticles using avocado seed extract and blue LED light. The resulting material exhibits excellent catalytic activity in the reduction of methyl orange and significant antibacterial activity against *Staphylococcus aureus*. This eco-friendly approach provides a sustainable alternative to conventional synthesis methods and offers potential applications in a range of industries, including wastewater treatment, environmental remediation, and healthcare. The ability to control nanoparticle size and dispersion through simple adjustments in the synthesis process further enhances the versatility of the biochar-AgNP composite, making it a promising candidate for future development in sustainable material science.

Reference:

Moreno, E., Gualle, A., Vizuete, K., Debut, A., Orejuela, L., & Ponce, S. (2024). Engineered Biochar: Novel Silver Loading onto Biochar Using Agri-waste Extract Assisted by LED Light for Catalyst and Antibacterial Capabilities. *Waste and Biomass Valorization*, 1-10.

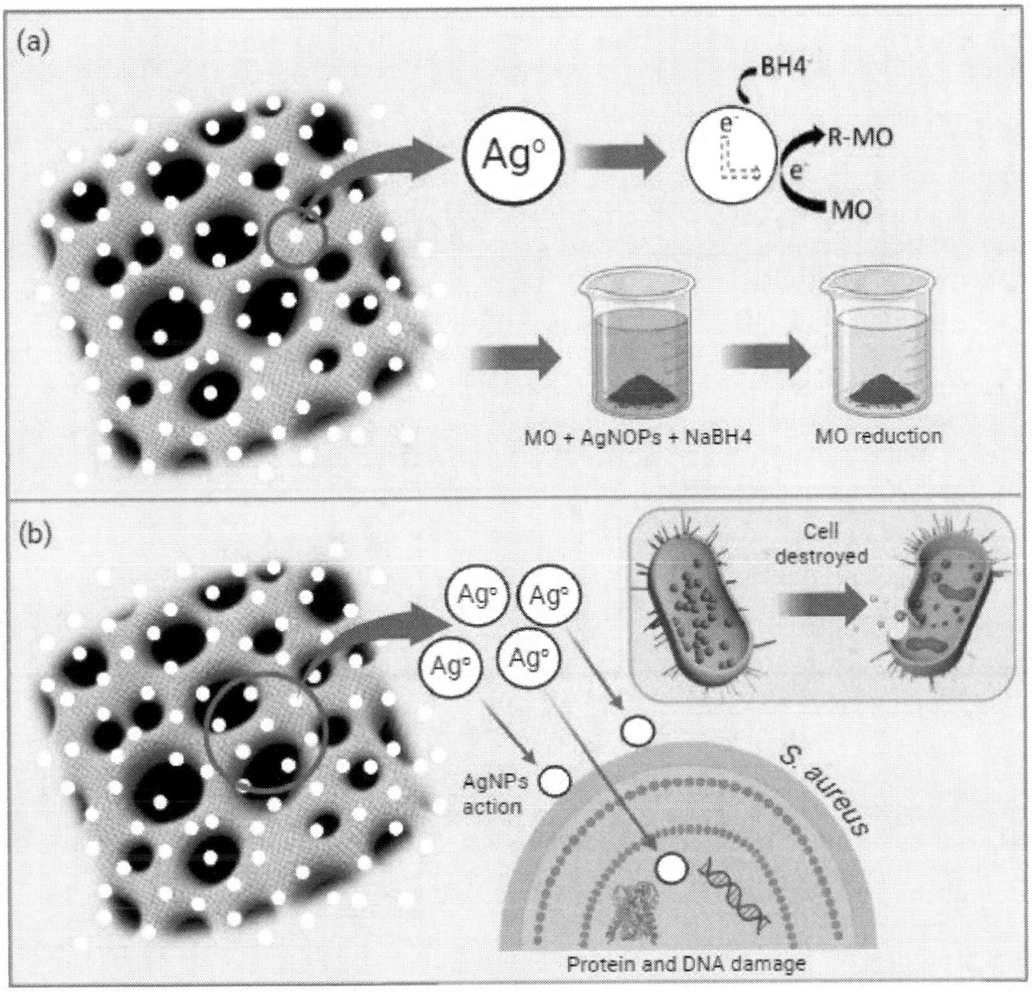
(a)
BH4⁻
Ag°
e⁻
R-MO
e⁻
MO
MO + AgNOPs + NaBH4
MO reduction
(b)
Ag°
Ag°
Ag°
Ag°
Cell destroyed
AgNPs action
S. aureus
Protein and DNA damage

Quantum Chemical Effects at the Iron Electrode in an Iron-Air Battery

Keyan O'Donnell

Iron-air batteries are a promising solution for mass grid storage due to their long-term energy storage stability and ultra-low material cost. However, the hydrogen evolution reaction (HER) during the charging cycle compromises battery efficiency (by reducing round -trip efficiency) and longevity. Modifying iron electrodes with sulphur presents a cost effective strategy to reduce the HER, leveraging sulphurs potential to alter the electronic environment of iron surfaces. This study employs state-of-the-art atomistic calculation methods, specifically the Solvated Jellium Method (SJM) developed in the Peterson lab, to investigate sulphur's impact on iron electrodes. In doing this, the viability of sulphur as an 'anti-catalyst' in reducing the material's work function will reduce the parasitic HER reaction to a commercially viable degree is studied. By advancing iron-air battery performance, this research promotes the integration of renewable sources into the grid, fostering a cleaner and more resilient energy infrastructure.

Low-Temperature and Low-Pressure Reaction for Unit Operations Laboratory Experiment

Andrew Sander, Gregory Neumann

We developed a Standard Operating Procedure (SOP) for the packed bed reactor at Rose-Hulman Institute of Technology for use in the Unit Operations Laboratory courses. Multiple reactions that would be possible under the conditions of the reactor were researched and then tested in the reactor. The limits of the reactor were temperatures less than 100°C and pressures less than 100 psig, with the reaction also being safe for undergraduate use. After testing the reactions and gathering example data, an SOP for that reaction and a project brief for faculty use was created. The first reaction that was tested was the combination of acetone and glycerol in an ethanol solvent while the reactor was packed with Amberlyst-15 acidic ion exchange resin to produce Solketal. The results from this reaction provided an average of 26% yield of the product with less than 2% deviation. Reaction conditions were changed to determine effects on conversion and yield. These tests resulted in expected trends based on reaction fundamentals and literature.[1,2] The second reaction that was tested was the dehydration of fructose over the same Amberlyst-15 catalyst to produce 5-hydroxymethylfurfural. The baseline results found were an average of 7.64% yield of 5-HMF product with a 2.01% deviation, while the average fructose conversion was 23.74% with a 3.76% deviation. To test the variability of this reaction, we changed two different points, the flowrate and the concentration of fructose. Testing the lower flowrate, we found that it converted nearly all of the fructose, but only produced a 5.50% yield of product. The results from increasing the concentration of fructose gave a lower yield of product, but overall, it produced slightly more moles than the original concentration experiments. One setback of this reaction with our setup is the temperature requirements. Our reactor uses a water bath heater, so we are unable to exceed 100°C. Due to heat loss across the reactor and tubing, the reactor struggled to reach 80°C, which hindered the conversion of fructose into the product. Another cause for the low yield of the product is its rehydration into levulinic acid and formic acid. We were able to realize the potential of Amberlyst-15 as a catalyst in the liquid-phase packed bed reactor for multiple reactions. These reactions were verified to meet the needs for the Unit Operations Laboratory.

References

1. Corrêa, Isabella, et al. "Continuous valorization of glycerol into solketal: From the fixed-bed adsorptive reactor to the simulated moving-bed reactor." *Industrial & Engineering Chemistry Research*, vol. 61, no. 11, 11 Mar. 2022, pp. 4017–4030, https://doi.org/10.1021/acs.iecr.1c04832.
2. Moreira, Miguel N., et al. "Solketal production from glycerol ketalization with acetone: Catalyst selection and thermodynamic and kinetic reaction study." *Industrial & Engineering Chemistry Research*, vol. 58, no. 38, 5 Sept. 2019, pp. 17746–17759, https://doi.org/10.1021/acs.iecr.9b03725.

Pre-Treatment of Used Cooking Oils and Further Epoxidation Using Amberlite IR-120 As Catalyst

Juan Chitiva Arteaga, Alvaro Orjuela, Laura R. Conde

Used cooking oils (UCOs), inevitable byproducts of food preparation in households, institutions, or industries, present disposal challenges as inedible materials. Predominantly composed of triglycerides, partial glycerides, and fatty acids, these oils retain residual triglycerides from virgin vegetable oils or fats with unsaturations that are active groups for further epoxidation. Oleochemical epoxides serve as essential components in epoxy resins and polyurethanes, offering an eco-friendly alternative to traditional first-generation epoxidized vegetable oils and mitigating UCOs mismanagement impacts. Despite the potential, exploitation of waste materials via heterogeneously catalyzed processes involves major challenges mainly related to deactivation of catalysts caused by the impurities in UCO. These impurities can adversely affect the conversion to oxirane ring, leading to lower yields and less efficient reactions. Studying the effects of these impurities on the epoxidation process is crucial to understand their impact and determine how they influence the overall kinetic performance and utility of UCOs in industrial applications. Then, this study aims to evaluate the pre-treatment process required to obtain a suitable UCO as raw material for further epoxidation using an ion exchange resin as catalyst (Amberlite IR-120) and to conduct a preliminary kinetic evaluation of epoxidation with in situ formed peracetic acid (i.e. Prileschajew reaction). Experimental procedures included the characterization and conditioning of UCO sourced from restaurants in Bogotá, Colombia, through filtration and neutralization. Subsequently, epoxidation tests were conducted to evaluate the effects of stirring speed and catalyst particle size to determine the adequate conditions for assessing the intrinsic kinetics of the reaction. Kinetic tests were carried out under recommended conditions and the conversion profile was analyzed to observe the performance of the resin in the presence of pre-treated UCO. The kinetic profiles were compared with those obtained using a refined oil feedstock of similar characteristics to advance in the understanding of the challenges of using second-generation raw materials for the production of oleochemical epoxides.

Automated Testing Platform for High-Throughput Electrochemical Catalyst Characterization

Shane Fuentecilla

One of the main bottlenecks in conventional combinatorial electrochemical catalyst characterization methodologies is the speed at which they can be screened. Incumbent methods also suffer from significant potential for human error. Here we describe an automated electrochemical testing platform allowing for the rapid screening of electrochemical catalysts. The platform is based around a customizable flow cell, which is positionally controlled via a low-cost XYZ control platform. The flow cell electrolyte is automatically replenished through a series of pumps and interfaced with a Biologic SP-50e potentiostat for data collection. Components of the platform are coordinated and controlled with a script written in Python, where all parameters can be managed through a unified user interface. This system was successfully validated through the testing and characterization of mixed-metal oxide catalyst OER activity for seawater electrolysis.

Stability of Doped NiO Catalysts for Ethane Oxidative Dehydrogenation

Ram Anthony Del Prado, Alyssa Hensley, Shuqiao Wang

Ethylene is a critical chemical building block in the petrochemical industry, while global ethylene consumption is expected to grow 6.1% every year to more than 200 million metric tons in 2020. However, the conventional industrial production method, steam cracking, is highly energy-intensive and leads to severe coke formation. Meanwhile, NiO-based materials have emerged as promising catalysts for an environmentally friendly alternative through ethane oxidative dehydrogenation (EODH), and the catalytic performance can be significantly improved by doping NiO with various elements (M-NiO).

To understand the nanoscale effects of dopant on the stability of M-NiO during EODH, the highly active (110) facet is investigated using density functional theory and ab initio phase diagrams. A systematic screening is conducted, examining simultaneous factors of dopant element (Al, Mo, Nb, Sn, Ti, V, W, Zr), dopant placement (surface or subsurface), oxygen vacancies (surface or subsurface), and surface oxygen species (O* or O2*) adsorption. The result reveals the dominant (most stable) structures for various M-NiO(110) systems under environmental conditions varying oxygen presence and temperature. Overall, this approach not only identifies the dominant structures during the EODH process but also provides a framework to accelerate experimental screening of M-NiO catalysts during working conditions.

Spin State of Single-Atom Alloy Catalysts and Its Effect on Chemical Properties

Collette Riviere, Shengjie Zhang, Matthew Montemore

Catalysts are crucial for developing energy efficient chemical processes and clean energy technologies. Single-atom alloys (SAAs), which are bimetallic alloys where one metal is introduced as isolated atoms into another metal, have emerged as promising catalysts for a number of important reactions. However, while the spin state (i.e., magnetism) of SAAs has been shown to play an important role in influencing their reactivity, this effect is not well understood. Thus, to harness the full potential of SAAs and effectively perform rational design, it is imperative to understand how their chemical properties are affected by their spin states. More broadly, SAAs exhibit unique electronic structures and spin states, making them excellent platforms for fundamental studies of the relationship between magnetism and chemical reactivity. Therefore, this research will involve a comprehensive computational examination of hydrogen adsorption on SAAs at various spin states, to determine the extent of spin's influence on chemical reactivity and the underlying causes behind it. Additionally, we will investigate why some SAAs exhibit nonzero spin states while others do not. This strong understanding of spin's role in SAA reactivity will aid in tailoring SAAs as catalysts for a wide range of applications.

Accelerating in Situ CO2 Mineralization: Detailed Exploration of Kinetics, Reaction Mechanisms, and Mineral Product Formation

Andrew Schwartz

CO2 mineralization offers a promising approach for long-term carbon capture and storage by converting CO2 into stable carbonate minerals. This study explores the optimization of CO2 mineralization under varying chemical conditions to enhance reaction kinetics and maximize carbonate formation. The use of amine-based accelerators, such as glycine, was investigated to increase reaction rates and improve the efficiency of mineral trapping processes. Experimental work involved systematically varying calcium chloride and accelerator concentrations, as well as examining the effects of temperature and solvent composition on reaction outcomes.

The results demonstrated that glycine not only accelerates CO2 mineralization but also acts as a pH buffer, improving the overall absorption efficiency of CO2. Analytical techniques including XRD, FT-IR, and TGA were employed to characterize the mineral products, confirming the formation of calcium carbonate in polymorphs such as calcite, aragonite, and vaterite. This work highlights the potential of amine-accelerated mineralization as a scalable and energy-efficient solution for CO2 sequestration, offering insights into improving the feasibility of mineral-based carbon capture technologies.

Computational Modeling of Catalytic Pathways for Nitric Oxide Reduction: Evaluating the Efficiency of $PtCu_3$ and $NiCu_3$ Catalysts

Nicole Szponar

Nitrogen oxides (NO_x), particularly nitric oxide (NO), are major contributors to environmental issues such as acid rain deposition and ozone layer depletion. The NO reduction (NOR) reaction is therefore essential for mitigating these harmful effects; also, if NO can be fully reduced to ammonia (NH_3), it would provide a route to this critical chemical feedstock. In this study, computational modeling was used to analyze the performance of different catalysts for NOR, and the thermodynamics of different competing pathways were examined. Using density functional theory (DFT), the structures of all NOR to NH_3 reaction intermediates and their Gibbs free energy were calculated, allowing for identification of the most thermodynamically favorable NOR to NH_3 pathways.

The nanoparticle catalysts Cu_4 and Pt_4 were first examined, with Cu4 demonstrating more favorable thermodynamics for NOR. Next, $PtCu_3$ and $NiCu_3$ nanoparticles were examined for potential improvement, as bimetallic catalysts can show synergistic effects. $PtCu_3$ demonstrated improved catalytic activity compared to individual Cu_4 and Pt_4 structures, because it maintained a steady decrease in energy across steps, making the NOR process more favorable. In contrast, Cu_4 and Pt_4 displayed sharp energy spikes, or barriers, though Cu_4 performs better than Pt_4 because it had only one of these barriers (NO $\rightarrow$ N) and a fully downhill pathway (NO $\rightarrow$ NOH). Pt_4, however, has barriers in both key pathways, making the reaction less favorable. $PtCu_3$ avoids these entirely, resulting in smoother energy transitions and making it the most efficient catalyst for NOR.

The $NiCu_3$ catalyst also showed a favorable energy profile, as it had a significantly steeper slope in energy. This suggests it could potentially serve as an efficient catalyst for NOR to ammonia. Though it has much lower energy requirements for its pathways, $PtCu_3$ still performs better. In $PtCu_3$, all pathways starting from NO are downhill, which is ideal. In contrast, $NiCu_3$ has one downhill pathway(NO $\rightarrow$ N), but the other includes a barrier, making $PtCu_3$ more favorable overall for NOR.

This study examines the importance of selecting catalysts that not only reduce NO efficiently but also minimize the required activation energy, leading to more sustainable processes. The findings provide insights into catalyst design strategies that could help address the environmental challenges posed by nitrogen oxide emissions. By improving the catalytic performance of mixed metal systems like $NiCu_3$, there is potential for developing more effective solutions for NO removal in environmental protection efforts.

Images for Reference:

Figure 1. Graph of PtCu3 with possible pathways

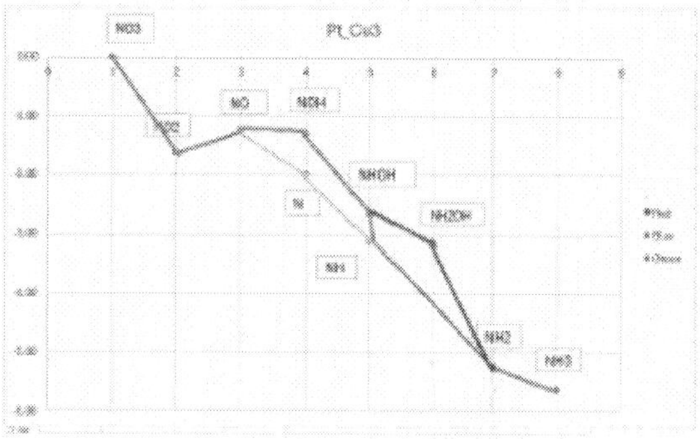

Figure 2. Graph of Pt4 with possible pathways

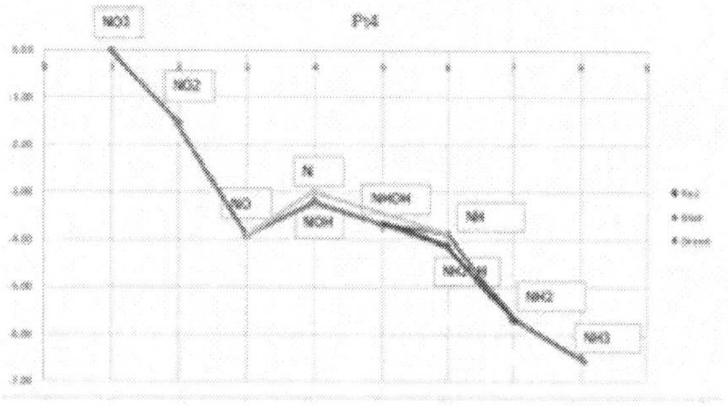

Figure 3. Graph of Cu4 with possible pathways

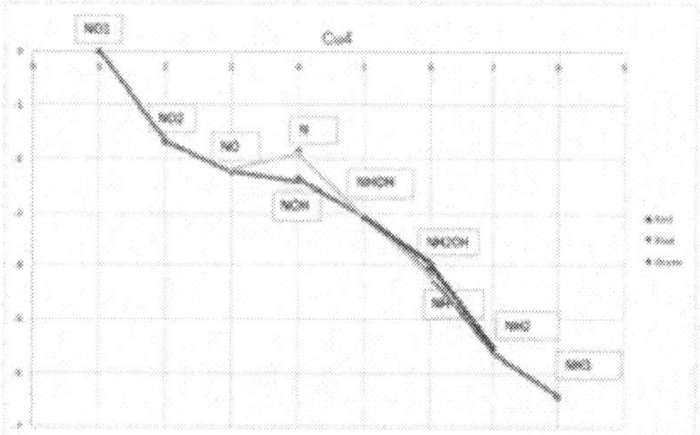

Figure 4. Graph of NiCu3 with possible pathways

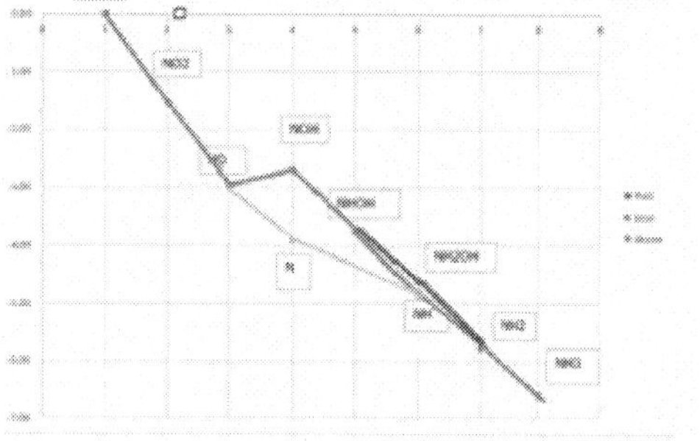

Evidence of CORE Coupling in Heterogenous Catalysts through Galvanic Current Analysis

Sam Miller

Cooperative Redox Enhancement (CORE) has emerged as a promising tool to enhance bimetallic catalytic activities in liquid-phase heterogeneous catalysis. Unlike conventional alloying, CORE occurs when catalyst metals are physically separated, polarizing each metal to either the oxidation or reduction side of the reaction concerned. It has been demonstrated that through this physical separation, the catalytic activity of the precious metal catalysts Au and Pd for the oxidative dehydrogenation(ODH) of 5-hydroxymethylfurfural (HMF) is increased more than when individual activities in separate monometallic catalysts are added. This has been confirmed in both electro- and thermochemical experiment setups. (***Catal. Sci. Technol.***, 2023,**13**, 47-55).

Even though the mechanism of CORE has been uncovered in recent articles, there is no direct evidence of electron flow between the two metals. Galvanic coupling tests, in which two differing monometallic catalysts that show CORE when coated bimetallically are immersed in the same solution on different electrodes and left to pass current between them, show electrochemical polarization between the active sites of each catalyst. Herein, we show clear evidence of the electron flow via galvanic coupling using ODH of HMF as a model reaction.

Controlling Single-Atom Sites Using Phosphonic Acids

Briella Riley, Zachary W. Meduna, J. Will Medlin

Olefin hydroformylation is a significant homogeneously catalyzed industrial process of aldehyde synthesis. Despite the efficiency of homogeneous catalysts, their recovery is a major and costly challenge, ultimately leading to the net loss of finite, expensive noble metals. Single-atom catalysts, composed of an atomically dispersed noble metal across a metal oxide support, combine the uniformity and defined active sites of homogeneous catalysts with the advantageous catalyst recovery of heterogeneous catalysts. This project was focused on the fundamental aspects of controlling single-atom sites using phosphonic acid self-assembled monolayers (SAMs) to steer the product selectivity of ethylene hydroformylation and hydrogenation. Phosphonic acids with differing tail group functionalities were deposited onto an atomically dispersed rhodium on titanium oxide catalyst. Catalyst surfaces were characterized by diffuse reflectance infrared Fourier transform spectroscopy (DRIFTS) both before and after CO adsorption to confirm the success of the phosphonic acid depositions and to probe rhodium site availability following SAM deposition. The simultaneous hydroformylation and hydrogenation of ethylene was utilized as a gas-phase probe reaction to evaluate the ability of phosphonic acid SAMs to steer single-atom rhodium product selectivity. Findings from this study show that, in comparison to the uncoated single-atom rhodium catalyst, the presence of phosphonic acid SAMs on the catalyst surface increased product selectivity of hydroformylation while decreasing that of hydrogenation despite some apparent rhodium site blocking caused by the SAMs. Additionally, it was found that aliphatic versus aromatic phosphonic acid SAMs, while causing similar rhodium site blocking, resulted in differing product selectivity, displaying the ability of phosphonic acids with varying tail group functionalities to influence single-atom rhodium sites.

Controlling Selectivity in Reactions for Sustainable Hydrogen Storage in Liquid Carriers

Genevieve Ansay, Matthew Edgar, Sara Ahsan, Siddarth Krishna

In order to fully integrate intermittent renewable energy sources into our energy infrastructure, new technologies for long-duration storage and transport of renewable energy are urgently needed. Hydrogen (H_2), produced from renewable electricity via water electrolysis, is an energy carrier that can aid in decarbonizing numerous industries, but storing pure hydrogen requires high pressures (>200 bar) and/or cryogenic temperatures (-250 C). One possible solution is liquid organic hydrogen carriers (LOHCs), molecules which can be reacted with hydrogen to produce hydrogen-rich counterparts that can be stored at ambient conditions, then later dehydrogenated to recover H_2. Viable LOHCs require moderate reaction energies and effective catalysts for the hydrogenation and dehydrogenation reactions to enable reversible H_2 storage. Past research indicates that nitrogen-containing heterocyclic aromatic molecules such as 'indoles' (Fig. 1A) can both store and release H_2 at mild conditions (< 200°C).[1]. One catalyst that has shown promise for nitrogen containing LOHCs are Pd nanoparticles anchored to high-surface-area, thermally stable metal oxide supports.[1] Reaction selectivity is of key importance because the LOHC must be essentially fully recyclable for subsequent storage cycles. However, the effects of LOHC molecular structure and catalyst surface structure on reaction selectivity remain poorly understood.

To address this, dehydrogenation reactions of 8H-Indole and 8H-N-Methylindole (8H-NMID) to indole and N-methylindole (NMID) were investigated with catalysts of Pd on alumina and silica supports. The reaction pathway for 8H-indole is shown in Figure 1A. Reactions were performed in a high-pressure batch reactor system for 3 hours at 180°C with a feed of 1% 8H-indole or 8H-NMID in dodecane. Products were identified using gas chromatography-mass spectrometry (GCMS).

Fig 1B shows that for two different LOHCs (indole, NMID), catalysts of Pd supported on SiO_2 or Al_2O_3 are active for dehydrogenation, forming a combination of partially and completely dehydrogenated products. However, the sum of product yields did not equal the observed reactant conversion, suggesting the presence of side-reactions (up to 40% losses for indole conversion over Pd/SiO_2). Subsequent control experiments mixing catalyst and reactant solution at ambient conditions revealed that silica and alumina supports were adsorbing indoles, possibly due to hydrogen bonding with OH groups on support surfaces. This accounted for a significant fraction of missing carbon, as displayed in the predicted carbon absorption bar of Figure 1B. A carbon balance is the percent carbon in a reaction feed also detected in the output. Since carbon atoms are not destroyed, values under 100% indicate undetected carbon products.

The extent of adsorption differs between 8H-indole and 8H-NMID (95 mg 8H-NMID vs 130 mg 8H-indole per gram SiO_2), suggesting steric hinderance from the 8H-NMID methyl group suppresses adsorption. These results show that improving selectivity in catalytic LOHC reactions requires tailoring LOHC structure and support properties to avoid parasitic processes such as support adsorption, motivating further research into the molecular factors governing LOHC-support interactions.

(1) Li, Y.; Guo, X.; Zhang, S.; He, Y. A Perspective Review on N-Heterocycles as Liquid Organic Hydrogen Carriers and Their Hydrogenation/Dehydrogenation Catalysts. *Energy Fuels* **2024**, *38* (14), 12447–12471. https://doi.org/10.1021/acs.energyfuels.4c01633.

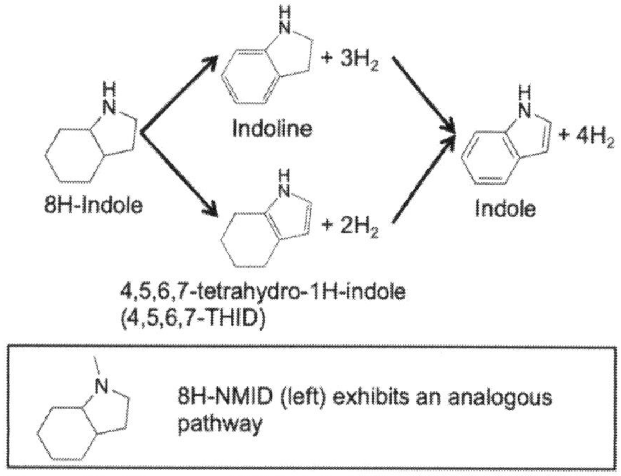

Figure 1A. Diagram of 8H-indole to indole dehydrogenation pathway and of 8H-NMID.

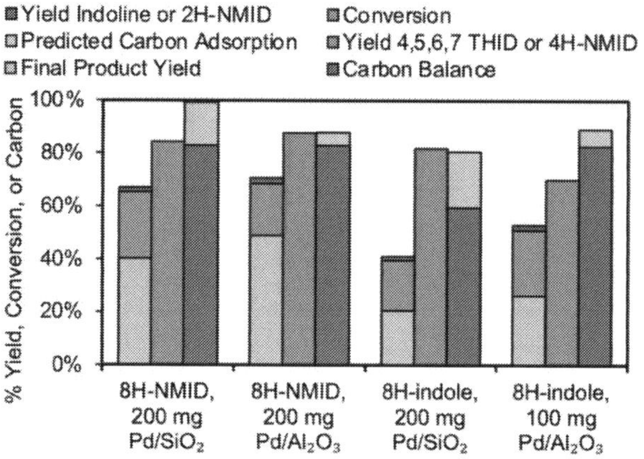

Figure 1B. Batch reaction studies of N-heterocycle dehydrogenation after 3 hours at 180 °C, with catalyst metal weight percent of 0.54-0.64%.

Photocatalytic Methane Coupling to Ethane over Heteropoly Acid Based Catalysts

Jack Sullivan, Nan Yi

Catalytic conversion of methane to more valuable chemicals, particularly fuels, is a promising process that offers economic incentive and environmental benefits. Industrial methane conversion currently faces multiple challenges, notably, high energy being required to maximize methane conversion. However, high reaction temperature leads to a random distribution of desired and undesired products. This study investigated a photocatalytic approach to convert methane to ethane and hydrogen through gold-modified heteropoly acid catalysts. By varying the morphologies of gold species, choosing different wavelengths, and altering pretreatment conditions, methane transformation was achieved as well as controlled selectivity to products, ethane and hydrogen. This work makes it feasible to develop a viable and economic approach to mitigate the environmental impact of methane while being a competitive source of revenue.

Optimizing Tungsten Powder Fluidization: Applications for Atomic Layer Deposition

Hermann Klein-Hessling Barrientos, Davis R. Conklin, Alan W. Weimer

Tungsten is a refractory metal which possesses unique characteristics: high melting point, high density, a low thermal expansion coefficient, high tensile strength, high thermal conductivity and favorable corrosion resistance. This set of attributes makes it ideal for high-temperature applications in hypersonics, cermet fuel elements in nuclear thermal propulsion and advanced radiation shielding. Traditional powder metallurgical methods of manufacturing tungsten parts do not scale well with highly customizable and complex geometries. Emerging alternatives like additive manufacturing (AM) methods of powder bed fusion, binder jetting and material extrusion, face limitations such as cracking and difficult post-processing. Introducing metal additives allows for favorable fine-tuning of enhanced properties to withstand these demanding applications. Incorporating metal additives is accomplished via atomic layer deposition (ALD), a gas-phase chemical process that utilizes alternating reactions to deposit conformal thin films upon a substrate. ALD is ideal for refractory metals, as melting them is impractical and energy intensive. Powder ALD typically relies on a Fluidized Bed Reactor, where the minimum fluidization velocity is the critical threshold for effective powder fluidization.

This work investigated the fluidization of several different sizes of tungsten powder: 1–5-micron, 12-16 micron and 50-70 micron. Fluidization of incredibly dense powders such as tungsten (19300 kg/m^3) proves difficult as the powder does not fall within the conventional Geldart classifications of particle diameter and densities. Literature on the subject is incredibly scarce, especially for particles that are smaller than 50-70 micron in size. The range explored was chosen based on previous literature and allowed for direct comparisons of fluidization behavior. While the smaller size range of 1-5 micron and 12-16 micron is most suitable for additive manufacturing applications and has no existing literature on its fluidization regimes or behavior. This work has compiled different thresholds of fluidization of this powder size with both mechanical agitation and the lack thereof. In addition to this, the effects of temperature during fluidization and that of powder bed height relative to reactor diameter (H/D) were investigated.

(H/D) ranges of 0.8-1.4 were explored and suggested that greater ratios resulted in more uniform fluidization with smaller bubble formation, greater pressure drops were also observed alongside greater minimum fluidization velocities. A temperature range of 200-400 °C was selected to account for the great majority of chemical precursors used in ALD for refractory metals. As predicted by the Ergun equation one would expect increased fluidization temperature to result in increased minimum fluidization velocities because of increased gas viscosity.

The findings on fluidization thresholds, temperature effects, and the impact of bed height (H/D) ratios extend our understanding of how to manipulate the processing conditions for more uniform powder dispersion, which is essential for enhancing the performance and precision of tungsten parts in high-temperature, high-stress applications. Importantly, the work addresses a gap in literature for finer tungsten powders (1-16 micron), which are pivotal for AM, while also laying the groundwork for the development of more efficient ALD techniques. Ultimately, these insights could advance the scalability and reliability of tungsten-based materials for aerospace, nuclear propulsion, and radiation shielding, enabling their use in increasingly demanding environments where traditional manufacturing methods fall short.

Grid-Connected Battery Energy Storage Systems for Power Modulation: A Holistic Assessment Framework

Mary Schilling, Ashley McCullough, Styliani Avraamidou

The United States currently relies on fossil fuels for energy production, as large-scale renewable energy use is not conducive to the current power grid infrastructure. The most prominent renewable sources—solar and wind—stochastically produce power and challenge the maintenance of constant frequency, and the matching of supply and demand on the electricity grid. As a result, more than 2,000 GWh of energy was curtailed by independent system operators (ISOs) in 2022 [1]. Therefore, energy must be stored during times of overproduction and sold back to the grid during times of underproduction and high demand to eliminate the need for fossil fuels. Battery Energy Storage Systems (BESSs) arise as an attractive grid supplement that may manage this stochastic power because of their storage performance, ability to modulate power, and ease of development compared to other Energy Storage Systems (ESSs). While different types of BESSs are currently being developed, estimating the economic and environmental impacts of their integration into the electricity grid becomes a challenge [2]. This work aims to address and balance the profitability and environmental impact of integrating different types of BESSs—such as Sodium-ion batteries (SIB) [3] and lithium-ion batteries (LIB) [4]—into the grid. To assess economic performance of BESSs in power markets, optimal bid-in strategies for the day-ahead market, real-time market, and frequency regulation market were found for all ISOs in the US. To assess environmental performance for the same BESSs, a lifetime cycle assessment (LCA) cradle-to-grave framework for power-modulating systems was created to quantify the environmental impact that the integration of batteries into the grid with the optimal bid-in strategies may have. Initial economic feasibility results highlight that participation in day-ahead markets can generate \$100,000-\$400,000/year/MWh capacity depending on the given ISO. The LCA framework will be applied to example SIB and LIB systems, and results will show a holistic analysis of the implementation of BESSs as power-grid supplements for sustainable energy retention.

Logical Proofs about the Genetic Code in Lean 4

Colin Jones, Tyler Josephson

The genetic code is at the center of biological engineering and modifying the genetic code allows us to regulate gene expression to generate biologics or engineer specific proteins. Genomic software facilitates this process by processing large datasets beyond the cognitive abilities of people. However, genomic software is vulnerable to human error during its translation from pseudocode to code. Eliminating bugs is pertinent to keeping the software robust and "correct". Traditionally, debugging has been done through writing up test cases and quizzing the software to see if it outputs the desired response. We propose a new approach through rigorous, mathematical proof. Test cases are vulnerable to human mistranslation like code, but proofs logically follow from strict axioms embedded within the software.

Lean 4 is a functional programming language that also facilitates computer-checked math proofs. By thinking of the genetic code as a function that maps 4^3 codon sets to 20 amino acids, we can define several properties about itself and prove them. Using Lean 4, we prove the redundancy and non-overlapping nature of the genetic code as a minimal working example for the uses of provable functions to biological engineering.

CFD Analysis of Mixing in the Transition Regime: User-Defined Functions and Viscosity Variations

Julesy Baker, David Foster

Mixing is a common process used in the manufacturing and processing industry. To optimize the mixing efficiency, it is critical to understand what is going on inside the tank. In a physical mixing tank, several flow regimes exist simultaneously. At the impeller, turbulent flow is likely to exist. In the bulk and the surface of the tank, the fluid can be in laminar flow. Between the two extremes, there will be a wide range of Reynolds Numbers, all falling into the transitional flow regime. The process for industrial mixing is typically modeled in the laminar or turbulent regimes, as the equations governing these regimes are well understood. [1-2] Due to the complexity of transition mixing, there is minimal modeling done in the regime despite it being a common working regime for many industries. The lack of accurate transition mixing models hinders these industries in fully understanding their processes. [1] Computational Fluid Dynamics (CFD) is used in modeling various physical systems to analyze velocity profiles, streamlines, volume fraction distribution, and numerous other fluid and flow properties.

A user-defined function is a C program function that can be dynamically loaded into CFD software packages, such as ANSYS Fluent. A UDF serves to enhance the standard features of code already apparent inside the ANSYS fluent solver code. In this case, a UDF was written to redescribe the shear-thinning viscosity behavior of the fluid, Carbopol, in the transition regime based off the Carreau-Yasuda model for non-Newtonian fluids. Carbopol is a non-Newtonian, shear-thinning fluid whose viscosity profile can be altered with change in pH and/or concentration. Two different viscosities, 7500 cP and 3350 cP, were used to investigate their dependence on flow patterns. The UDF was incorporated into the two different viscosities with two different ANSYS fluent models: The Spallart-Allmaras Model and The Transition Shear-Stress Turbulence Model. Different mixing regimes were achieved by changing the speed of the impeller. Without the UDF, both RANS models struggled to characterize the flow and shear-thinning viscosity behavior for Reynold's numbers that existed within the transition regime. The UDF worked to better represent the dispersion of dye and the shear-thinning of the Carbopol throughout the entire mixing tank.

References

[1] Oldshue, J.Y. "A Guide to Fluid Mixing.", 2003.

[2] Kresta, Suzanne M. Advances in Industrial Mixing: A Companion to the Handbook of

Industrial Mixing. Wiley, 2016.

Unbiased Characterization and Classification of Coffee

Ryan Koes, Katsuyuki Wakabayashi

Coffee quality assessment traditionally relies on certified sensory specialists who are not easily accessible. The reliance on subjective evaluations can lead to inconsistencies and inaccuracies in coffee assessment, making it difficult to achieve reliable and standardized evaluations. This study leverages machine learning to objectively classify brewed coffee by brand, class, and roast level using physicochemical characterization. 112 brewed coffee samples from six different coffee brands ranging across commodity, specialty, and premium classifications, at varying roast levels, were subject to pH, color, turbidity, potassium ion, and total dissolved solid measurements. We utilized Support Vector Machine (SVM), Decision Tree, and Logistic Regression models to predict coffee brand, class, and roast level. SVM, enhanced by hyper-parameter tuning, achieved 100\% accuracy using 10-fold cross-validation, outperforming Decision Tree and Logistic Regression models. Key features essential to high-accuracy classification were pH, turbidity, color, and potassium ion content. Our findings demonstrate that machine learning models can accurately classify quality, brand, and roast level, offering a reliable and cost-effective alternative to traditional subjective methods. This approach may benefit manufacturers in maintaining consistent quality and aid consumers in identifying their preferred coffee profiles.

Adsorption Column Design for Solvent-Based Plastic Recycling

Priscilla Lee, Yoel Cortes-Pena, Charles Granger, Tianwei Yan, Victor M. Zavala, George W Huber

Multilayer plastic films used in food packaging are designed with several layers of polymers that improve the product shelf life, transportation, and aesthetic appeal. The presence of many polymers in multilayer plastics makes it impossible to mechanically recycle them, so they are ultimately either landfilled or burned. A potential solution to this recycling challenge is the solvent targeted recovery and precipitation (STRAP) process. In this process, each polymer layer is chemically dissolved within a selective solvent and recovered by precipitation (Fig. 1). The resins produced from STRAP can then be reused in the production of plastics.

The Chemical Upcycling of Waste Plastics center is currently developing STRAP technology to process municipal solid waste, producing plastic resins from the multilayer plastics and ethanol from the biogenic material. However, municipal solid waste contains impurities and additives like minerals and dyes that need to be separated from the solvent to improve the quality of the resin products. It may be possible to implement adsorption to purify the solvent for reuse. Adsorption is a purification process where solute particles adhere to an adsorbent through van der Waals forces (Fig. 2). An adsorption column, packed with activated carbon adsorbent, can thus remove impurities from the solvent.

The goal of this work was to investigate the potential economic and environmental impact of an adsorption process to purify the solvent. To this end, we developed an adsorption column model in BioSTEAM, an open-source software in Python for the design, simulation, and analysis of chemical production processes, and implemented the adsorption column in a STRAP model for processing municipal solid waste. We then conducted uncertainty and sensitivity analyses to determine the factors driving the economic and environmental impact (Fig. 3). The adsorption column was modeled based on first principles of mass transfer and adsorption equilibria and kinetics. Key parameters in the design of the adsorption column include the mass-transfer transfer coefficient, equilibrium constant, cycle time, solution flow rate, and solute concentration. Adsorption experiments were conducted to inform the adsorption equilibrium and mass transfer parameters.

A preliminary techno-economic analysis and life cycle assessment suggest a minimum selling price of $1.39/kg-resin and global warming potential of 38 g-eq-CO2/L-ethanol (Fig. 3). Additionally, our sensitivity analysis showed that the hydrolysis yield and processing capacity were the most impactful in the system while adsorption column parameters had less bearing. These results suggest that adsorption may be a cost-effective method to purify the solvent in the STRAP process.

References:

Azizian, S., & Setareh, E. (2021). Adsorption isotherms and kinetics. In M. Ghaedi (Ed.), *Adsorption: Fundamental Processes and Applications* (pp. 445-509). Elsevier. https://doi.org/10.1016/B978-0-12-818805-7.00011-4

Gabelman, Alan. (2017). Adsorption Basics: Part 1. *CEP*, July 2017, pp. 48-53. https://www.aiche.org/resources/publications/cep/2017/july/adsorption-basics-part-1

Seader, J. D., Henley, E. J., & Roper, D. K. (2016). *Separation process principles: With applications using process simulators* (4th ed.). Wiley.

Figure 1.

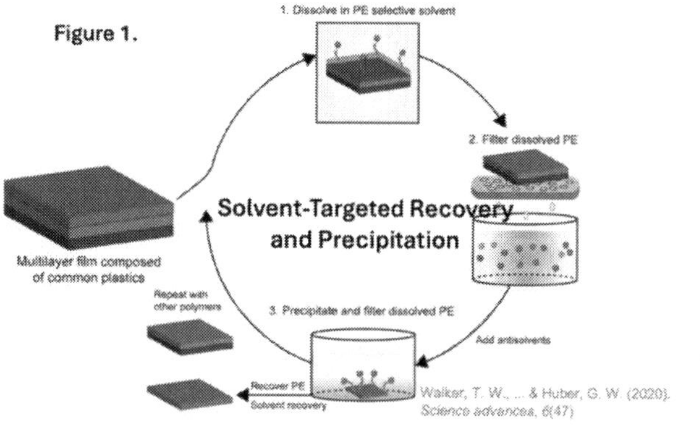

Figure 2.

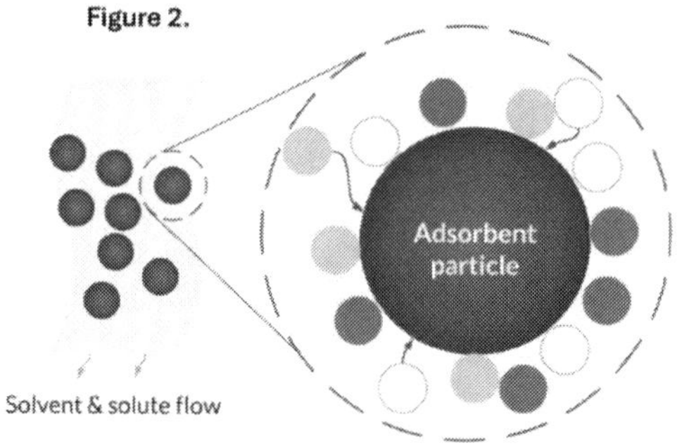

Figure 3.

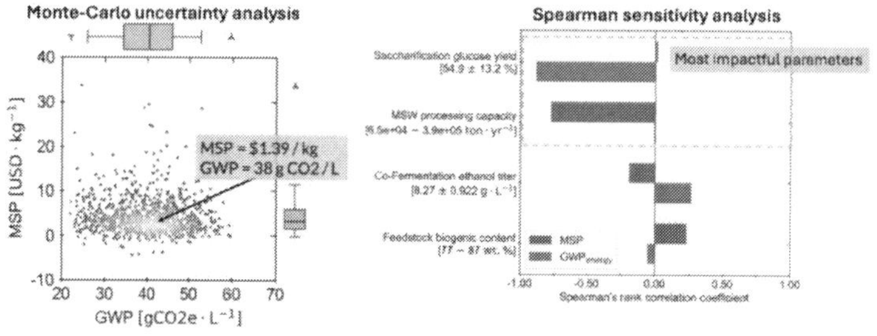

Leveraging Machine Learning for Life Cycle Assessment: Predicting Environmental Impacts and Scaling Sustainable Chemical Production

Milo Barkow, Matthew Conway, Marcella McMahon, Jahnvi Patel, Brendan Weil, Robert Hesketh, Kirti Yenkie

Life cycle assessment (LCA) is a systematic analysis of potential environmental impacts of raw materials, products, or services during their entire life cycle. Initial design stages are inherently complex and often lack comprehensive information, posing challenges for effective LCA evaluations. Machine Learning (ML) emerges as a valuable solution to address these challenges. ML algorithms, particularly Artificial Neural Networks (ANN) and XGBoost, prove effective in predicting environmental impacts of new chemicals with limited data. This study focuses on comparing ML models trained on data sets consisting of a feature set (provided inputs) and a label set (predicted outputs). The feature set is comprised of both thermodynamic properties and molecular descriptors of the chemicals. The label dataset consisted of four endpoint and twelve midpoint impact assessment metrics. These assessment metrics are used in determining a chemical's impact on human health, resource utilization, ecosystem quality, and climate change. The effectiveness of the ANN algorithm was tested using a dataset of 500 points, divided into training, testing, and validation sets.

Current and future work aims to expand this research towards technology scale-up analysis, focusing on the transition from lab-scale studies to industrial-scale production of greener chemicals. By examining case studies in sectors such as pharmaceuticals and specialty chemicals, we are collecting detailed data on costs, emissions, global warming potential, and energy consumption. Additional case studies will be simulated in programs such as ASPEN/SuperPro followed by Simapro to fill in missing data if necessary. These metrics will serve as inputs for algorithms to identify common scalability indices for environmental emissions.

Closed-Loop Materials Discovery for Multi-Principal Element Alloys

Jarett Ren, Maitreyee Sharma Priyadarshini, Eddie Gienger, Paulette Clancy

Advanced functional materials have widespread applications in energy, health, and space exploration, but their discovery is limited by a large combinatorial space that is expensive to characterize experimentally or computationally. One such material class is multi-principal element alloys (MPEAs), which are alloys consisting of three or more elements in significant amounts. These alloys are desirable for their excellent mechanical properties such as high hardness, strength, and thermal stability, but synthesizing and testing them is costly. To address this problem, we used PAL 2.0, an algorithm we previously developed that utilizes a physics-based prior in concert with Bayesian optimization to efficiently search the material space. We demonstrated the effectiveness of this method through closed-loop materials discovery of high-hardness MPEAs with experimental collaborators at the Johns Hopkins University Applied Physics Laboratory. Based on known MPEA data, we used PAL 2.0 to recommend a list of high-hardness alloys, some of which the experimentalists then manufactured and tested. Using these results, we retrained PAL 2.0's models and derived new recommendations, thereby continuing the cycle. To expand the search space beyond confirmed MPEA compositions, we developed an alloy generation algorithm based on k-means clustering to further identify novel and previously undiscovered alloy compositions with high hardness. Based on recommendations from PAL 2.0, we manufactured thirteen new MPEAs. While our initial 350-alloy dataset contained only four alloys with high hardness (Vickers hardness value > 1000), we were able to identify five more high-hardness alloys within just three feedback cycles. Among those five alloys, two had exceptionally high Vickers hardness values of 1263 and 1269. This demonstrates that, using a closed-loop approach, PAL 2.0 can greatly accelerate the discovery of new materials with optimal properties from a complex, high-dimensional material space.

Figure 1: Manufactured MPEA recommendations. Vickers hardness values are shown for each recommended MPEA that was manufactured. The color and shape of each data point denote the closed-loop cycle during which the corresponding alloy was manufactured. Two manufactured alloys with particularly low hardness were omitted.

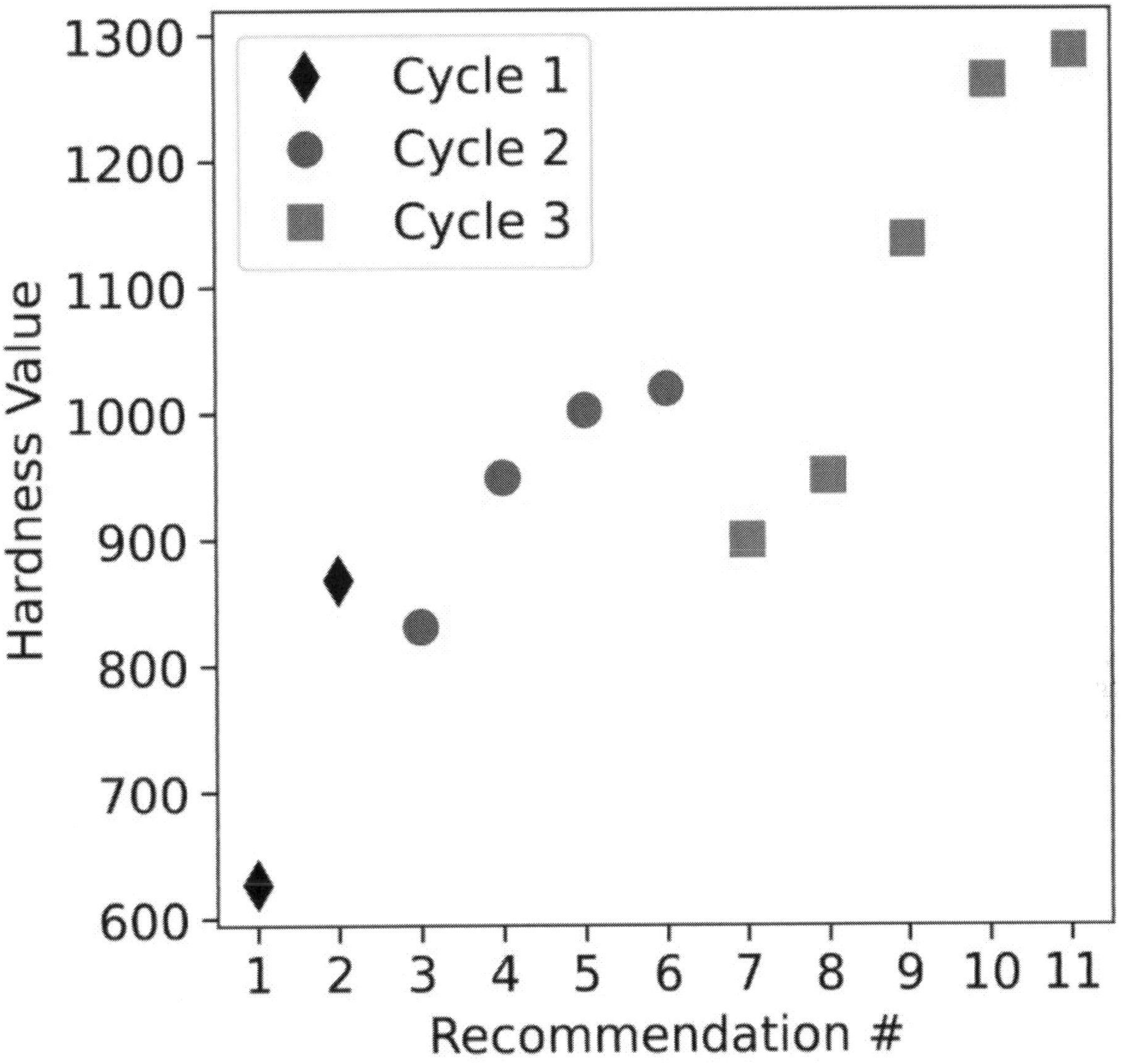

Cycle 1
Cycle 2
Cycle 3
Hardness Value
1300
1200
1100
1000
900
800
700
600
1
2
3
4
5
6
7
8
9
10
11
Recommendation #

Virtual Nodes for Zeolite Property Prediction

Nga Vu, Mingrou Xie, Rafael Gomez-Bombarelli

Zeolites are crystalline, nanoporous materials that are used in industrial applications in catalysis, separation, and ion exchange. However, their synthesis and properties are complex topics that require significant study. Recently, there have been many approaches to understand zeolite with machine learning. Task-relevant virtual nodes have been shown to enhance machine learning model robustness - that is, the model's ability to produce accurate predictions when faced with data or conditions outside its training set - by incorporating additional valuable information(1). In the context of zeolites, Voronoi nodes—points in three-dimensional space that locally maximize the distance from surrounding host atoms in a crystal structure—can serve as task-relevant virtual nodes for understanding zeolites. These nodes can represent zeolite cages—nanoscale void spaces within the zeolite's crystalline framework formed by interconnected rings of silicon and oxygen atoms. These cages are crucial for understanding the docking process between zeolites and organic structure-directing agents (OSDAs) in zeolite synthesis(2), as they determine the size and shape of OSDA that can be accommodated within the zeolite structure. These Voronoi nodes show potential for improving model robustness of predicting the binding energy between zeolite and OSDA for zeolite synthesis. This study aims to explore the effectiveness of Voronoi nodes in predicting other zeolite properties. We will utilize the zeo-1 dataset (3) to generate Voronoi nodes for zeolite structures and assess the robustness of the MACE (Higher Order Equivariant Message Passing Neural Networks for Fast and Accurate Force Fields) model(4) when trained with these Voronoi nodes for energy, force, and stress prediction. Our research aims to investigate whether Voronoi nodes can enhance our understanding of the structure-property relationships in zeolites, and offer new insights into their complex behavior and characteristics.

1. Okabe R, Chotrattanapituk A, Boonkird A, et al. Virtual node graph neural network for full phonon prediction. Nat Comput Sci. 2024;4(8):522-531. doi:10.1038/s43588-024-00661-0

2. Schwalbe-Koda D, Gómez-Bombarelli R. Supramolecular recognition in crystalline nanocavities through Monte Carlo and Voronoi Network algorithms. *The Journal of Physical Chemistry C.* 2021;125(5):3009-3017. doi:10.1021/acs.jpcc.0c10108

3. Komissarov L, Verstraelen T. Zeo-1, a computational data set of zeolite structures. Sci Data. 2022;9(1):61. doi:10.1038/s41597-022-01160-5

4. Batatia I, Kovacs DP, Simm G, Ortner C, Csányi G. MACE: Higher order equivariant message passing neural networks for fast and accurate force fields. Adv Neural Inf Process Syst. 2022;35:11423-11436.

Computational Studies of Phase Transformations in Sulfur Cathodes for Li/S Batteries

Arianny Putera, Perla Balbuena, Francisco Ospina-Acevedo, Saul Perez Beltran, Joseph Loa, Gerard Real

Lithium/Sulfur (Li/S) batteries are promising high energy density technologies because of the high theoretical specific capacity of S cathodes. However, the complex chemistry of the lithiation (discharge) reactions where Li is stored in Li-S compounds, and the de-lithiation (charge) reactions where Li leaves the Li- S compounds yielding plain sulfur, face significant challenges. Sulfur (S8) is stable at room temperature in a crystal rhombic configuration known as the α-phase. However, other S allotropes can exist as metastable phases at room temperature, such as β (monoclinic) and γ (amorphous). The monoclinic phase can become stable at higher temperatures. The transformation between stable and metastable phases is of interest because each crystalline S structure is also associated with a specific electronic configuration. Since the cathode material requires good electronic conductivity, converting the S electrode material into specific phases that may be better conductors of electrons is of high interest.

Here we use classical molecular dynamics (MD) and a temperature ramp in a low T range to investigate sulfur phase transitions at solid-solid interfaces, and in a higher T range we evaluate solid to liquid transitions such as b → disordered → liquid, or a → b → liquid. During the temperature ramp, we select trajectory frames to perform theoretical X Ray diffraction (XRD) analyses that can help determine the establishment of a new phase.

Predicting Phase-Specific Entropy of Polymerization Using Machine Learning to Accelerate the Design of Recyclable Plastics

Hunter Lee

Today's plastics are difficult to recycle, and therefore plastic waste accumulates in landfills, oceans, and other natural habitats. It is important to design novel intrinsically circular polymers (ICPs) that can be generated from renewable feedstocks with the hope of eliminating the end-of-life problem and pollution of today's plastics. These novel ICPs can undergo a process called chemical recycling through which the quality of the starting polymer is maintained as it is broken down into its building blocks for reuse, allowing the material to theoretically be recycled infinitely. In order to expedite the design of such materials, we are developing a machine learning model that can predict a thermodynamic indicator of recyclability for a polymer-reaction environment pair. This machine learning model is trained on a dataset assembled for the first time of phase-specific entropies of polymerization with features engineered to represent the complexities in entropy of polymerization, a property that is notoriously difficult to capture accurately using molecular modeling methods. This model will be used to rapidly screen the vast molecular search space created by hybrid pathways for monomer candidates that have thermodynamic properties suitable for sustainable chemical recycling.

Developing Support Vector Machine Models to Enable Precision Oncology

Cameron McMahan, Atefe Alimirzaei, Chris A. Kieslich

In the growing world of cancer research and treatment research, machine learning is growing alongside it. Our goal is to develop a machine learning model to aid in treatment selection with a focus on an individuals' genetic sequence to guide the way. Data is taken from the 50% growth inhibition data set (GI50) from the NCI-60 database. This database includes 60 cell lines each with multiple cancer types. A support vector machine (SVM) is applied to find the most important features of drug compounds for each of the cell lines; the cell line with the most entries, A549-ATCC, is the subset used for training and testing. The drug compounds are encoded through the Simplified Molecular Input Line Entry System (SMILES). These strings are used with a python package known as RDKit to create Morgan fingerprints. The fingerprints serve as the data that the SVM is trained on, as each bit in a fingerprint represents a substructure of the compound. The compounds were labeled as active or inactive based on the concentration used in the GI50 trials. The kernel of the SVM is made by clustering the compounds by calculating the similarity scores using the Tanimoto algorithm. Finally, an in-house function, MISTIC - Model Informed Feature Selection Through Importance and Contribution, has been used to guide the feature selection process with the aim of accurate drug-response prediction.

Principal Component Analysis in Prostate Cancer Research

Molly Smith, Hyeju Song, Christopher Kieslich

Prostate cancer is the most common type of cancer in the United States, excluding skin cancer. Prostate cancer is also a very diverse cancer, which creates obstacles in regard to treatment, study, and prediction of the disease. Computational modeling is useful for finding information about the disease but can have barriers when it comes to price, time, and accessibility. When utilizing the large datasets found in prostate cancer studies, there is also a large computational cost. Principal Component Analysis aims to provide accurate results while lowering the barrier between the problem and the cost. PCA,here primarily using Python software, condenses large datasets while maintaining the overall variance.This lowers the computational burden. PCA is then combined with a feature selection model to try and extract valuable data. Gene expressions derived from prostate tumors contain information that can be run through a PCA and feature selection model. The model has the potential to reveal if some patients are more at risk for relapse of the disease than others. Up to thirty percent of men were found to relapse after prostate tumor removal, indicating an urgency for developing prediction methods. The tumors were removed from patients in various stages of prostate cancer, allowing for high variability of data. Some patients reached complete remission after removal while others unfortunately experienced reemergence of the disease. The study indicates that there is an underlying correlation between patients who relapse and abnormalities found in gene expressions. No singular gene has been found sufficient for prognosis, but using PCA and the full gene expression dataset can reveal related genes. PCA graphs the results in direction of most variation and exposes the most important features, or principal components. These principal components are used as axes and show the relationship between the original data points and principal components. This method can be coded in several computer languages, making PCA an accessible tool. The goal of this experiment is to show how PCA can be an accurate step in cancer diagnosis.

Use of Computational Modelling to Derive Structural Insights of SFB Protein P3340

Tausif Tamim, Abhishek Sharma, Radmila Janjusevic

Segmented Filamentous Bacteria (SFB) are gram-positive gut bacteria distinguished for their roles in developing Th17 and Th22 immune cells. These immune cells are well known for their robust response against pathogens and infections and the role of SFB in immune response is well studied. SFB_NYU_P3340 (P3340) is a surface protein of SFB possessing high expression in the human gut, where SFB binds with intestinal epithelial cells. While the amino-acid sequence of P3340 is well known, the exact structure remains elusive. AlphaFold 3, released in May 2024, can be used to predict the structure of P3340 using artificial intelligence and deep learning, but the accuracy of this prediction requires further testing and optimization. To bridge this gap, the protein simulation package OpenMM can be employed to run simulations of P3340 and computationally assess results, including the stability of the protein structure under standard force fields. In this study, OpenMM was used to find a stable structure of P3340 and simulate P3340 under variable solvents conditions. In-silico analysis was conducted on the P3340 simulations, finding important data such as conformational energy minimization, radius of gyration, density, contact distance to assess the stability of P3340 at various instances of simulating the protein. P3340 has been produced in our lab from synthetic DNA that was codon-optimized for E. coli expression, but with low yields. Our experiments have shown that the use of small-molecule additives is promising in increasing the solubility of P3340, which is characteristically insoluble, thus increasing the yield of P3340 during purification. OpenMM was also used to simulate P3340 in the presence of different additives and the solubility of P3340 was computationally measured using various solubility parameters. In summary, we have used machine learning enabled protein structure determination and molecular simulations to well characterize the three-dimensional structure of P3340 and explore how we can improve yields in our processes. In the future, we aim to use our models to explore mechanistic questions about this particular protein.

Uterine Cancer Image Analysis with Convolutional Neural Networks

Mason Brown, Nellone Reid, Joshua Young, Alexander Brogan

Endometrial uterine cancer is the fourth most common cancer in women in the United States. Unlike most cancers, its rates have rising incidences and mortalities. Although Black women are less likely to be diagnosed with the disease, they are more likely to die from it. This discrepancy can be partially credited to the fact that Black women are often diagnosed at a later, less treatable stage. Studies attribute this diagnostic discrepancy to a variety of factors, including a lack of accurate and precise diagnosis during early stages of the cancer. A possible solution is using artificial intelligence and machine learning (AI/ML) models. Machine learning techniques are shown to be excellent at diagnosing cancer through the analysis of data and images such as X-rays, magnetic resonance imaging (MRIs), and ultrasounds. However, a major limitation of machine learning techniques is that if the models are trained on bad or biased data, they will reproduce that bias in their evaluations. Similarly, if diagnostic devices, such as sensors, are designed without an understanding of cell behavior differences based on ethnicity, the device will also have biases in its outcomes. Convolutional neural networks (CNNs) are an advanced form of ML that show promise in medicine, accurately performing diagnoses and medical image analysis. A key advantage of CNNs is less required manual image preprocessing compared to other ML techniques. Although, CNNs require large datasets to be trained from scratch, repurposing pre-trained CNNs and artificially expanding the dataset significantly cut down the required data collection. To address racial disparities in patient survival rates, this study developed convolutional neural network models that predict type of endometrial cancer in medical images, using pre-trained machine learning algorithms and ultrasound images. Analyzing the change in the accuracy of models as a function of race will determine if there is a bias.CNNs - ResNet50, DenseNet-201, VGG16, AlexNet, and Inception ResNet-V2 - were able to discern the race of the patient and the type of cancer with accuracies of 98.31%, 97.06%, 97.92%, 93.18, and 96.67% respectively, demonstrating the power of CNNs in medical diagnosis. Training the model on a racially biased dataset led to a notable decrease in test accuracy on the underrepresented races. ResNet50 could discern the type of cancer 86.42% of the time when there were an equal number of black and white patients in the training set, but the accuracy dropped to 61.22% when trained on only white patients. The other models displayed a similar drop-off in performance. This signifies a need to ensure a racially diverse training set to avoid introducing biases into the diagnostic process. Using the source data, we were able to 1) identify machine learning models that can recognize uterine cancer through the analysis of existing scans at high accuracies; and 2) identify racial disparities in diagnosing endometrial uterine cancer via ultrasound imaging and analysis. The methodology used by this study is applicable to other diseases diagnosed with medical imaging and can assist with identifying and solving racial disparities.

VLE-ML: Prediction of Vapor-Liquid Phase Equilibrium Based on Molecular Descriptors with No Code Machine Learning

Nicolas Iglesias Giraldo, Alejandro Valencia Virguez, Pablo Ortiz Herrera, Andres Fernando Gonzalez Barrios

The recent rise of machine learning, particularly through sophisticated models like deep learning, has revolutionized problem-solving and introduced new methodologies in chemical engineering. This study aims to predict vapor-liquid equilibrium (VLE) diagrams using machine learning models. Unlike previous studies which rely on physical models, this research focuses on predicting VLE diagrams for various molecules at different pressures and temperatures without assuming any physical model. By employing a detailed experimental procedure and evaluating several models, we sought to correlate molecular descriptors (obtained from the RDkit library) with VLE diagrams (specifically with vapor compositions) using codeless tools such as RapidMiner.

The research followed these experimental steps: First, extensive data collection was conducted using robust databases like CHERIC, peer-reviewed scientific articles, and the NIST ThermoData engine in Aspen Plus. We collected experimental data for 562 compound pairs, which amounted to 27,349 unique equilibrium datapoints.

To enhance model performance, dimensionality reduction techniques like Principal Component Analysis (PCA) and feature selection by correlation were applied. This reduction decreased the input features from 208 to 53 while retaining 80% of the original data variance. The data was then split into training (80%) and test (20%) sets which were split by compound pairs to preserve independence. Batch cross-validation was employed to optimize the models and ensure further generalization capacity.

Several models were trained and optimized, including Linear Regression, Random Forest, XGBoost, and Artificial Neural Networks (ANNs). A grid search was conducted to optimize the hyperparameters of each model. The models were evaluated using root mean square error (RMSE), mean absolute error (MAE), and the coefficient of determination (R^2). Additionally, the results were compared with the UNIFAC model, a commonly used method for VLE diagram prediction.

The results indicated that Artificial Neural Networks (ANNs) achieved the best performance among the evaluated models. While the UNIFAC model was efficient for common compounds, it exhibited significant errors for rare molecules. Linear Regression had the worst performance, indicating the problem's non-linearity. Random Forest showed poor performance, whereas XGBoost demonstrated a marked improvement over Random Forest. Neural Networks outperformed all models, effectively handling the problem's non-linearities.

This research demonstrates that machine learning models, especially neural networks, can substantially improve VLE diagram prediction. Moreover, using codeless tools like RapidMiner facilitated the implementation and optimization of these models, making the process accessible to the academic community.

Towards Greener Solvents: A Machine Learning (ML) Pipeline for Sustainable Alternatives

Rohan Datta

Organic solvents are widely used in industry but pose significant environmental and health risks, with many classified as ecotoxic or carcinogenic. Solvent selection guides (SSGs) were developed as a tool to help scientists responsibly and safely select solvents. However, current SSGs are limited in scope and rely heavily on experimental data, making the process of finding "greener" solvent alternatives slow and costly. In this work, we present a computational pipeline that 1) predicts the 'greenness' of any solvent and 2) suggests potential replacements. Trained on the GlaxoSmithKline Solvent Sustainability Guide–which scores solvents according to their environment, health, safety, and disposal impact–our Gaussian Process Regression (GPR) model can predict the greenness score of any solvent using only its SMILES. Incorporating Hansen Solubility Parameters (HSP), the pipeline can suggest replacements that are both greener and "similar" to the target solvent. This method provides a rapid, cost-effective approach for discovering greener solvents, with potential applications in green polymer synthesis and sustainable chemical process design.

Predicting Amyloid Fibrillation through Transfer Learning

Ethan Eschbach

Drug discovery is a complex multi-optimization problem that involves balancing the biological activity and developability of target compounds. Recent advances in generative machine learning models have contributed to the streamlining of these processes but lack precise control over target biophysical properties—many of which can be harmful to the development of a peptide therapeutic. Amyloid fibrils are a form of aggregate characterized by ordered stacks of peptide, which form a fibrous structure. In the discovery process, these aggregates are difficult to metabolize and are often less potent than free molecules. In this work, we develop a generalizable model that can predict amyloid fibrillation through protein language model (pLM) embeddings. We utilize ESM2—a pLM—to generate latent embeddings for sequences of interest, which are then passed to our model. Experimental data for fibrillation is limited and the largest public datasets consist of sequences that are much shorter than therapeutically relevant peptides. We explore transfer learning preprocessing strategies that allow us to effectively generalize to new sequence lengths, including mean-pooling and a modified convolutional neural network with attention weights—which is referred to as light attention (LA). Processed embeddings are passed to a standard multilayer perceptron (MLP) and predictions are scored against labelled data. These architectures demonstrate high predictive power when evaluated on two publicly available datasets and serve to expedite the development of peptide-based pharmaceuticals.

Machine Learning and AI Based Models for Modeling Optimization of Photocatalytic Conversion of CO_2

Antonio Franco Aguado, Ankita Juneja

Photocatalytic conversion is a chemical reaction process in which a photocatalyst is activated by an energy light source (sunlight or artificial). The activated photocatalyst can reduce CO2 into different valuable by-products that can be used in pharmaceutical, chemical and fuel industries. The imminent urgency of climate change has caused an increase in the efforts to develop sustainable methods for reducing atmospheric CO2 level, however, the complexity of the photocatalytic process influenced by numerous features and parameters need of the advanced modeling techniques to optimize the reaction conditions and improve the efficiency. This study presents a novel framework that integrates machine learning (ML) and artificial intelligence (AI) models to predict the outcome of photocatalytic conversion of CO2. By levering extensive dataset created by obtaining large amount of data obtained from experiments and peer reviewed articles we employ supervised learning algorithms to model the relationships between input parameters—such as catalyst composition, reaction conditions, and light source characteristics—and output parameters, including type of primary product, yield and productivity. Our approach utilizes models, such as Extreme Gradient Boosting (XGBoost), Artificial Neural Networks (ANNs), and Generative Adversarial Networks (GANs), enhanced with hyperparameter tuning and regularization techniques to achieve high predictive accuracy. The models are further validated using performance metrics like Mean Squared Error (MSE), cross-validation techniques, and R2 score to ensure robust predictions

Developing Magnetic and Nonmagnetic Machine-Learned Interatomic Potentials for Gold-Promoted Nickel Catalysts

Jennifer Gonzales-Pasion, Isabella Furrick, Mitchell Wood, Alyssa Hensley

As a result of fossil fuels being burned to produce energy, the emission of greenhouse gases has increased and are the leading cause for climate change. An alternative to this practice that shows promise in the transportation section is the use of hydrogen alkaline fuel cells, a clean and renewable energy source. Despite this, the widespread use of these fuel cells is hindered by its reliance on scarce and expensive noble metal catalysts (e.g. Pt). Previous work indicates that sustainable and cheaper Au-promoted Ni-based catalysts show promising catalytic activity for the hydrogen fuel cell. As catalytic reactions occur at the nano-scale, experiments cannot fully resolve key information about the working catalyst, such as atomic-level surface reconstruction and active site formation. To tackle this issue requires the use of atomistic molecular dynamics to uncover the mechanisms of this evolving catalytic activity. However, two challenges exist: (1) current computational modeling techniques are unable to reach the necessary length- and time-scales and (2) due to nickel displaying ferromagnetic properties, there is an open question on how magnetism affects model accuracy. To address these challenges and be able to capture the multiscale performance and structure of NiAu catalysts operating within hydrogen fuel cells, we developed two machine learned interatomic potentials (ML-IAPs) for NiAu catalysts. These two potentials differ in that one is magnetic while the other is nonmagnetic. This was done to assess the impact magnetism has when predicting accurate catalytic structures of NiAu. The methodology for building an accurate machine learned interatomic potential first involved the development of training data consisting of both domain and beyond domain knowledge structures. The acquired training data was then converted into descriptors using the Fit Spectral Neighborhood Analysis Potential (FitSNAP) software. The hyperparameters and group weights were optimized using a genetic algorithm called the design analysis kit for optimization and tera-scale applications (Dakota). The goal of the Dakota optimization was to minimize prediction errors for the energies, forces, and a set of known objective functions (e.g. vacancy energy). Once the potential was deemed optimized it was then applied to various surface and nanoparticle NiAu structures to run molecular dynamics simulations at varied temperatures using Large-scale Atomic/Molecular Massively Parallel Simulator (LAMMPS). The current magnetic and nonmagnetic potentials demonstrated low errors and successfully modeled the structures mentioned earlier. Both potentials were applied to surfaces of pure Ni and Au at the (100), (110), and (111) facets, displaying stability and no anomalies at a temperature of 300 K. Additionally, the magnetic potential was also applied to a NiAu nanoparticle and was tested at two separate temperatures of 300 K and 1000 K. During these tests on the nanoparticle, the structure maintained its overall shape and did not display irregularities. These results show that both potentials were able to accurately predict the provided configurations and can further be tested on different structures and reaction conditions. Overall, this study signifies a promising step towards building accurate ML-IAPs for NiAu catalysts and advancing towards cost effective hydrogen fuel cells.

Machine Learning Driven Quantitation of Viable Capsids for Gene Therapy Applications

Juhee Park, Lilianna Gutierrez, Anne Robinson

Viral vectors are a promising treatment option for a range of genetic disorders and diseases due to their ability to effectively deliver therapeutic genes to specific cells. However, large-scale production remains challenging, as viral vectors must not only be efficient but also safe to administer to patients. Currently, transmission electron microscopy (TEM) is the standard for high-fidelity quality control, allowing trained technicians to directly visualize and count viral capsids. This manual process of counting viable capsids is both time-consuming and prone to human error, which poses a challenge, especially when scaling up the production of potentially life-saving therapies.

To address this bottleneck, this study explores the application of machine learning algorithms to automate the quantitation of viable capsids in gene therapy applications. The first step involved preprocessing data by creating image patches from original TEM images, with each patch containing a single capsid. Using the Computer Vision Annotation Tool (CVAT), all capsids were labeled as 'full,' 'partial,' or 'empty,' representing a simplified classification of capsid fill values. Following the labeling step, a balanced dataset of these annotated capsid image patches was created and subjected to thorough data analysis, where meaningful features were extracted and applied to various machine learning algorithms. During feature engineering, the introduction of certain features—particularly one that compares the symmetry of image patches—yielded promising results, although further investigation is required to confirm its significance. Ultimately, this study aims to reduce the time and error rate associated with TEM-based manual quantitation of capsids, thereby streamlining the production of gene therapies.

Low-Cost Automation of Steam Distillation for Essential Oil Extraction: A Practical Approach Using Arduino Technology

Gustavo Perez, Josue Loro, Max Dueñas, Odell Liñan, Osvaldo Carrion

This study presents the development of a cost-effective automation system for the steam distillation process used to extract essential oils, utilizing Arduino technology. The system was designed to address common operational challenges, including maintaining water levels, regulating temperature, and preventing spills. Key components of the setup include servomotors, solenoid valves, various sensors, and an Arduino UNO microcontroller. Testing demonstrated improved process control and operational efficiency. The Arduino-based system successfully regulated the water flow and condensate levels, showing that accessible and affordable technology can enhance industrial processes, particularly in the context of small-scale or pilot-level operations. This project provides an example of how low-cost automation can optimize essential oil extraction processes, making such advancements feasible even for smaller operations.

Predicting High-Density Slurry Concentration with the Use of Variable Pathlength Model for Raman Spectroscopy

Rui Matsubara, Steven Crouse, Rupanjali Gurprasad, Martha Grover

Raman spectroscopy has been widely used as a feedback tool for process control, especially when quantifying suspended solids, such as silica in the nuclear waste. Quantification of the Raman spectra has historically followed a linear relationship between solid concentration and the Raman spectra intensity; however, when quantifying suspended solids in a slurry, the spectra-to-concentration relationship has been reported to become non-linear in previous research. In the literature, this issue was addressed by applying several data-driven methods. However, there still lacks a first-principal model to understand what the mechanism leading to the observed non-linear behavior. To fill this gap, this research focused on Raman intensity formula and pointed out that the change in traveling distance of the emitted laser from Raman apparatus, laser path length, could be the descriptor causing the non-linearity. The suspended solid can block the laser from traveling deeper if the suspended solid concentration is significant, causing laser path length to vary.

To verify our hypothesis that path length varies with suspended solid concentration in a slurry, a Monte Carlo simulation algorithm was constructed. Laser path length was calculated via the Monte Carlo simulation from inputs of suspended solid concentration and particle size distribution (PSD). The adjusted, non-linear Raman spectra intensity formula was applied to predict the spectra intensity of solids with four silica samples of varying PSDs, and this predicted value was compared to the experimentally observed intensity that was obtained under the same conditions. The simulation results suggested that the laser path length diminished with increasing concentration at any PSDs, and the predicted model based on variable path length had a strong agreement with NJ60 and NJ6 silica samples, with the coefficient of determination of 0.822 and 0.908, respectively. However, the other two silica samples (NJ0 and NJ00N) had no agreement with the proposed model, with -1.243 and -0.717 coefficient of determination, respectively. The difference of model performance on these four samples can be highlighted by the difference in PSDs among the four samples; NJ60 and NJ0 have median radii around 300–400 micrometers, while NJ0 and NJ00N both have median radii of around 600 micrometers. These results suggested that the developed variable laser path length model may predict the Raman spectra intensity more accurately at the smaller solid PSD ranges, and for the solid samples that have similar size as laser interrogation length (1000 micrometers for this research), the model did not describe the experimental data. From these conclusions, further investigation should be made on the best model for quantifying suspended solids with large particles, and possibly for quantifying smaller particles where the size to be smaller than 100 micrometers.

Automatic Generation of Reduced-Space Optimization Formulations of Process Systems for Faster Deterministic Global Optimization in Julia

Joseph Choi, Matthew Stuber

Optimization plays a critical role in engineering because of its ability to address decision-making problems. For problems involving nonconvexity, which is inherent to many chemical engineering systems of interest, deterministic global optimization solvers are required to guarantee a certificate of optimality. However, global solvers are ill-suited for large-scale applications because of their worst-case exponential runtime with respect to problem dimension. This has prompted research into the generation and use of reduced-space optimization formulations, where a model is algebraically transformed from a high-dimensional space into its intrinsic dimension, to reduce computational cost and therefore broaden the scope of problems for which global solvers can be applied.

In this work, we detail the development of user-defined functions that provide a bridge between an easy-to-use open-source modeling framework and a robust open-source deterministic global optimization solver, namely EAGO.jl, within the Julia programming language. Specifically, we utilize ModelingToolkit.jl's intuitive modeling interface and structural simplification features along with Symbolics.jl's function compilation tools to automatically generate reduced-space models as numerically usable functions, which can then be used as EAGO-compatible equality constraints in JuMP models.

We then apply our methods to formulate and solve a process design optimization problem with the objective of minimizing capital and operational costs, where we demonstrate improvements in convergence times and reductions in memory allocations compared to its full-space formulation. These advantages are further exemplified through the provision of our novel approaches to automatically generate reduced-space models by using powerful and highly accessible open-source tools, advancing the applicability and availability of global solvers for optimizing large-scale nonconvex chemical engineering systems.

Data-Driven Surrogate Modeling and Optimization for Ammonia Production

Thisali Wadunambu Arachchige Dona, Md Masud, Yuhe Tian

Solving large-scale flowsheet optimization using first-principles unit operation models remains a significant challenge for conventional mathematical programming methods. This complexity arises due to the high computational cost and difficulty in capturing non-linearities and interactions within interconnected process units. Therefore, a new framework is needed to holistically represent an entire process system by employing data-driven surrogate models, which offer sufficient modeling accuracy while significantly reducing numerical complexity. By simplifying the representation of individual process units, surrogate models facilitate the optimization of complex flowsheets, allowing for the identification of both optimal structural configurations and corresponding operating conditions.
In this study, ammonia production is used as a case study to demonstrate the feasibility and effectiveness of the proposed approach. Surrogate models are constructed using data generated from process simulators, effectively replacing traditional first-principles unit models. An integrated software platform is developed for this purpose, combining the capabilities of Aspen Plus for process simulation, MATLAB for input-output data sampling, and Python for surrogate modeling using neural networks. A comprehensive superstructure is formulated to consider multiple process routes and technologies for hydrogen production (methane steam reforming, water electrolysis), nitrogen production (air separation units, membrane separations), ammonia synthesis (Haber-Bosch reactor, microwave-assisted reactor), and downstream separations (distillation, membrane separations). These alternative flowsheet designs are systematically evaluated against performance metrics such as cost, sustainability, and productivity, enabling a holistic assessment of different configurations.

A Process Operability Approach for Safety-Critical Chemical and Energy Production

Theodore Malencia, Beatriz Dantas, Austin Braniff, Fernando Lima, Yuhe Tian

Effective management of exothermic reactions, particularly those involving volatile substances, is crucial in chemical process industries to prevent catastrophic events such as thermal runaways. Thermal runaway occurs when a reaction generates heat that accelerates the reaction further, leading to uncontrollable temperature increases. To mitigate such risks, this study employs a comprehensive risk management approach combining steady-state operability, dynamic operability, and control strategies.

Steady-state operability focused on assessing the achievability of desired operating regions by specifying the range of input variables, known as the Available Input Set (AIS), defined by upper and lower bounds of design and/or operational variables. Input-output mappings generate the Available Output Set (AOS), which represents the outputs that the system can achieve. By the identification of the desired operating boundaries, known as Desired Output Set (DOS), the inverse mapping enables the identification of their respective Desired Input Set (DIS), providing a feasible operating range that ensures desired risk levels. Dynamic operability introduces a time-dependent component, assessing system responses to disturbances. This approach parallels steady-state operability but incorporates variations in heat transfer characteristics over time, allowing for the evaluation of how the feasible ranges of input variables adjust in response to disturbances. The assessment is conducted through both open-loop and closed-loop simulations. In the closed-loop scenario, a Proportional-Integral (PI) controller is employed to manage the system.

This methodology is demonstrated on two safety-critical systems: a Continuous Stirred-Tank Reactor (CSTR), modeled after the T2 Laboratories accident, and an electrolyzer used for hydrogen production from water for clean energy applications. The findings of this study offer a valuable framework for enhancing the safety and efficiency of these critical processes.

Using CFD Simulation to Analyze the Repaired Coarctation of the Aorta

Matthew Bradley, Dominic Lippa, Olivia Dillenbeck, Jason Mandell, David Foster

Coarctation of the aorta (CoA) is defined as a stenosis of the descending thoracic aorta (DAo) [1] which results in narrowing of the aortic arch. Current clinical guidelines recommend intervention when there is significant aortic narrowing, [2] and when a significant pressure gradient is measured during cardiac catheterization. The arch contains four outlets, three of which distribute oxygenated blood to the upper half of the body, while the fourth delivers to the lower body. If left untreated, serious heart conditions can arise later in life due to the hindrance in blood flow to both the upper and lower extremities. Patients with untreated CoA remain at high risk due to complications including increased mortality, hypertension, stroke, and decreased exercise capacity. [3] This condition affects approximately 6-8% of the population. [4] Today, surgical techniques to repair arches include extended end-to-end anastomosis which involves removing the constricted part of the aorta. Depending on the size of the constricted area, the ends of the artery can be reconnected simply by stitches or inserting a graft. Earlier techniques involve aortoplasty, which uses synthetic materials such as balloons or stents to reinforce the artery wall.

This project aims to model the flow through the aortic arch using computational fluid dynamics (CFD) to analyze the complex flow patterns and pressure gradients in patients pre and post CoA repair with the goal of providing insight to the flow patterns to better understand the effect of the repair, as well as contribute to future repair strategies. CFD is a branch of mechanics that creates simulations for a variety of applications. Under an IRB (Internal Review Board) approved study, clinical data was supplied by the University of Rochester Medical Center. This data included aortic arch anatomy in MRI and CT scan format, along with flow data and heart rates from cardiac MRI studies, and blood pressures from outpatient visits. These scans were then mapped onto a 3D model and imported into ANSYS, the CFD program used to conduct all flow modeling. Once an unrepaired arch was modeled, a virtual surgery was performed on the arch at two common cutting points in end-to-end anastomosis. The blood flow was then modeled again through the newly repaired arch to assess the change and quality of the flow. This will allow surgeons to be better informed on the quality of the surgery they are performing, and how to better assess the optimal snipping points in the arch. All simulations were conducted with normal blood parameters for density and viscosity, and governed using the Continuity and Navier-Stokes equations for turbulent flow under the SST k-omega flow model.

References

(1) Jonas R. Comprehensive Surgical Management of Congenital Heart Disease. 2nd ed. CRC Press; 2014.

(2) Stout KK, Daniels CJ, Aboulhosn JA, et al. 2018 AHA/ACC Guideline for the Management of Adults With Congenital Heart Disease: Executive Summary: A Report of the American College of Cardiology/American Heart Association Task Force on Clinical Practice Guidelines. Circulation. 2019;139(14).

(3) Kempny A, Dimopoulos K, Uebing A, et al. Reference values for exercise limitations among adults with congenital heart disease. Relation to activities of daily life--single center experience and review of published data. European Heart Journal. 2012;33(11):1386- 1396.

(4) Doshi, Arpan R, and Sathish Chikkabyrappa. "Coarctation of Aorta in Children." Cureus vol. 10,12 e3690. 2018.

Impact of L- and D-Peptide Isomer Mixtures on Self-Assembly in Tissue Engineering: A Molecular Dynamics Study

Justin Kim, Kyle Lampe, Rachel Letteri, Phillip Taylor

Tissue regeneration is a critical challenge that necessitates the design of dynamic materials. Such design requires the material to mimic the extracellular matrix (ECM), by understanding features such as proteolytic remodeling, cell adhesion, and viscoelastic properties. Recent studies have focused on the design of ECM-mimetic, water-swollen hydrogels, due to the flexibility of their mechanical and biochemical properties. Specifically, our group has previously developed Rapidly Assembling Pentapeptides for Injectable Delivery (RAPID) hydrogels, composed of the peptide sequence KYFIL (lysine–K, tyrosine–Y, phenylalanine–F, isoleucine–I, and leucine–L), and has used turbidity, oscillatory rheology, and X-ray diffraction to examine the effects of amino acid stereochemistry (i.e., L- and D-amino acids) on peptide self-assembly and hydrogel mechanical properties. We used atomistic molecular dynamic simulations to probe the molecular mechanism governing the stereocomplexation-directed self-assembly of KYFIL. Five peptide systems with varying L:D ratios, solvated with physiological ion concentrations, were conducted under constant pressure and temperature using NAMD and GROMACS software. Simulations trajectory files revealed that pure L- and D-KYFIL systems formed larger but fewer clusters of self-assembled peptides, while stereoisomer mixtures formed a larger number of smaller clusters. Hydrogen bonding and pi-pi stacking had similar results with the L- and D- mixtures and pure peptide systems. Lastly, secondary structure analysis indicated the formation of β sheets and random coils in all systems. Overall, varying stereochemical ratios significantly impacted the assembly mechanism of KYFIL systems, influencing the size of the cluster, secondary structure, and hydrogen bonding, in qualitative agreement with previous experimental observations. Future investigations will include coarse-grained simulations utilizing GROMACS and MARTINI, to simulate hydrogels at larger length scales and longer timescales.

Investigating the Effect of Dispersity on the Conformation and Dynamics of Semiflexible Polymer Melts – Insights from Molecular Dynamics Simulations

Dominick Szabo, Taofeek Tejuosho, Janani Sampath

Synthetic polymers inherently possess a distribution of molecular weights due to the statistical nature of their polymerization path, which is typically described by dispersity (Đ). Mathematically, Đ is the ratio between the weight average molecular weight M_w and number average molecular weight M_n. Studies have shown that both high and low dispersity can offer complementary characteristics, significantly impacting the polymer's processability and its mechanical properties. However, the design space for manipulating this parameter is vast and performing experiments is time consuming. An important feature of polymer melts is the degree of entanglement, or the topological constraints that polymer chains exert on one another. By utilizing coarse-grained molecular dynamics simulations, we study the structure and dynamics of chains spanning the unentangled to marginally entangled regime as a function of dispersity, making comparisons with that of chains in a monodisperse melt. We specifically use the bead – spring polymer model which has been demonstrated to capture essential physics of polymer melts. In order to control the degree of flexibility of the chains, which in turn affects observable macroscopic properties, we employ the three-body bending potential. First, we investigate the structure by computing the static structure factor and radius of gyration to benchmark these quantities with their behavior in the corresponding monodisperse melts to assess any deviation from ideal chain behavior. Next, we examine the effects of heterogeneity in surrounding test chain lengths on melt entanglement using primitive path analysis. Lastly, we analyze the mobility of specific test chains through the mean squared displacement to gain insights into molecular mechanisms that may contribute to the diverse relaxation processes in a polymer melt. In summary, this work provides the necessary foundation for an improved understanding of the behavior of chains in synthetic polymer melts to guide the predictive design for future applications.

Machine-Learning the Near-Infrared Fluorescence of DNA-Stabilized Silver Nanoclusters with Quantified Uncertainty and Explainability

Kyrstin Datanagan, Stacy Copp, Cory Simon

Single-stranded DNA can stabilize silver nanoclusters with bright fluorescence colors (AgN-DNAs). AgN-DNA properties depend strongly on DNA sequence. Due to the programmability of DNA sequence, it is possible to synthesize sequence-selected silver nanoclusters with diverse structures and, thus, properties. In particular, there is recent interest in the design of AgN-DNAs with optimal optical properties for bioimaging in the near-infrared (NIR) tissue-transparency window.

Building on previous work by the Copp group at UC Irvine, we train and test an ensemble of gapped k-mer SVR's (support vector regressors) to predict, with quantified uncertainty and explainability, the NIR fluorescence intensity of an AgN-DNA from its DNA sequence. Our training data constitutes ~2,400 examples of AgN-DNAs and their measured NIR emission intensities. The gapped k-mer kernel scores the similarity between a pair of DNA sequences according to matchings of gapped k-mers between them; a length-k DNA subsequence with wildcard positions. We expect our machine learning model to be useful for computationally-designing AgN-DNAs for bioimaging, circumventing a time- and cost-intensive trial-and-error search in the lab for these promising bioimaging agents.

Acoustic Enhanced Machine Learning Modeling of Li Metal Batteries

Benjamin Schwartz

Higher energy density batteries are crucial for efficient energy storage. Anode-free Li-metal batteries have high energy-density, and we used mechanical information from the batteries to model their capacity degradation. Electrochemical and acoustic data were collected as the cells underwent formation and cycling. Ultrasound transmission through the cell provided acoustic data, with modulations in the received waveforms containing *operando* mechanical information. Raw waveforms or derived acoustic variables were used as features to train a random forest regressor with capacity data as the labels. Our model, using the derived variables, showed a mean R^2 of 0.998 across 33 cycles, accurately inferring the capacity for each cycle from the same cycle's acoustics. This result indicates the modeling power of the acoustic information. Furthermore, we can perform highly accurate future predictions. This displays how chemo-mechanical acoustic data can predict future electrochemical degradation. Future predictions become more accurate with the derived variables, indicating that our variables successfully extract the valuable information from the raw waveforms. Highly predictive models, based on low-cost, *operando*, mechanical measurements provide valuable insight into battery degradation, aiding in the development of high-energy-density batteries.

Deciphering Chondroitinase ABC Thermal Stability Via Molecular Dynamics

Yvette Olivas Biddle

Chondroitinase ABC (chABC) is a 115 kDa bacterial lyase that has shown significant potential as a therapeutic for treating spinal cord injuries[1–5]. Specifically, chABC breaks down glial scarring by degrading the side chains of chondroitin sulfate proteoglycans and has been shown to promote spinal cord plasticity, neuronal regeneration, and axonal sprouting in animal models[1,3,4]. However, the impact of chABC is limited as it completely loses its activity within a few hours in dilute conditions at or above 37°C (human body temperature), thereby necessitating continuous intrathecal administration every two weeks over the course of two to six weeks to maintain therapeutic efficiency[6]. This rapid degradation poses a serious barrier for therapeutic application despite the potential of chABC. While past work performed by the Webb and Gormley labs identified incredibly promising copolymers for the stabilization of chABC, the mechanism of the stabilization is not yet clearly identified[7-9]. The experimental results do not provide insight as to how the copolymers identified in the previous research endow thermodynamic stability, or how they will be impacted by other components necessary for therapy. Additionally, it must be considered that while these copolymers may provide stabilization in ideal conditions, their affects may be impacted by further chemical modification of functionalization that may be necessary for the optimization of drug-release. Therefore, it is necessary to perform further experimentation to elucidate the mechanisms of stabilization, allowing for deeper understanding and more informed implementation of future steps for therapeutic application. Molecular dynamics (MD) simulations were selected to elucidate the molecular-level interactions that lead to enhanced thermodynamic stability of chABC when in conjunction with copolymers. By conducting simulations of solely chABC in a solvated system, a baseline has been established of the temperature at which the in-silica version of this enzyme is expected to begin to degrade. At around 400K chABC in this system may begin to lose function, and at 460K and above it begins to entirely lose its original structure. This information is critical for future determinations of the stabilizing effects of the co-polymers as they are introduced to the system, as now there is a baseline of an un-stabilized system for comparison. Further work for this project currently underway has included introducing both polymers that are known to have stabilizing effects, as well as polymers that are known to not be stabilizing for similar 100 ns production MD simulations.

Citations

1. Bradbury, E. J. et al. Chondroitinase ABC promotes functional recovery after spinal cord injury. Nature 416, 636–640 (2002).
2. Barritt, A. W. et al. Chondroitinase ABC Promotes Sprouting of Intact and Injured Spinal Systems after Spinal Cord Injury. J. Neurosci. 26, 10856–10867 (2006).
3. Caggiano, A. O., Zimber, M. P., Ganguly, A., Blight, A. R. & Gruskin, E. A. Chondroitinase ABCI Improves Locomotion and Bladder Function following Contusion Injury of the Rat Spinal Cord. J. Neurotrauma 22, 226–239 (2005).
4. Chau, C. H. et al. Chondroitinase ABC enhances axonal regrowth through Schwann cell-seeded guidance channels after spinal cord injury. FASEB J. 18, 194–196 (2004).
5. Rosenzweig, E. S. et al. Chondroitinase improves anatomical and functional outcomes after primate spinal cord injury. Nat. Neurosci. 22, 1269–1275 (2019).
6. Tester, N. J., Plaas, A. H. & Howland, D. R. Effect of body temperature on chondroitinase ABC's ability to cleave chondroitin sulfate glycosaminoglycans. J. Neurosci. Res. 85, 1110–1118 (2007).
7. Gormley, A. J. & Webb, M. A. Machine learning in combinatorial polymer chemistry. Nat. Rev. Mater. 6, 642–644 (2021).
8. Kosuri, S. et al. Machine-Assisted Discovery of Chondroitinase ABC Complexes toward Sustained Neural Regeneration. Adv. Healthc. Mater. 11, 2102101 (2022).
9. Tamasi, M. J. et al. Machine Learning on a Robotic Platform for the Design of Polymer–Protein Hybrids. Adv. Mater. 34, 2201809 (2022).

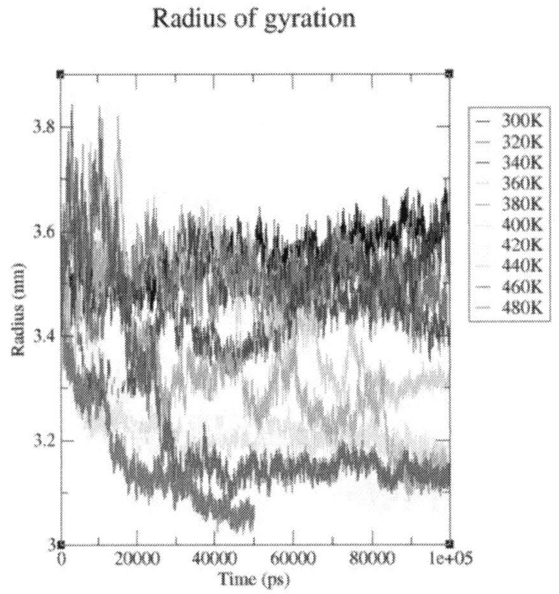

Figure 1: Radius of Gyration

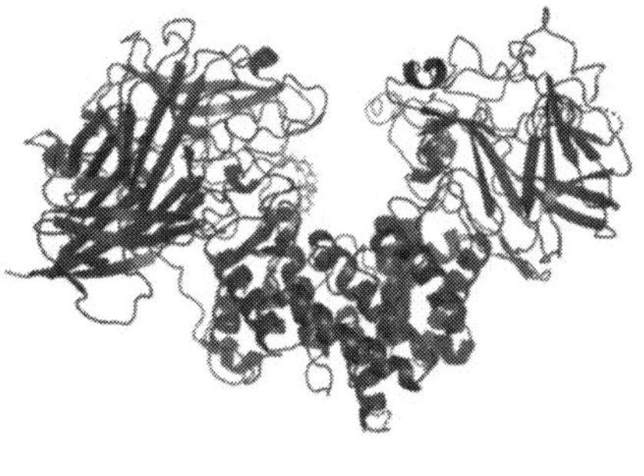

Figure 2: chABC in initial position with key active residues in orange

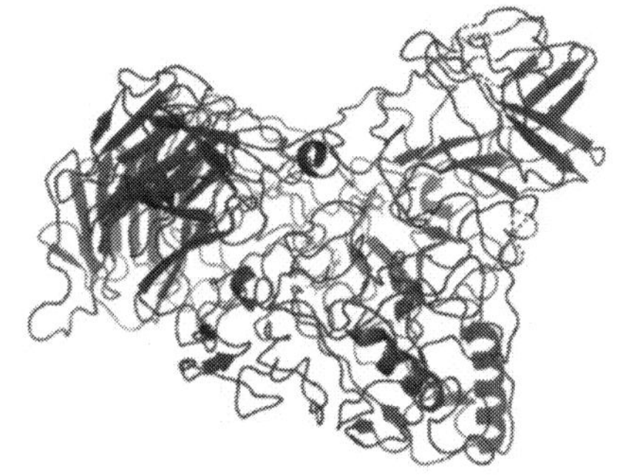

Figure 3: chABC after 100 ns at 460K with key active residues in orange

Development of Algorithms for Quantum Material Measurements Under Extreme Conditions

Christopher Stines, Abhishek Singh, Prakash Regmi, Yun Suk Eo

Measuring the physical properties, including electrical transport, thermal conductivity, and thermoelectricity, of quantum materials requires the use of specialized hardware and software. The Quantum Design Dynacool system can reach temperatures as low as 0.35 Kelvin with the use of a Helium-3 system and magnetic fields as high as 9 Tesla. To accurately measure the thermal conductivity and thermoelectricity, reliable thermometry readings of the material are required in addition to a carefully designed thermal puck that is free from unwanted external heat transfer. Reliable thermometry is difficult in extreme conditions because measurements are susceptible to noise and thermal drift. Also, the typical duration of a thermal conductivity or thermoelectricity measurement run can be as long as days or weeks, and therefore the data acquisition must be automated. In addition, the thermal puck design requires a thorough understanding of thermal properties at these extreme conditions to avoid external conductive heat transfer.

We report our current development of the thermal puck hardware and measurement software required to collect thermometry data. When determining the thermal conductivity and Seebeck coefficient of a material, it is essential that thermometer measurements have reached stability. The software is used to collect the temperature data from these thermometers and ensure data stability. One challenge is determining a stability condition when a large outlier in the data is present. Many different filtering algorithms were considered. The chosen algorithm to be used is an exponentially weighted moving average filter with an arbitrary filter weight. The software must be able to ensure data accuracy and precision before accepting a set value while simultaneously controlling the measurement hardware.

Another major aspect of the software is the control of the Dynacool system and the lock-in amplifiers that measure the output voltages of the thermometers. Control of the Dynacool system is required to set the sample environment such as the temperature and magnetic field. A single measurement run can range from days to weeks, so constant human monitoring is not viable. As a result, automatic control of the Dynacool and lock-in amplifiers is required. The thermal puck and control software have both been developed simultaneously to meet requirements of the existing hardware and needs of the lab. Some materials to be tested are also being synthesized in the lab. These materials include $FeGa_3$, $CrSb_2$, and $FeSb_2$.

Life Cycle Sustainability Assessment of Hydrogen Energy Systems: Framework and Case Study

Raul Ortiz Ibarra, Marco De Sousa, Betsie Sara Monserrat Montaño Flores, Rahul Kakodkar, Efstratios Pistikopoulos

The Sustainable Development Goals (SDGs) developed by the United Nations emphasize the necessity for progress towards peace, prosperity, and sustainability by 2030. There is an incentive for businesses to progress towards achieving these goals, as reputational risk represents a significant market value. Progress toward these SDGs requires the use of a Life Cycle Sustainability Assessment (LCSA), which is the accumulation of a Life Cycle Assessment (LCA), Life Cycle Costing (LCC), and Social Life Cycle Assessment (SLCA).

This work presents a framework for a Life Cycle Sustainability Assessment of hydrogen systems. The LCC and LCA of hydrogen have been heavily researched. In the case of the Social Life Cycle Assessment (SLCA) of hydrogen, this work seeks to support the establishment of a consensus on social impact categories, the SLCA database, and methodology. The medium risk hours (mrh) metric of quantifying social indicators is presented to be the best option for hydrogen systems. It allows businesses to improve the quality of their outputs by emphasizing the "workers" stakeholder group [1]. Other potential metrics that are assessed include social costing and value-added but are dismissed as environmental or economic indicators [1]. A fine line between environmental, social, and economic indicators needs to be drawn. Green hydrogen systems such as wind-powered electrolysis have been sought after due to their low emissions quantified by an LCA. However, the reduction in emissions often comes at the cost of increased prices and decreased social welfare [2]. A combination of different hydrogen systems including green hydrogen and conventional hydrogen is proposed towards a sustainable hydrogen system, which can be assessed through an analysis such as the Life Cycle Sustainability Assessment (LCSA).

References

[1] B. Monique, G. Norris, C. Norris Benoit, Shdb documentationv5, The Social Hotstpots Database (2022).

[2] A. Valente, D. Iribarren, J. Dufour, Comparative life cycle sustainability assessment of renewable and conventional hydrogen, Science of the Total Environment 756 (2021) 144132.

Revealing Microplastic-Water Interactions in the Atmosphere Via Coarse-Grained Simulations

Zach Ostrand, Thomas E. Gartner III

Microplastics have emerged as a significant environmental concern due to their global presence. Recently, microplastic particles have been discovered in the atmosphere, though their atmospheric assembly behavior and possible impacts on climate processes is largely unknown. In this work, we study the potential effects of atmospheric microplastics on climate using molecular dynamics simulations. In particular, we aim to understand how the presence of microplastic particles may affect water phase behavior and droplet nucleation, relevant to cloud dynamics and precipitation. We pair the ML-BOP coarse-grained water model with typical bead-spring polymers to provide a preliminary look at water-microplastic interactions. First, we show that ML-BOP, despite being designed to model water's condensed phases, reasonably predicts water's liquid-vapor phase behavior. We then simulate ML-BOP in the presence of polymers of varying hydrophobicity to evaluate the polymers' impact on water phase behavior and liquid droplet nucleation. Understanding such interactions is crucial for predicting climate trends and developing strategies to mitigate the global impact of microplastic pollution.

Optical Molecular Recognition from Chemical Reaction Mechanism Images

Ching Ting Leung, Yufan Chen, Hanyu Gao

To support the development of AI models in helping organic chemistry research, optical chemical structure recognition (OCSR) systems are designed to build databases by extracting molecular structure information, typically in the form of molecular graphs or SMILES, from chemical molecule images. While many chemical reaction images only included reactant and product molecules, it is not uncommon for organic chemical reaction images to include additional information such as partial charges and curved arrows indicating electron movements. These noises are structurally not differentiable from the essential information in a molecule such as atoms and bonds and cause significant difficulties in OCSR tasks.

We propose a novel pipeline that combines an image-segmentation model and an OCSR model in dealing with information extraction from chemical reaction mechanisms, which typically contain rich information about electron flow. The image-segmentation model is used for data pre-treatment to identify and remove curved arrows indicating electron movement, followed by the OCSR model in molecular identity recognition. In addition, there are no existing benchmarking datasets designed specifically to target chemical reaction mechanisms. We created a dataset of molecule images and structural Molecular identifiers of Molecular images in Chemical Reaction Mechanisms (SMiCRM). It consists of 453 molecular images, with mechanistic features, such as curved arrows and partial charges. They are labeled with their Simplified Molecular Input Line Entry System (SMILES) and their Structural Data Files (SDFs). Comparing the performance of the molecular recognition accuracy using the proposed pipeline and the performances of using only the OCSR model on the dataset collected, the proposed pipeline's performances significantly improved from 12.09% to 64.5% in exact SMILES matching and from 15.36% to 90.97% in Tanimoto similarity in recognizing the identity of mechanistic molecules of curved arrows removed. We also used this approach for in complete reaction mechanism image parsing and demonstrate positive performance improvements.

In conclusion, the research proposes an autonomous and effective procedure for collecting molecular-level information for chemical reaction mechanisms and highlights the further avenues for improving chemical information extraction methodologies.

Explicit pH and Temperature Control of Complex Pharmaceutical Bioprocesses

Dylan Miller, Austin Braniff, Yuhe Tian

With the recent impetus on advanced biotechnical processes in the pharmaceutical industry, it is critical to explore methods that can improve the efficiency and performance of the real-time operation for these sophisticated processes. The advanced control of these systems represents a challenging task due to the highly complex dynamics. The majority of in-use control techniques for pharmaceutical production use single-input single-output (SISO) and error reactive methods which can lead to suboptimal operations. The use of Model Predictive Control (MPC) provides an approach for optimally controlling systems. However, in pharmaceutical bioprocesses, the systems may be too complex to model accurately, or intricate models may lead to computational inefficiencies in the online implementation of MPC. Alternatively, multi-parametric model predictive control (mp-MPC) can solve the control problem offline to obtain optimal control laws as linear piecewise affine functions of system parameters (e.g. process states, setpoints, etc.). Understanding the effects of pH and temperature in pharmaceutical processes is critical in safely and reliably controlling these systems to produce high-grade pharmaceuticals. To facilitate the application of model-based controls, system models need to be validated in their ability to capture the effects of manipulating system variables while predicting accurate operational trajectories.

In this work a model for the batch lactic acid fermentation production is adopted and validated from literature for its robust and generalizable formulation featuring dependencies on temperature, pH, and substrate concentration. In validating this model, utilizing parameter values taken from literature, accuracy errors are present. This is rectified by employing non-linear parameter regression techniques to minimize modeling errors. To this end, we present a new set of regressed modeling parameters that more accurately fit the experimental data. The high-fidelity model with optimized parameters is linearized to enable to use of mp-MPC, thus avoiding computational complexity involving the elaborate system model. An explicit MPC problem is formulated utilizing the linearized models to maintain the pH and temperature of the biotechnical batch reactor at desired setpoints by manipulating the addition of acids or bases and the heating/cooling utility flowrates to the reactor. Control studies are presented and validated against the true nonlinear system to demonstrate the efficacy of advanced control techniques on these pharmaceutical bioprocesses. The improved high-fidelity model derived in this work and the application of mp-MPC stand as foundational steps for optimizing the operations of future large scale industrial implementations of pharmaceutical bioprocesses.

Physics-Based Simulation of HSV Surface Glycoproteins to Inform Vaccine Design

Anna Broerman, Hannah Padgette, Kayla Sprenger

Herpes Simplex Virus (HSV) infects a significant portion of the global adult population. Though this virus may remain dormant after acute infection, it has been known to resurface and have serious potential side effects. HSV is an enveloped, double-stranded DNA virus that enters host cells via endocytosis. HSV cell entry depends on four essential surface glycoproteins, with glycoprotein D (gD) mediating the crucial receptor-binding step. While the dynamics of this process have been studied with respect to nectin and HVEM as essential receptors, there are no studies involving the *in silico* dynamics of glycoprotein D binding to its initial co-receptor, 3-O-sulfated heparan sulfate (HS). Due to the ability of HSV to establish latency, it can evade the immune response, resulting in the lack of any effective licensed vaccine. Studying the molecular dynamics of the gD binding interaction with heparan sulfate, and how it may influence following protein-protein interactions, will potentially inform a structurally based vaccine design.

To address this, we modeled the initial state of gD using AlphaFold3, finalizing glycan locations for structural accuracy. Molecular dynamics simulations of the gD monomer in solution, performed with GROMACS, resulted in a stable structure, providing a promising basis for further study. Additional optimizations, such as hydrogen mass repartitioning, were explored to enhance simulation capabilities. Next, we will simulate the gD dimer in complex with heparan sulfate in an intracellular environment to assess any conformational changes induced by secondary receptor binding. The results of this simulation will be essential in uncovering the individual mechanisms involved in the viral cell entry process, to ultimately inform structure-based HSV vaccines.

Structure-Sensitivity of Nitrate Reduction on Bimetallic Surfaces

Aaron Bal, Eli Shopbell, Deep M. Patel, Luke T. Roling

Excess nitrate (NO_3^-) leaching from agricultural runoff poses a significant environmental hazard to the nitrogen cycle and the quality of drinking water. Electrocatalytic reduction of NO_3^- is a sustainable route to utilize renewable electricity as a sustainable solution by converting nitrate into environmentally benign nitrogen gas (N_2) or value-added chemicals like ammonia (NH_3). Despite its overall promising outlook, identifying desired catalyst surface properties to promote both N-O dissociation (favored at strong-binding metals) and N-N/N-H bond formation (favored at relatively weak-binding metals) poses a significant challenge to engineer efficient electrode materials for synthesizing the desired product(s).[1] Bimetallic surfaces with a strong binding metal (Pd, Pt) and a weaker binding metal (Ag, Au, Cu) provide binding sites with varying electronic properties to possibly accelerate both classes of reactions and allow tuning of the product selectivity by optimizing the surface composition and distribution.[2,3]

This presentation will use density functional theory (DFT) calculations to develop structure-sensitive scaling relations of key reaction intermediates involved in NO_3RR to NO, N_2, and NH_3 on (111) and (100) facets of Pd, Pt, Cu, Au, and Ag bimetallic surfaces. These scaling relations correlate the adsorbate binding energy to be accurately represented as a function of the binding energy of the metal site (BE_M) with mean absolute errors less than 0.07 eV.[4] This enables us to derive thermodynamic selectivity towards NO, N_2, and NH_3 as a function of a binding site (BE_M). These scaling relations will be further combined with Brønsted-Evans-Polanyi (BEP) relations to calculate activation energies as a function of computationally inexpensive BE_M. Finally, we plan to incorporate the structure-sensitive scaling relations and modified BEP relations in microkinetic modeling to derive the correlation between turnover frequencies and determine the selectivity of desired products as a function of BE_M. This correlation can be used to predict the structure of more active and selective materials for NO_3RR to NH_3 through inverse design.

References

- Liu, J-X. et al., *ACS Catal.* **9**, 7052-7064 (2019)
- Liu, H. et al., *ACS Catal* **11**, 8431-8442 (2021)
- Chun, H.-J. et al., *ACS Catal.* **12,** 1394-1402. (2022)
- Choksi, T. et al., *J. Phys. Chem. Lett.* **10**, 1852-1859 (2019)

Efficient and Sustainable Farming through Artificial Intelligence

Minlu Wang-He

One of today's greatest challenges is feeding the world's growing population sustainably. It is crucial to have efficient and sustainable farms that can maximize crop yield with minimum environmental impact and resource consumption, such as fertilizer and water usage. To achieve this goal, one critical and arguably the most important factor is to have an effective soil microbiome. The soil microbiome is the complex community of microorganisms inhabiting the soil that play many vital roles in the ecosystem. An effective microbiome can offer benefits to its host, including plant growth promotion, nutrient use efficiency, and control of pests and phytopathogens. For example, beneficial bacteria, such as *Rhizobiales*, can help plants capture nitrogen and other nutrients, improve soil fertility, and provide disease protection. On the other hand, harmful bacteria, such as *Xanthomonas*, can cause plant diseases, produce toxins, etc.

Based on this common knowledge, we further hypothesize that the composition of a soil microbiome can be a good indicator of crop yield. If this is true, information on soil microbiomes can be used to help farmers plan out how to allocate their resources and grow food in a more sustainable fashion.

In this work, using the data reported in [1], we tested different machine learning (ML) algorithms to identify the correlation between the microbiome composition and crop yield, as well as to identify the key microbes that contribute to crop yield. The dataset included soil samples with microbial counts of 630 different species from 1,344 plots, together with sweet corn yields on those plots. Four ML algorithms were investigated: Multiple Linear Regressor, Multi-layer Perceptron Regressor, K-Neighbors Regressor, and Decision Tree Regressor. Among them, Multiple Linear Regressor is a linear method, while the rest are nonlinear methods. The model input is microbiome composition or bacteria count, and the model output is the crop yield (measured in kilograms of crops per hectare of land area). Three different metrics were utilized to evaluate the prediction performance of different methods: coefficient of determination (R^2), mean absolute error (MAE), and relative squared error (RSE). We found that a Decision Tree Regressor (DTR) model was the most accurate model, which delivered the highest R^2 and the smallest MAE and RSE.

This result confirms that soil microbiome composition can serve as a very good indicator for crop yield, and a good ML model can serve as the linkage. The ML model can provide guidance on resource allocation and farming management practices, for example, using biochar to increase the abundance of the microbes that have positive impacts on crop yield.

References:

[1] S. Nielsen *et al.*, "Comparative analysis of the microbial communities in agricultural soil amended with enhanced biochars or traditional fertilisers," *Agric Ecosyst Environ*, vol. 191, pp. 73–82, 2014.

CFD Modeling of Non-Newtonian Power Law Fluids in the Transition Regime

Courtney Palmeri, David G. Foster

Mixing is an essential operation used in the manufacturing and processing industries, and modeling the flow it creates throughout the tank is crucial to gain the level of understanding needed to optimize its efficiency. Flow regimes are characterized by their Reynolds number, a dimensionless quantity expressing the ratio of the inertial force to viscous force in the flow. For stirred mixing tanks, Reynolds numbers below 30 indicate the laminar regime and Reynolds numbers above 1000 suggest the turbulent regime. Governing equations exist to model flow in the laminar and turbulent regimes; however, this is not the case for the range of Reynolds numbers falling between these regimes, called the transition regime. In real mixing tanks, all flow regimes occur simultaneously. Turbulent-like flow tends to exist closer to the impeller while laminar-like flow tends to exist closer to the baffles; between these exists the complex transition flow. This project aims to create one model that can accurately model flow throughout all regimes using Carbopol, a non-Newtonian fluid that exists in many shampoos, dishwashing liquids and laundry detergents as the test fluid. Carbopol at viscosities of 3,355 cP, 7,500 cP, and 10,000 cP were modeled and compared in both a labatory mixing tank and using ANSYS Fluent, a commercial Computational Fluid Dynamic (CFD) software package. To model the mixing patterns in ANSYS, multiphase transient simulations that show the volume fraction distribution of dye released into the mixing tank were analyzed using different viscous models under the laminar-transition, transition, and transition-turbulent regimes for each starting viscosity. Blend times for the transition, transition-turbulent, and turbulent regimes were also calculated for the 10,000 cP Carbopol simulations using torque data from mixing Carbopol in the laboratory. The viscous models tested were the Scale-Adaptive Simulation (SAS) model, the Spalart-Allmaras Model (SAM), and the Detached Eddy Simulation (DES) model. The results suggest that the CFD simulations match well with the laboratory results and contribute to the further understanding of mixing in the transition region.

Development of Courses for Nuclear Chemical Engineering Education: Uses of Radioactive Materials in Science

Raul Velazquez, Benedek Szalai, Eric Palmer

Radioactive isotopes hold immense potential to contribute to societal advancement. To leverage New Mexico's strategic proximity to nuclear research and waste management facilities, an undergraduate general education course titled 'Discovery and Use of Radioactive Materials' has been developed. This course aims to provide students with a comprehensive understanding of the role isotopes play in scientific research and environmental monitoring. Foundational concepts in chemistry and physics will be covered to ensure a robust grasp of radioactive isotopes, with additional focus on the production, lifecycle, supply chain, and economic considerations of commonly utilized isotopes. The course will also emphasize the global applications of isotopes, such as Cesium-137 in industrial thickness monitoring and Uranium-235 in energy production. By fostering an informed understanding of the benefits and risks associated with radioactive isotopes, this course seeks to empower students to engage with critical issues in public health, energy policy, and environmental conservation, ultimately promoting scientific literacy and responsible decision-making in these areas.

Redesigning Heat Exchanger for Enhanced Air Compression in a Global Industry Leader's Separation Process

Camila Velandia, Jimena Zesati, Camila Franco

Water scarcity is one of the most critical global issues facing our world today, significantly affecting various sectors, including industrial processes. This urgent challenge has widespread implications for economies, societies, and the environment. To effectively mitigate the economic impact of this situation and safeguard the company's reputation in an increasingly competitive market, it is crucial to implement meaningful and strategic modifications to existing equipment. In this context, a comprehensive redesign of the heat exchanger was sought for a leading company in the production and supply of industrial gases, with the aim of optimizing the air separation process, particularly in the critical area of air compression. Notably, this redesign was conducted under the assumption of using only oxygen, effectively treating nitrogen and argon as if they were not present in the mixture.

The analysis of the current heat exchanger was conducted using detailed operational statistical data and flow information provided by the training partner. This thorough examination included calculating key factors such as global coefficients, the allowed fouling factor, and local heat transfer coefficients. The design modifications focused primarily on improving these critical coefficients by optimizing overall thermal resistance, Reynolds number, fluid velocity, and pressure drop in both the tubes and the shell.

The implemented changes successfully achieved a remarkable 26% increase in oxygen mass flow, significantly enhancing overall heat transfer efficiency. This substantial increase will allow the company to store and transport more oxygen, resulting in greater profits and improved operational capabilities. During the redesign process, essential regulatory standards were carefully considered to ensure that the results remained within the ranges established by heuristic rules. Additionally, relevant norms from associations such as TEMA were thoroughly researched to identify the specific parameters that the new design needed to meet.

Furthermore, the design changes resulted in a 7.4% reduction in water mass flow, thereby fulfilling two of the three objectives set by the company. These advancements not only improve operational efficiency but also strategically position the company to tackle the pressing challenges of water supply in the industrial sector. Ultimately, this approach contributes to sustainable practices, economic growth, and a positive corporate image in a world that increasingly values environmental responsibility. By focusing solely on oxygen in the redesign, the company can further streamline its processes and enhance productivity in the air separation industry.

Development of Courses for Nuclear Chemical Engineering Education: Review of Radioisotopes Used in Medicine

Eric Palmer, Catherine Brewer

Radioisotopes are an important part of modern medicine and are likely to play a larger role in some areas of medicine in the future. There are many factors that go into the selection of radioisotopes for medical practices: radioactive dose, time of exposure, chemistry, and half-life of the isotope. A review of radioisotopes in the medical field was done in preparation for a new undergraduate general education course, entitled, "Discovery and Uses of Radioactive Materials." The course aims to educate students about the opportunities, advantages, and disadvantages of using radioactive materials in everyday life, energy, medicine, defense, and science. For the medical applications literature review, articles and secondary resources were identified using Google Scholar, PubMed, Mayo Clinic, and then references within those resources. The medical field uses short-lived isotopes for imaging, diagnosis, and treatment of many different medical conditions such as cancer, chronic heart failure, and rheumatoid arthritis. Among the types of treatments using radioisotopes are external beam radiation therapy, brachytherapy, radiopharmaceuticals, targeted alpha therapy, radioembolization, and radio-synovectomy. There are many advancements being made in the field of radiation medicine that will affect how disease is treated in the future.

Blend Times with Dual-Impellers in a Stirred Vessel

Maria Logothetis, Adam Peru

In recent studies, researchers have made significant discoveries about the effects of baffles on the blend time in stirred tanks. In this study, the focus is on comparing, baffled and unbaffled vessels. Baffles change the direction of the flow of fluid from rotational motion to turbulent motion; to improve mixing. On the other hand, a vessel without baffles is usually considered to have undesirable fluid flows, leading to vortex formation, gas entrainment, segregation, and longer blend times. However, industries use unbaffled vessels for many process operations. The reasons cited for using unbaffled vessels are primarily concerned with cleaning the vessel between batches and the potential for stagnation in high-viscosity fluids, particularly when properties change over the processing time. Previous work has shown that the flow patterns are dramatically different between unbaffled and baffled vessels. There is a familiar turbulence throughout the vessel at high Reynolds numbers with the baffled configuration. Without the baffles the fluid segregates into two zones; a swirling inner zone near the shaft and a more turbulent outer zone. Two blend times were identified in using unbaffled tanks, a short blend time for the outer zone and a much longer time for the inner vortex-centered zone which defines the overall blend time for the vessel. At a low Reynolds number, the baffled blending is in thew transitional flow regime between turbulent and laminar with a combination of eddies and striations. The last region of the tank to become completely mixed is located behind the baffles. Unbaffled blend times are shorter than baffled and as the Reynolds number decreases. The vortex length is reduced with decreasing Reynolds number. Most of the blend time studies were reported on vessels with a single impeller. In an industrial setting, vessels may have a liquid height higher than the diameter of the vessel. In this vessel configuration, it is common in the industry to use multiple impellers placed on a single shaft. This research study examines the effect of dual impeller systems on blend time in both baffled and unbaffled tanks.

Exploring Cultural Capital Among Black Engineering Students at Minority Serving Institutions

Cheery Chukwukelu, David Horton Jr., Jeanette Jarvis, Jerrod Henderson

Despite ongoing efforts to increase the representation of Black students in engineering, participation among Black students remains low. While past research on Black students in engineering has highlighted the challenges they experience in higher educational institutions, a growing amount seeks to use assets-based perspectives to highlight their experiences. Although a significant number of Black engineering students graduate from MSIs, there has been less published research on their experiences compared to research in PWI settings. We draw and build upon the work of researchers who highlight how engineering students possess cultural wealth that allows them to succeed in engineering.

Using community cultural wealth (CCW), we explored how five Black engineering undergraduate students at Historically Black Colleges & Universities and Hispanic Serving Institutions cultivated and used their cultural capital. We used a qualitative approach that included semi-structured interviews. The interviews were analyzed using an iterative process of inductive and deductive coding. The six forms of cultural capital (linguistic, resistant, familial, social, aspirational, and navigational) served as *a priori* codes for deductive analysis.

We developed two themes that describe how students built, supported, and used their cultural capital, which included *Relationships as the Bedrock for Success* and *Intersections of Cultural Capital in Overcoming Challenges*. Our research findings revealed that familial and social relationships are crucial for early STEM engagement, emotional support, building aspirations, and developing needed skills for navigating engineering. Black engineering students were also sources of cultural capital for their peers, reinforcing their cultural wealth. To overcome challenges, Black engineering students used intersections of their cultural capital. These students employed combinations, layers, and multi-directional aspects of their cultural capital to persist and become successful in engineering. This study emphasizes the necessity for research acknowledging differences in Black students' experiences at HSIs and HBCUs and the importance of diversifying faculty and their approaches to engaging with students. We contribute to the body of research that uses antideficit approaches to examine the experiences of Black engineering students at MSIs.

Introducing Pollution Prevention and Life Cycle Analysis through Popcorn

Marcella McMahon, Emma Padros, Kirti Yenkie, Robert Hesketh, Brendan Weil, Alec Guerra, C. Stewart Slater, Barnabas Gao, An Pham

An important aspect of sustainability is preventing environmental pollution, yet many consumers overlook the implications of food packaging. A significant concern is that many paper and cardboard food containers are coated with 'forever chemicals.' These chemicals create a barrier against oil but complicate recycling and reuse, resulting in increased waste. Based on the curriculum, chemical engineers are well-equipped to handle this challenging problem.

One important tool for chemical engineers in evaluating sustainability is Life Cycle Analysis (LCA). This analytical method assesses the environmental impacts of a process from beginning to end, incorporating frameworks like cradle-to-gate (from raw materials to production) and cradle-to-grave (from production to disposal). Through LCA, engineers can evaluate material and energy flows, allowing them to select environmentally friendly materials, minimize energy use, and enhance process efficiency. The insights gained from LCA help in designing processes that minimize environmental impact and promote responsible resource use. Topics like pollution prevention and sustainability are not commonly covered in school curricula. Our aim is to engage and educate students on these topics to bring awareness to this cause as well as to motivate them to pursue engineering as a career so they shall develop the skills and knowledge to tackle such large-scale global problems.

We achieved this using a popular snack like popcorn. Four methods for popping corn were compared: a Hot Air Popper, Microwave, Movie Theater Machine, and a stove with a Jiffy Pop container. The Hot Air Popper and Movie Theater Machine allow users to pour in kernels directly, opposed to the Microwave and Jiffy Pop which relies on prefilled bags and aluminum containers. While the convenience of microwave popcorn is appealing, our experiment highlighted the importance of evaluating each method's environmental footprint. During the experiment, students used a wattmeter to measure the energy consumed by each technology. They then compiled data to perform a streamlined LCA, assessing metrics such as energy used, time taken, waste produced, as well as taste. This practical application of LCA helped them determine which method was more sustainable. Additionally, the experiment included a discussion on the science behind popping popcorn. Students learned about the three primary heat transfer principles which were conduction, convection, and radiation. In the Hot Air Popper and Movie Theater Machine, heat is transferred through convection as hot air circulates around the kernels. Jiffy Pop used heat transferred from a hot surface through the aluminum pan that heats the oil. Finally the oil heats the kernels until the explode or pop. In contrast, the Microwave uses electromagnetic radiation to excite water molecules within the popcorn, generating heat that causes the kernels to burst. Understanding these scientific principles enriched the students' overall learning experience. At the conclusion of the experiment, students evaluated all four methods, considered the waste generated, energy used, and tasted the popcorn produced from each technique. This reinforced their learning about LCA, sustainability, and pollution prevention. Introducing these important chemical engineering concepts early in their academic careers will invite them to consider sustainable practices in their futures.

Assessing the Impact of 'People-Oriented Recitation Problems' on the Interest of First-Year Intro Chemical Engineering Students

Gabriel Mendez-Sanders, Joanne Beckwith Maddock

Chemical Engineering (ChemE) curricula often don't demonstrate to students how core engineering concepts are applied in professional practice [1]. This disconnect may negatively impact students' interest and intention to pursue ChemE, particularly in their first year. The purpose of this study is to determine the effects of a recitation intervention on students in a R1 private university's introductory ChemE course. The researchers define a recitation problem as 'People-Oriented' if it is introduced by a current faculty member or alum who relates the problem's content to their field of work. The study's primary research questions are to determine if 'People-Oriented Recitation Problems' impact students': 1) interest in ChemE and specific topics therein?; 2) intention to pursue ChemE? To answer these questions, mixed methods will be used, analyzing data from students' academic performance, a pre- and post-semester survey, and student focus groups. The quantitative survey questions are modeled after the STEM-CIS instrument [2] and apply Social Cognitive Career Theory as a theoretical lens to understand students' career interest, choice goals, and outcome expectations. If 'People-Oriented Recitation Problems' are shown to increase students' interest in and intention to pursue ChemE, it may be a good learning activity to improve student retention.

[1] New Directions for Chemical Engineering. Washington, D.C.: National Academies Press, 2022. doi: 10.17226/26342.

[2] M. W. Kier, M. R. Blanchard, J. W. Osborne, and J. L. Albert, "The Development of the STEM Career Interest Survey (STEM-CIS)," Res Sci Educ, vol. 44, no. 3, pp. 461–481, Jun. 2014, doi: 10.1007/s11165-013-9389-3.

A Generalized Antoine Equation to Estimate Vapor Pressure Predicted By the SRK Equation

Elizabeth Cuneaz, Michael Misovich

The Soave-Redlich-Kwong (SRK) equation of state may be used to predict vapor pressure for any substance, but the calculations require a complex algorithm. A generalization of the simple Antoine equation replicates vapor pressure predictions of the SRK equation. This is done by introducing two similarity variables, adjusted temperature denoted by η and adjusted pressure denoted by Ψ. The goal of this project was to further develop this generalized Antoine equation. The first improvement used a computer code written in the Maple® programming language to search for generalized Antoine constants (A, B, and C) which more closely fit the SRK vapor pressure predictions. Four distinct sets of constants fit the vapor pressures within 0.05 percent for adjusted temperatures from 1.0, the critical point, down to 0.15. The second improvement was finding a single equation for each of the Antoine constants as a function of the upper limit of a user-specified adjusted temperature range. Using ranges of width 0.20, each of the constants was given by a single polynomial equation for adjusted temperatures as low as 0.40.

Understanding the Experiences of Undergraduate Students in STEM Research across Neurodiversity

Cassie Lafleur, Sarah Young, Mariah Arral, Elise Baribault, Julianna Gesun, Jeffrey Halpern

In this poster, we explore the varying experiences of undergraduate students engaged in STEM research, with a focus on neurodivergent individuals. Neurodivergent individuals often face marginalization and stigma to the point of feeling compelled to 'camouflage' or 'mask' their behavior to fit in with their neurotypical peers. To gain a deeper understanding of these experiences across neurodiversity, we developed a quantitative student survey, which was refined using cognitive interviews before being distributed to STEM undergraduates. Our survey allowed us to begin to understand the experiences and perspectives of undergraduate students in STEM research across neurodiversity, focusing on areas of communication, compensation, and interest in graduate school education. We have been able to assess and compare the initial responses between neurodivergent students and their neurotypical peers on these topic areas, providing hypotheses behind any potential disparities. We plan to further investigate the reasonings behind these differences through qualitative interviews so that we may provide recommendations for faculty to improve equity within their research laboratories.

Association Modeling of Ethanol + Chloroform + Dioxane

Adil Kolah, Carl T. Lira

Recently Killian et al. (https://dx.doi.org/10.1016/j.fluid.2021.113299, https://dx.doi.org/10.1016/j.saa.2022.121837, https://dx.doi.org/10.1021/acs.iecr.3c00968) have demonstrated that infrared spectroscopic data of alcohols in hydrocarbon exhibit cooperative hydrogen bonding that is inconsistent with the widely used Thermodynamic Perturbation Theory 1 (TPT1) association model implemented in the SAFT equation of state. Experimental spectroscopic data are more accurately modeled with Resummed Thermodynamic Perturbation Theory (RTPT) of Marshall and Chapman (https://dx.doi.org/10.1063/1.4834637). RTPT uses a smaller association strength for the dimer and a single larger association strength for all subsequent oligomers. Application of RTPT to phase equilibria was also demonstrated by Killian et al. The model has been published only for a single associating species (alcohol) in an inert solvent. This work extends both TPT1 and RTPT to an inadequately studied ternary mixture with both self- and cross-associating species (ethanol + chloroform + dioxane) and compares vapor-liquid equilibrium (VLE) modeling results from both to experimental VLE data from Gonzalez et al. (https://dx.doi.org/10.1021/je00034a018).

Vapor-liquid equilibria are calculated using $y_iP = x_i\gamma_iP^{SAT}{}_i$, where γ_i is the activity coefficient calculated by contributions as $\ln(\gamma_i) = \ln(\gamma^{association}{}_i) + \ln(\gamma^{combinatorial}{}_i) + \ln(\gamma^{attractive}{}_i)$. $\ln(\gamma^{association}{}_i)$ is the deviation caused by molecular association of molecules, $\ln(\gamma^{combinatorial}{}_i)$ is the deviation due to sizes of molecules, and $\ln(\gamma^{attractive}{}_i)$ is the empirically adjusted residual term for dispersion and polarity. This work focuses on modeling the association contribution to the activity coefficient for each component and the impact of the RTPT or TPT1 assumption when the same cross-association strengths are used to model the experimental data. The association contribution differs significantly for the two models. Spectroscopic data for mixtures are needed to determine which model is most appropriate.

Chemicals Deemed Non-Flammable Are More Dangerous Than They Appear

Ashley Sheehan

Knowing how to safely heat flammable chemicals is crucial for chemical engineers aiming to minimize fire hazards in plants and surrounding communities. Engineers heavily rely on information provided in a chemical's Safety Data Sheet (SDS) to determine the safest manufacturing processes. However, unforeseen safety hazards can arise when the SDS contains inaccurate information. Some heavily halogenated hydrocarbons are labeled as non-flammable because they appear not to have a flash point. Instead, an outgassing phenomenon occurs, creating an extreme flame rather than the expected blue flash upon ignition during flash point testing. This project aims to demonstrate that these hydrocarbons are flammable and will flash above ambient temperature. Numerous trials of five chemical species were conducted to test for flashing within a flammability chamber. All five tested chemicals flashed at relatively low temperatures (most below 42 °C). These results indicate that each of the five chemicals should be considered a flammable or combustible liquid, depending on the chemical's flash point. Based on these findings, it is recommended that the flash point of all hydrocarbon-based chemicals is tested. If a liquid chemical is determined to be flammable or combustible, its SDS should be updated accordingly.

Semi-Continuous *Ex Situ* Carbon Dioxide Mineralization in Produced Water for Calcite Production

Quinn Bennett

Flue gas, a high concentration point CO_2 waste stream created across many industries, offers an opportunity for capturing CO_2 at rates much greater than direct air capture. In this work, flue gas is contacted with produced water, a brine waste produced from hydraulic fracturing (fracking) that contains high concentrations of Ca^{2+} ions, to mineralize CO_2, forming calcium carbonate ($CaCO_3$) [1]. This mineralization route simultaneously sequesters CO_2 into a valuable commodity product, mitigates two common waste streams, and can utilize "off-the-shelf" technologies including common crystallization reactors for mineralization and a tandem chlor-alkali cell to provide hydroxide. Herein the design and operational conditions of the crystallizer are studied, and the chlor-alkali cell is simulated with a concentrated NaOH stream.

Crystallizer parameters including CO_2 concentration in flue gas, temperature, and reaction pH were studied in a two-level design of experiments. The $CaCO_3$ precipitation rate and the material properties of the collected product were characterized. Rates as high as 16.59 ± 0.37 $\mu mol.L^{-1}.s^{-1}$ were sustained in a bench scale (1 L) crystallizer allowing full recovery (~100% yield) of Ca^{2+} as $CaCO_3$. The reactor design also influences the crystal structure of the collected precipitate. Examples of designs that will produce a majority aragonite (orthorhombic) structure as opposed to the more stable and valuable calcite (trigonal) structure will be shown. Effects of reactor parameters on particle size were also studied. Overall, this study of CO_2 mineralization utilizing flue gas and produced water is aimed at providing the necessary information to scale this process to a pilot scale (200 L crystallizer) capable of mineralizing >10 kg CO_2/day as well as informing techno-economic analyses of this promising process.

[1] Zhu, B.; Wilson, S.; Zeyen, N.; Raudsepp, M. J.; Zolfaghari, A.; Wang, B.; Rostron, B. J.; Snihur, K. N.; von Gunten, K.; Harrison, A. L.; Alessi, D. S. Unlocking the Potential of Hydraulic Fracturing Flowback and Produced Water for CO2 Removal via Mineral Carbonation. Applied Geochemistry 2022, 142, 105345. https://doi.org/10.1016/j.apgeochem.2022.105345.

Modulation of Hydrophobic Organic Contaminant Toxicity to Soil Protists Via Co-Exposure with Microplastics

Mehr Chhatre, Naomi Adler, Christopher Hawxhurst, Leslie M. Shor

With the increasing demand for disposable and low-cost goods, the use of plastics and thereby the amount of microplastics in the environment have been growing. Microplastics absorb contaminants, thus, as microplastics are transported through the environment they carry absorbed contaminants with them. The impact of this pollution displays itself all throughout the environment, affecting the various microorganisms of the soil ecosystem, including soil protists. Protists regulate the soil bacterial population through grazing and transport which in turn also modulates soil fertility and plant productivity. Soil protists are also a significant food source for larger microorganisms, and as hydrophobic contaminants are known to bioaccumulate up the food chain, these contaminants can ultimately reach higher order predators such as humans while beginning at soil protists. Here we show the effect of microplastics to modulate the toxicity of a model hydrophobic organic contaminant (HOC) to a model organism. We measure lethal as well as sub-lethal toxicity endpoints including protist growth rate and motility. Naphthalene was selected as a model HOC because its hydrophobicity (quantified as Log K_{ow}) is in the range of many common HOCs likely to sorb to microplastics under environmental conditions, also because naphthalene can be readily quantified in aqueous solution by optical methods, as well as its known toxicity to soil protists. *Colpoda Steinii*, a large soil ciliate, was selected as the model soil protist. The microplastics (MPs) selected are small spheres of polyethylene (size ranging between 35 and 45 micrometers), the most commonly used plastic in the world, being lightweight and versatile. The lethal endpoint of the protists when exposed to the contaminants was determined, by utilizing an aqueous solution of naphthalene as well as an aqueous solution of naphthalene treated with MPs. Replicate chambers were set up with equal number of viable protists, and then various concentrations of the two naphthalene solutions. The protists were left in these solutions for 24 hours, after which the fractions of living protists were determined, and plotted to measure the percentage of protists killed versus total naphthalene added (both MP-free and MP-containing) to demonstrate the effect of toxicity from co-exposure of HOC with microplastics. Going further, sub-lethal effects of lower naphthalene concentrations will be studied, such as the impact on growth rate and protist motility. Through modeling the impact of HOCs on the soil ecosystem, specifically on soil protists, we can better understand their detrimental effects on the environment, particularly on their conjunction with MPs, an environmental contaminant not yet fully understood. Thereby ultimately being able to predict and mitigate these effects on a larger environmental scale.

Chitosan-Based Hydrogel Beads for Encapsulation of Bacteria for Bioremediation of Chloride Solvents

Aira McDaniel, Celia H'Almeida, Eleanor Cole, Jaida Izen, Andrea Valle

Chlorinated compounds, such as 1,1,1-trichloroethane (TCE) and 1,4-dioxane, are found in many groundwater aquifers that supply drinking water to the US. TCE was used commonly in the dry cleaning of clothes and degreasing of engines on many Air Force bases. These contaminants are linked to causing long-term health complications in humans. Although remediation techniques like pump and treat exist, such methods are simply not effective enough to manage the spread of these compounds. These contaminants can be broken down by bacteria through a process called cometabolism. Rhodococcus rhodochrous ATCC 21198 is a bacterium capable of degrading these specific chlorinated compounds. A community of bacteria can be created by immobilizing them inside of hydrogel beads. The ultimate goal is to place the beads into a perforated column in a groundwater aquifer for passive bioremediation. However, the beads will be stacked on top of each other and must withstand the weight of the bead pack. Based on the chitosan coating techniques of Sarma et al., our research examines if chitosan-coated sodium alginate beads would be stronger than sodium alginate beads alone, or if stronger hydrogel beads could be made by using only chitosan. The chitosan beads also needed to be tested to determine if they are habitable for the bacteria.

Therefore, the engineering objectives of the research project were:

(1) Develop a method to increase the modulus (stiffness) of current sodium alginate

beads using a chitosan coating.

(2) Develop a method to produce chitosan beads with a higher modulus (stiffness) than

sodium alginate beads.

(3) Immobilize ATCC 21198 bacterium in chitosan beads and test for their viability,

since chitosan is known to have antimicrobial properties.

These trials showed that not only is chitosan weaker than sodium alginate, it is also not a viable method of bacteria immobilization. This could be attributable to the more acidic pH from the addition of glacial acetic acid, and the antimicrobial properties of the chitosan used in the beads.

Uranium Absorption Onto Biochars

Michael Fry, Shermal Fernando, Eric Palmer

Heavy metal and radionuclide contamination waste (such as that from runoff from former uranium mines) seriously threatens clean water sources in several regions, including the Four Corners area of the Southwest. This study evaluated the potential of biochar derived from waste biomass as a low-cost absorbent to remove uranium from contaminated water. Biochars from the pecan, cattle, and hemp industries were tested in batch adsorption studies focusing on the effects of initial uranium concentration, solution pH, and contact time. Hemp residue-derived biochar has significantly higher absorption than those derived from the pecan wood and cattle manure. Studies conducted been at a concentration of 1ppm, with a pH of 3-4, for time intervals as long as 72 hours. In future work, batch testing will be conducted at different concentration and pH ranges; In addition, the uranium loaded biochars will be combusted to analyze radionuclide fractionation between gas/vapor and ash, providing insights into the potential for radioactive waste volume reduction without regeneration of the absorbents

Interactions of PFAS with Common Surfactants and Dissolved Solids

Rohini Ghosh, Tyler Durkin, Suchol Savagatrup

As a result of their use in a variety of consumer products and industrial processes, the widespread contamination of per- and polyfluoroalkyl substances (PFAS) in the water supply has been linked to adverse impacts on both human health and the environment. This has necessitated the development of rapid and on-site detection and remediation efforts. In the context of environmentally relevant systems, contaminant PFAS molecules exist in tandem with a variety of other pollutants, some of which may also be surface active or involved in electrostatic interactions. In this project, we sought to characterize how the interactions between PFAS molecules and other environmental contaminants, particularly surfactant molecules and metal ions, alter the interfacial behavior of PFAS. Disentangling the interactions within these systems is necessary for the development of a sensing device which can accurately distinguish and quantify PFAS contamination. We observe that the interactions follow either synergistic or antagonistic behaviors, depending on the structural factors of the compounds involved. We identify several key factors that dictate such behaviors, including (1) headgroup charges, (2) length of the hydrophobic tail group, and (3) chemical identity of the hydrophobic tail group. Our results demonstrate how structural characteristics dictate the interaction of PFAS with other compounds and brings into focus potential challenges sensors may have in accurately differentiating between PFAS and other contaminants.

Treatment of Cyanobacterial Harmful Algal Blooms (cHABs) Using Microbial Fuel Cells

Nada Shetewi, Amanda Simson, Radmila Janjusevic

Cyanobacterial harmful algal blooms (cHABs), driven by toxin-producing cyanobacteria, are increasingly prevalent due to climate change factors, such as rising water temperatures and nutrient pollution. A major concern associated with cHABs is the production of cyanotoxins like microcystins, with microcystin-LR (MC-LR) being among the most toxic and widespread, produced primarily by *Microcystis aeruginosa*. Microbial fuel cells (MFCs) have emerged as a promising method for the on-site treatment of freshwater contaminated by cHABs. MFCs employ electrochemically active bacteria (EABs) to convert organic matter into electricity, leveraging microbial communities found in environments like soils, sediments, and aquatic systems. These bacteria perform anaerobic oxidation of organic matter, transferring electrons to an electrode in a closed circuit. Given the nutritional content of cyanobacterial biomass, specifically *Microcystis aeruginosa*, it serves as an effective substrate for EABs, supporting both energy production and water treatment in MFC systems. However, MC-LR secretion by *M. aeruginosa* can inhibit critical microbial processes, diminishing MFC efficiency in both electricity generation and bioremediation. Recent advancements have incorporated biodegradable co-substrates to boost EAB growth and enhance MC-LR degradation within MFCs. This research investigates the interactions between *Microcystis aeruginosa* and EABs within MFCs, specifically targeting strains and cyanobacteria found in Central Park Lake. The study evaluates the potential of MFCs to effectively treat organic matter and degrade MC-LR, aiming to enhance bioremediation strategies for cHAB-contaminated waters.

Removal of PFAS from Water Using Zeolites

Abigail Peters, Charles Ponge, Mark Shiflett

Per- and Polyfluoroalkyl Substances (PFAS) are man-made chemicals, found in almost all water sources around the world: including drinking water, and are detrimental to our health. Accurate PFAS analysis is challenging due to contamination from their prevalence in most products. The latest EPA regulations (April 10, 2024) require all public water facilities to remove PFAS to a specified level by 2029. PFOA and PFOS need to be below 4.0 ppt, while PFHxS, PFNA, and GenX need to be below 10 ppt, and PFHxS, PFNA, HFPO-DA, and PFBS mixtures need to have a hazard index below 1. Their removal from water using an effective and sustainable method is currently unavailable. Project DISCOVER focuses on PFAS removal using zeolites. Zeolites are alumina-silica structures that have molecular selectivity through diffusion. Compared to other methods such as granular activated carbon, ion-exchange resins, and membranes, zeolites are easily reusable. In this study, zeolite beta CP814E* was tested to see its effectiveness at PFAS removal, as well as CP814E* modified for ion-exchange of the hydrogen with sodium and potassium. All materials were HDPE or PP to ensure no PFAS contamination and the reusable materials were washed with methanolic ammonium hydroxide. Zeolites were placed in water, spiked with varying concentrations of PFAS; K-PFOS, PFOA, 6:2 FTS, PFHxS, PFHxA, PFBS, and PFBA. The samples were vortexed, shaken for at least 24 hours, and centrifuged followed by a 50% dilution with methanol. After centrifugation, a final sample was spiked with internal standards. PFAS identification/quantification used LC/MS-QQQ. This was repeated for kinetic data collection, using syringe and syringe filters to sample in under 30 seconds. The study showed that zeolite BEA can adsorb K-PFOS and PFOA, with more than 90% removal. However, PFBA is most challenging to adsorb, with less than 5% removal. The kinetic data shows the adsorption process goes to its equilibrium concentration in under 30 seconds for K-PFOS and PFOA. The porous material identified could be scaled up for large scale pump/column separation.

Ion Exchange Chromatography for Enhanced Phosphorus Removal from Wastewater

Ria Patel, Morwan Osman, Arturo Coronel, Milena Zeru, Sebastian Oakes, Jennifer Le, Emiri Tsubouchi, Gabe Miguelino, Ga-Eun Kim, Nancy Le, Jameson Draney, Emerson Parker, Osama Edobaa, Chloe Pracharktam, Michelina Fiteny

Addressing excess phosphorus in wastewater is crucial, as eutrophication threatens aquatic ecosystems and biodiversity through the creation of harmful algal blooms and hypoxic zones. Effective management of phosphorus pollution is vital for preserving water quality, sustaining biodiversity, and supporting livelihoods reliant on healthy ecosystems. While various removal technologies have been investigated, there remains a critical need for improved phosphorus management systems in the San Diego area. Here we demonstrate that continuous flow ion exchange columns efficiently remove phosphorus from simulated wastewater solutions. The columns packed with cationic ion exchange resin beads utilize electrostatic attraction to filter out phosphate anions. As wastewater flows through the chromatography column, the phosphate anions engage in an exchange reaction with the ions present in the resin. We analyzed the removal process using ultraviolet-visible spectroscopy, calibrated to measure phosphate concentrations in both the initial wastewater solution and the column effluent. The integration of custom-designed, 3D-printed components significantly enhanced the column's efficiency, resulting in reduced trace phosphate levels in the effluent. After the ion exchange process, the cationic resin can be regenerated using a weak base solution, facilitating the recovery of isolated phosphate anions for reuse as a fertilizer source. This streamlined ion exchange column presents a scalable solution for phosphorus removal from wastewater, contributing to a sustainable approach to mitigating eutrophication and minimizing waste.

Cohesin-Dockerin-Based Multi-Enzyme Assembly for Efficient PET Depolymerization

Kira Scotia Kenney, Brian Ortiz, Siddhant Gulati, Qing Sun

Efficient depolymerization of recalcitrant polymers like cellulose and plastics requires the synergistic performance of multiple enzymes. For the complete depolymerization of polyethylene terephthalate (PET) plastics into their constituent monomers, two enzymes- PETase and MHETase- are required. Cellulosomes, which are multi-enzyme complexes on the cell surface of many anaerobic bacteria, serve as inspiration for directed assembly of multi-enzyme clusters. Functional assembly of enzymes onto designer cellulosomes has shown improved performance over free enzyme mixtures.

To achieve complete depolymerization of PET, we aimed to assemble a multi-enzyme cluster using highly-specific cohesin-dockerin interactions observed in cellulosomes. In this work, we fused PET-degrading enzymes FAST-PETase and MHETase enzymes with dockerins from three distinct cellulolytic bacteria - Clostridium cellulolyticum, Clostridium thermocellum, and Ruminococcus flavefaciens. These fusion proteins were expressed in Escherichia coli and purified using Histag purification. Enzyme activity was subsequently tested on a soluble pNPA substrate and a solid PET film substrate, and all three enzymes were observed to be active.

These enzymes could then be successfully attached to cohesin domains on a trifunctional scaffold in displayed on the surface of a Saccharomyces cerevisiae cell. It was observed that the enzymes retained activity after binding. This work is an important step in designing multienzyme complexes for PET depolymerization.

Mycoparasitic Marvels: Trichoderma's Role in Sustainable Pest Solutions

Shravani Gajendragad, Shruti Dhawade, Suraj Mali, Ishwari Khade, Shubhangi Rastogi

India is an agriculture based country. About 65% of the country's population is engaged in agriculture and its allied activities. With the exponential increase in the population necessity for exponential productivities have forced us towards indiscriminate use of pesticides and fertilizers resulting in disturbed soil composition, fertility and microflora. As the population is increasing exponentially it is necessary that the yield also parallelly increases. Amid the growing challenges of global warming, agriculture has been the most affected sector. The growing climate change, use of chemicals, affects the quality of the yield. Another setback is the degrading fertility of soil. So it is necessary to find an alternative to improve the quality and quantity of yield while protecting the fertility of the soil. Disturbed microflora has given birth to new pests and diseases posing a significant threat to agricultural productivity. The most common agricultural pests are Cotton bollworm *(Helicoverpa armigera),* Tobacco whitefly *(Bemisia tabaci),* Two-spotted spider mite *(Tetranychus urticae)* etc. Chemicals such as glyphosate, acephate, DEET, propoxur, metaldehyde, boric acid, diazinon, Dursban, DDT, and malathion are being widely used to combat infestations of these organisms. Although these chemicals are effective, they ultimately find their way to the human gut and animals causing serious health concerns. This resulted in the need to explore novel alternatives to these substances. Recent advancements in the field of biotechnology has led to the use of microorganism based fertilizers and pesticides. It is both environmentally sustainable and an effective method. *Trichoderma* is a genus of fungi renowned for its significant contributions to agriculture and environmental management. These remarkable microorganisms play a vital role in sustainable farming practices by effectively controlling plant pathogens, enhancing soil health, and promoting plant growth. Their ability to suppress diseases and improve nutrient uptake makes them indispensable allies in the quest for environmentally friendly agricultural solutions. Studies conducted to date have demonstrated that *Trichoderma* can directly control insect pests through mechanisms such as parasitism and the production of insecticidal secondary metabolites, antifeedant compounds, and repellent metabolites. Consequently, the use of *Trichoderma* in agriculture proves to be effective not only against plant pathogens but also against insect pests, highlighting its potential as a sustainable alternative for future agricultural practices. This study is based on development of low cost production medium and formulation to be applied directly in the field or on seeds to help build agri-economics.

Sawdust As a Biopolymer for Packaging

Shirin Mohamed

With the increasing demand for sustainable and eco-friendly materials, there is a growing interest in repurposing agricultural waste for various applications. Sawdust, a byproduct of wood processing industries, is abundantly available and poses environmental challenges when not properly managed. This research explores the feasibility of utilizing sawdust as a renewable and biodegradable resource for 3D printing and molding applications. We investigated the solubility of our binder, ratio of sawdust, effect of plasticizer. Alongside, this research looked at the consistency, mechanical properties, and blowing capacity of the products from the furnish of binding agent and sawdust was monitored.

Characterization of Marine Bacteria for Bioplastic Degradation in Ocean Environments

Diana Sullivan, Tiana Rohe, Anne Meyer

Plastic pollution is one of the biggest environmental issues, specifically its impact in ocean environments. Nearly 1-2 million tonnes of plastic waste ends up in the ocean every year, harming the environment by disrupting ecosystems, breaking down into microplastics, and releasing toxic chemicals. Bioplastics offer a viable alternative to traditional petroleum plastics because of their ability to biodegrade. However, most bioplastics require specific conditions, such as warm, low oxygen environments, which are not present in marine environments naturally, in order to properly break down. Poly(3-hydroxybutyrate) (PHB) has been discovered as a naturally occurring polymer that can be used as an alternative plastic material. It is produced by bacteria as an energy storage molecule and can be metabolized in the Krebs cycle, therefore does not produce any harmful waste. Moreover, some strains of bacteria have developed the ability to break down PHB produced by other cells. Research has shown that synthetic biology can be used to take advantage of the PHB degrading properties of these strains to accelerate the rate of degradation of ocean plastics. Thus, they continue to be investigated to find possible applications and uses for a sustainable plastic alternative.

PHB-degrading bacteria strains were collected on submersed sample discs off the coast of California in the Santa Barbara Basin and the UCSB Lagoon. These bacteria strains were then analyzed and characterized in order to determine further specific applications. Plate readers were used to determine the growth rate for each strain. A 3D bioprinting technology developed in the Meyer Lab was used to create hydrogel based "biostickers" that allow for stabilization and longevity of the cells, as well as controlled analyses and applications of these strains to PHB products. Additionally, PHB degradation experiments via clear-zone analysis were conducted with the biostickers on PHB-powder agar plates at different temperatures to determine the rate of degradation. To determine the best conditions for performing long term experiments, humidity conditions were varied. The results showed various growth patterns which will allow for different applications of the bacteria, such as use for PHB degradation at different ocean depths or intended lifespan of PHB products. Furthermore, storing the degradation tests in a box with a beaker of autoclaved water has proven to be the best set up to prevent plates from drying out.

Magnetic Nanocomposites for the Remote Activation of Sulfate Radicals for the Removal of Rhodamine B

Marissa Nicholson, Pranto Paul, J. Zach Hilt

This project explores the potential of utilizing the exposure of a magnetic nanocomposite system to an alternative magnetic field (AMF) to remove organic pollutants from water and investigates the effects of sulfate radical activation on removal efficacy. When exposed to an AMF, magnetic nanocomposites become locally energized, and the associated heating can degrade contaminants. This research addresses the need for sustainable water treatment methods due to the potential threats of organic hazards on the environment and human health. The project aims to determine the effects of AMF and sodium persulfate (SPS) exposure on contaminant removal. The magnetic nanocomposite system consisted of a polyacrylamide-based hydrogel loaded with iron oxide magnetic nanoparticles (MNPs). Rhodamine B (RhB) dye was used as a model organic contaminant, allowing for degradation to be measured quantitatively using UV-vis spectroscopy. Results indicated that RhB removal by the nanocomposite without SPS was proportional to the concentration of MNPs in the nanocomposite. With the addition of SPS, 100% of the RhB was removed within 20 minutes. The removal was attributed to the activation of sulfate radicals by the MNPs and the heating caused by AMF exposure. This magnetic nanocomposite system shows potential as a unique water treatment method for environmental contaminants.

Cellulose Based Bio-Adsorbents for Removal of Chromium (Cr)

Sable Phillips

Increased use of heavy metals in production has caused a rise in pollution due to metal-based industries such as mining metal, smelting, and foundries. Chromium specifically has been found to have adverse health effects such as DNA damage, neurotoxicity, and cancer. Our goal is to remove chromium from water by using an environmentally friendly adsorbent based on biochar. All the biochar adsorbents prepared are cellulose based. First, cellulose is converted to biochar using hydrothermal carbonization which uses water as a green solvent. Biochar is then impregnated with potassium hydroxide (KOH) at varying concentrations. This impregnated biochar is then activated at 700, 750, 800, 850 Celsius and is labeled as one-step activation. A two-step activation is also prepared, by first taking the cellulose biochar and activating it at 1000 Celsius in an inert atmosphere (nitrogen). Once thermally activated, chemical activation of this biochar then follows the same procedure as the one step activation The adsorption capacity of each one of these different bio-adsorbents was then tested by setting up adsorptions at a concentration of 1.5g/L of the bio-adsorbent in chromium solutions (50, 100 and 200 ppm) both with an adjusted pH of 2 (the experimentally determined point of zero charge, PZC where theoretically adsorption should work best) and a non-adjusted pH (at neutral pH). Once the adsorption is complete after 24 hours, the solution was analyzed with a flame atomic adsorption spectrometer (FAAS). It was observed that as activation temperature increased, adsorption capacity of all bio-adsorbents increased. Comparison between the one step and two step activation, it was seen that at neutral unadjusted pH, two step activation produced better results than 1-step. However, at the PZC (pH 2), the reverse was true: 1-step activation did significantly better than the two step activation. When comparing the two methods at pH2, one-step activation fares better than 2-step at all chromium concentrations.

Sustainable Copper Extraction: Transforming Iron Waste into Jarosite for Improved Vanadium Recovery

Briana Franco

The shift towards renewable energy is increasing the demand for batteries and, consequently, the need for efficient extractions of key elements such as copper, lithium, and nickel. Hydrometallurgy, which uses aqueous solutions to extract metals, often involves leaching to selectively dissolve desired metal compounds while leaving the impurities in solid form. Managing iron waste to prevent environmental contamination while maintaining leach cycle efficiency is a continuous industry challenge. To address this, the West Lab is investigating a novel leaching process for copper extraction that aims to enhance waste management and maximize the recovery of vanadium– a valuable and costly reagent used in the leaching process. The proposed iron control method converts iron from the post leach solution (PLS) into an iron-sulfate solid, Jarosite. Adapting an iron-jarosite synthesis process from zinc hydrometallurgy, the optimal pH, temperature, and cation conditions were identified to maximize jarosite precipitation and minimize vanadium content in the solid form. A 100mL batch reactor was established with synthesized PLS to develop an effective and repeatable separation procedure at lab scale. Analytical techniques, including Inductively Coupled Plasma (ICP) and Powder X-Ray Diffraction (PXRD), confirmed a K-Jarosite purity of 92.6% and a vanadium recovery of about 80% in solution. Future work aims to explore additional metallurgical methods for improving vanadium and iron separation, reducing environmental impact by decreasing vanadium losses and establishing a stable iron waste product for safe disposal or repurpose in other industries, such as steel production.

Hydrogel Coatings for Urban Heat Resilience

Colby Snyder, Russell J. Composto

Human activities and urban infrastructure elevate city temperatures significantly above those of surrounding rural areas. This process, known as the urban heat island effect, exposes city residents to dangerously high temperatures in summer. We take a dual approach to treating urban heat, partnering with the Philadelphia Office of Sustainability (OOS) to pair fundamental materials research with an applied pilot project. Dark-colored surface coatings that absorb a high fraction of solar radiation are a major cause of extreme heat, and materials that prevent solar heat from entering the environment can passively cool buildings and cities. However, existing cool surface coatings are limited by the atmospheric IR transparency window, which can be blocked on humid or cloudy days. A promising alternative is evaporative cooling, which circumvents atmospheric transparency issues by absorbing moisture overnight and releasing it when heated by sunlight. For this study, we formulated a biocompatible hydrogel consisting of sodium carboxymethylcellulose (NaCMC), polyethylene glycol (PEG), and citric acid (CA). The cooling power of the coating was quantified in laboratory and outdoor environments, showing surface temperature reductions exceeding 10 °C. Beyond the laboratory, a partnership with the Philadelphia OOS to evaluate a new cool pavement coating is described. This work uses materials science to guide public policy, bridging the gap between academic research and climate solutions.

Exotic Fruit Residue Hydrochars: Sustainable Materials for Circular Engineering Applications in Environmental Remediation, Electronics and Agriculture

Camily Rivadeneira, Erick Mendizábal, María Fernanda Lecaro, Yuliana Pullas Gallardo, Victoria Alomía, Luis M. Prócel, Andrea C. Landázuri

The excessive use of non-renewable materials and products derived from fossil fuels has intensified the need to find alternative and sustainable routes to replace them with bioresources in recent decades. These bioresources are by-products or residues from second-generation biorefinery processes. Hydrochar has emerged as a promising solution due to its properties and applications. This carbonaceous material, generated from the hydrothermal carbonization (HTC) process, retains exceptional properties such as high carbon content, stability, capacity to absorb and eliminate contaminants, and high surface area. These properties make it a suitable candidate for various environmental and agricultural applications, contributing to the main objectives of circular and green engineering.

In Ecuador, the primary feedstock for hydrochar comes from diverse biomass waste materials, including sugarcane bagasse, oil palm fiber, peel and waste, as well as banana, plantain, rice, and fruit waste. This research focuses on potential hydrochars derived from various exotic fruit residues such as mango endocarps, moringa husks, palm residues, mangosteen, and blackberry bushes. The raw material undergoes preliminary analysis to define operating conditions for the hydrothermal process to produce hydrochar and HTC liquor.

The final product is subjected to different chemical analyses, including morphology characterization using BET and SEM to determine porosity and assess the hydrochar's capability to adsorb various molecular species. Additionally, the samples are analyzed using Fourier Transform Infrared spectroscopy (FTIR) and X-ray diffraction to identify functional groups, structural changes, and crystalline compounds. Finally, a semiconductor device parameter analyzer measures voltage retention capacity at different frequencies.

Initial results indicate that the generated materials are suitable for insulation, dielectric applications, and potential agricultural purposes. Furthermore, the study highlights the influence of fruit type, hydrothermal carbonization conditions, and post-treatment processes on the electrical characteristics of hydrochars.

In conclusion, hydrochars from exotic agroindustrial residues show potential as an innovative and sustainable alternative for waste management practices, promoting circular engineering and bio-economy. This research contributes to the development of eco-friendly materials with diverse applications, addressing the growing need for sustainable solutions in various industries.

Mechanochemical Depolymerization of Poly(ethylene terephthalate) - Rigorous Characterization of a Single Collision

Lauren Mellinger, Kinga Golabek, Carsten Sievers

Understanding the mechanisms involved in alkali-depolymerization of poly(ethylene terephthalate) (PET) is essential in optimizing this polymer recycling process. Complete depolymerization of PET into disodium terephthalate (Na_2TPA) and ethylene glycol (EG) is achieved by ball milling it with sodium hydroxide (NaOH). This depolymerization can be conceptually described by the shrinking core model, where the NaOH reacts with the interface of the PET particle, achieving full monomer yield after 20 minutes.[1] To better understand this process, a detailed study of the single model collision was conducted by impacting pure PET and NaOH coated PET with a milling ball.

The experimental setup included a height-adjustable drop collision system which utilized an electromagnet to release the steel ball. PET films of varying thicknesses were coated with a suspension of NaOH in methanol and dichloromethane, leaving ~5 mg of NaOH on the film surface. The ball was dropped from different heights, where the standard height of 1.07 m correlated to a ball velocity of approximately 4.58 m/s. This value was within the range of collision velocities calculated for the upcycling of PET in a ball mill operating at 30 Hz.[1] The diameter and depth of the crater formed from the impact were examined with an optical microscope. Additionally, the extent of structural and chemical changes were analyzed using spatially-resolved Raman spectroscopy. Spectra were measured at random points on the unimpacted film and across the diameter of the crater of impact.

The first band analyzed in the spectrum was 1292 cm^{-1}, which is characteristic of the ester group in PET. The area reduction of this band was measured to determine the PET conversion from a single collision. For the NaOH coated film at the standard drop height, a conversion of roughly 25% was observed, with lower conversion in the outer regions of the crater. On the uncoated PET films, the conversion inside the crater did not exceed 4%.

Another important spectroscopic band was the carbonyl vibrational band at ~1730 cm^{-1}, which combines several subcomponents representing phases of the PET chain. The peak at 1726 cm^{-1} represents the crystalline state, where the carbonyl groups in the terephthalate units are coplanar with the benzene ring and glycol units are in the *trans* conformation. The peaks at 1721 cm^{-1} and 1733 cm^{-1} correspond to the amorphous state of PET, where terephthalate units are completely disordered and glycol units are in *gauche* conformation.[2]

The peak areas were calculated for each film to establish a crystallinity basis ratio in PET before and after impact. In the uncoated films, there was a significant increase in the amorphous phase after the collision. Conversely, the NaOH coated films retained a higher crystallinity, implying that the NaOH layer prevents significant amorphization. This trend was observed across varying film thicknesses and different ball drop heights.

Overall, this characterization of the single collision allowed for a detailed understanding of the amorphization of the sample. This new insight into the process of depolymerization of PET is important in improving the efficiency of PET recycling at a commercial level.

References

1) W. Tricker, G. Samaras, K.L. Hebisch, M.J. Realff, C. Sievers, Hot spot generation, reactivity, and decay in mechanochemical reactors, Chemical Engineering Journal 382 (2020). https://doi.org/10.1016/j.cej.2019.122954

2) S. Fleming, K.L.A. Chan, S.G. Kazarian, FT-IR imaging and Raman microscopic study of poly(ethylene

terephthalate) film processed with supercritical CO2, in: Vib Spectrosc, 2004: pp. 3–7. https://doi.org/10.1016/j.vibspec.2003.10.003.

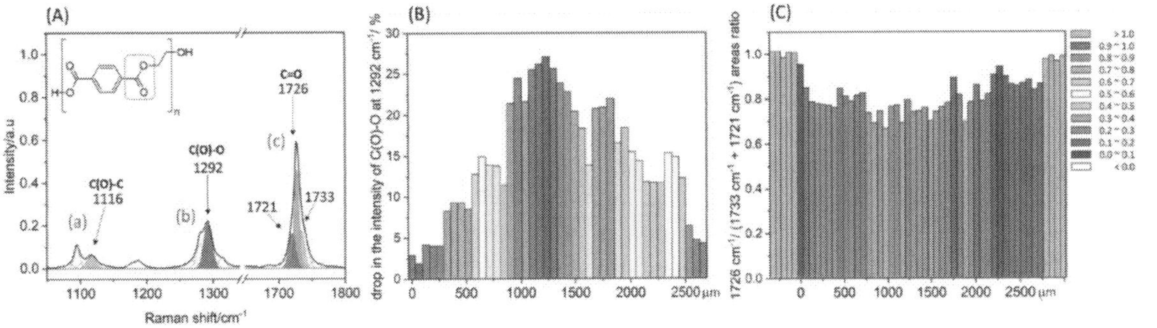

Figure 1. Raman spectra with bands of interest for identifying PET conversion and crystallinity (left), conversion of polymer determined by 1292 cm^{-1} band area across the crater diameter (middle), and crystalline to amorphous domains ratio of PET across the diameter of the crater (right).

Nanobubble Applications in Graphene Adsorption of Hydrophobic Pollutants: Towards Wastewater Treatment on the International Space Station

Madison McCarthy, Kenneth Mensah, Manisha Choudhary, Sonia Moavenzadeh, Onur Apul

The National Aeronautics and Space Administration (NASA) aims to improve the wastewater treatment system efficiency on the International Space Station by preventing the accumulation of dimethylsilanediol (DMSD), part of the siloxane chemical group, in the filters. Adsorption has proven to be a potentially effective approach. Meanwhile, recent literature has indicated that nanobubbles in water can improve the hydrophobic interactions between hydrophobic adsorbents like carbon and organic pollutants such as per- and polyfluoroalkyl substances (PFAS), thereby enhancing the pollutant removal efficiency. However, it is not yet established whether this nanobubble-enhanced adsorption phenomenon applies to DMSD. Because DMSD is highly volatile and difficult to measure, a proof-of-concept adsorption study was conducted with phenanthrene (PNT), a model organic hydrophobic pollutant, as a preliminary step towards DMSD adsorption studies. Adsorption tests were performed using graphene, graphene oxide, and reduced graphene oxide and different adsorption matrices viz degassed water and nanobubble water. The presence of NBs decreased the removal of phenanthrene by all three graphene derivatives. Therefore, a potential mechanism is that NBs and PNT compete for the adsorption sites on the graphene and due to the higher concentration of NBs, less PNT is adsorbed to the graphene. However, blank experiments performed without graphene demonstrated that PNT volatilization occurred in degassed and deionized water but was prevented in NB water. These data suggest a second potential mechanism: PNT has a higher affinity for NBs than graphene and their interactions alter the surface chemistry of the NBs such that they no longer adsorb to graphene. Further experiments will be conducted to determine the mechanism for interactions between PNT, NBs, and graphene and to understand why PFAS and PNT behave differently in a NB-graphene adsorption system. This research is crucial for developing technology that can manage spacecraft waste efficiently and with limited resources, allowing NASA to continue expanding its space exploration.

Effect of Keratin on Adsorption of Copper Ions Using Sodium Alginate

Jonathan Ackley, Catherine B. Almquist

Wastewater containing heavy metals are a byproduct of industrial processes in a wide variety of sectors, not limited to construction, electronics, and mining. Heavy metals are toxic, and not only pose a threat to human health, but to the health of ecosystems and the environment. Many different methods to remove heavy metals from waste streams have been developed to remove heavy metals from water, including chemical precipitation, membrane filtration, ion exchange, and adsorption, among others. Of these methods, adsorption is often considered an economically viable and environmentally-friendly method to remove heavy metals from solution.

One type of adsorbent is a biosorbent, which is made from organic compounds to adsorb metal ions in wastewater. Two bioadsorbents that have received recent attention are adsorbents made with alginate, a biopolymer derived from seaweed, and keratin, a protein found naturally in hair, wool, and feathers, among other sources. These two biopolymers are high in carboxyl and hydroxyl groups, which bind metal ions, allowing for the facilitation of adsorption of metal ions. Additionally, these materials are sustainable, cheap, nontoxic, widely-available, and biodegradable.

This project investigated the adsorption of copper (Cu) ions as a representative heavy metal from water on keratin-alginate composite bioadsorbents. The effects of keratin to alginate mass ratios were investigated with consideration to both adsorption capacity and adsorption kinetics. It was found that bioadsorbents with a keratin /alginate mass ratio were detrimental to adsorption capacity. Pure sodium alginate bioadsorbents had the highest adsorption capacities in this study for copper ions of up to 138 mg / g. The kinetics of adsorption of the bioadsorbents made in this study were comparable to each other.

Validation of Sea Surface Latent and Sensible Heat Fluxes in Three Global Reanalysis Products Against Saildrone Observations during 2023 Atlantic Hurricane Season

Cheryl Reuben

Many devastating hurricanes, such as Katrina, Harvey, and Maria, have seen a 30+ kt increase in the maximum sustained wind speed in a 24 hour period. This is known as rapid intensification and is challenging for models to predict. One key factor in rapid intensification is the energy transfer from the ocean to the atmosphere. While observations from systems such as saildrones, buoys, and aircrafts can be used to observe the air-sea energy exchange, they are irregular in space and/or time. Reanalysis procedures assimilate all available observations and synthesize them into products that are uniform in both space and time. However, model errors can impact these products. Before reanalysis products are used as proxies for in situ observations, they should be validated, especially under extreme conditions such as hurricanes. This study aims to validate sea surface latent and sensible heat fluxes from three commonly used global reanalysis products (NASA MERRA2, ECMWF ERA5, NOAA CFSR2) against saildrone observations taken during the 2023 Atlantic hurricane season. Hourly averaged saildrone data were paired with reanalysis data from each product. Discrepancies between saildrone observations and reanalysis outputs were mostly within one standard deviation with sporadic large spikes. The largest discrepancies are found under hurricane conditions. CFSR2 demonstrated a systemic negative bias for latent heat fluxes and large negative and positive errors for sensible heat fluxes. These results suggest heat fluxes from MERRA2 and ERA5 are a reliable representation of air-sea interactions in the Atlantic Ocean except under hurricane conditions. Away from hurricanes, the results are consistent with those from a similar study conducted for the Pacific sub-Arctic Ocean.

Surface Tension Measurements of Organic Gas-Exposed Hanging Droplet Aerosol Mimics

Joanna Kuan, Michael Haines, Bruno Loyola, Joseph Woo

Proof-of-concept measurements have indicated that the surface tension of a suspended water droplet will shift when in the presence of evaporated organic compounds in the air surrounding it, with shifts that would be greater than expected of the relatively small amounts of gas that are being transferred into the droplets. This study aims to take a more systematic approach to this phenomenon by exposing hanging droplets of water and ammonium sulfate solution to ethanol and acetone, which are both known to easily dissolve in water and can affect solution surfaces.

Property Characterization and Continuous-Flow Measurements of Dark and Irradiated Methylglyoxal-Ammonium-Sulfate Aqueous Aerosols

Maia Merriman, Erin O'Leary, Katherine Pierre-Louis, Joseph Woo

Methylglyoxal (MGLY), a commonly-observed carbonyl-containing volatile organic compound, has been observed to undergo a variety of shifts in composition and physical properties (e.g. UV-visible absorbance, surface tension, etc.) when processed with nitrogen-containing compounds and/or ultraviolet radiation. However, comparisons between aged and irradiated aerosol mimic solutions containing both MGLY and ammonium sulfate (AS) under bulk and small, hanging droplet (10μL) solutions have yielded discrepancies in SOA product compositions when exposed to UV light. As MGLY and proposed oligomeric products are expected to exhibit surfactant properties, it is proposed that SOA pathways that would not otherwise be possible are occurring at gas-droplet interfaces where surfactant SOA concentrations are locally high. This work expands on this comparison by assessing the relative yields of irradiated MG/AS reaction products under bulk, μL-scale hanging-droplet, and polydisperse aerosol measurements. UV/visible absorbance and ESI-MS data are reported. Effective amounts of oligomerization in irradiated MGLY/AS aerosol solutions are found to vary depending on duration of exposure to UV light and type of UV radiation applied (i.e. UV-A vs UV-C radiation.)

Investigating Cryodesalination As a Low-Energy and Affordable Solution to Freshwater Scarcity

Ravneet Kahlon

The 21st century has seen numerous breakthroughs across a wide multitude of fields including, but not limited to, science, technology, mathematics, and medicine. As beneficial as these breakthroughs have been, the means by which they were achieved have devastating impacts on the climate. Climate change is nothing short of a global catastrophe, and the consequences of such are exacerbated in low income and rural communities. This creates an unprecedented disaster where the technologies created to combat climate change are unable to aid those most vulnerable to its impacts. One of these issues is the lack of freshwater access in less affluent regions of the world. The solution to such is desalination; however, current methods are expensive, require immense infrastructure, and are highly energy intensive. To address this dilemma, a new method of desalination must be studied, one that is affordable, requires little to no infrastructure, and is energy efficient. CryoDesalination has the potential to meet all of these requirements by utilizing the simple fact that salt water will freeze at a slightly lower temperature than freshwater. As the water approaches freezing, the ice formed initially will be fresher than the brine left behind. The simplicity in theory is matched by the simplicity in materials, utilizing an Igloo cooler as an insulated housing, copper cups as a vessel for heat transfer, and acetone as a coolant. The salt water is transferred into copper cups and placed in an acetone bath within the Igloo Cooler. Dry ice is periodically dropped into the acetone to facilitate the cooling and reach the desired freezing point quicker. Recent tests have shown roughly a 25% decrease in salt content between the freshwater and initial salt water feed in an hour. Larger decreases were achieved in the past, albeit using a model with more expensive and energy intensive materials. The current results are promising; however, the cooling rate and overall freshwater yield are not enough to be scaled for large quantities of feed salt water. Future iterations need to be automated to improve accuracy of measurements and allow for more consistent separation of freshwater from the salty brine.

Advancing Water Treatment: Comparing Atomic Layer Deposition and Vapor Phase Infiltration for Polymer Membrane Modification

Prasi Desai, Bezawit A. Getachew

Clean water is essential for all communities, both for sustenance and energy needs. Membrane technologies are a growing solution for water treatment due to their high separation efficiency and low energy consumption. However, as these technologies evolve, there is still a need for membranes with improved selectivity, reduced fouling, and stability in harsh conditions.

Atomic Layer Deposition (ALD) is a thin-film deposition technique used to generate atomically precise layers of deposited materials. In industry, this method has primarily been used in the semiconductor and energy conversion space. More recently, ALD has been applied to engineer membranes for water treatment. Vapor phase infiltration (VPI), an ALD-derived method, allows materials to infiltrate into polymer chains. As a relatively new technique, its distinction from ALD remains under-defined. So far, VPI has been used to enhance materials' mechanical strength and fine tune their physical properties (conductivity, internal strain, etc.), and has found success in creating organic-inorganic hybrid materials. However, its use in water treatment membranes remains underexplored. This review seeks to clarify the distinction between ALD and VPI and highlight VPI's potential benefits for polymer membrane modification in water treatment.

To assess current work done by ALD and VPI for water treatment membranes, we conducted a systematic literature review to research ALD and VPI's use in modifying polymer membranes. Our search was restricted to polymer membranes specifically as this is an industry standard and is compatible with the described processes. We examine how these techniques enhance membrane properties like flux, hydrophilicity, pore size, and fouling resistance.

ALD has demonstrated efficacy in modifying polymer membranes, particularly through metal oxide coatings that reduce organic fouling and enhance water flow. However, balancing hydrophilicity and pore size with ALD remains a challenge. Specifically, it's difficult to ensure ALD coatings are thick enough to enhance flux but not so thick that they reduce pore size and water flow. This is one area where VPI could be a promising alternative, as it addresses these challenges by integrating materials into polymer chains, enhancing durability and hydrophilicity without compromising pore size.

Although both ALD and VPI allow for precise membrane modification, ALD's limitations in long-term membrane durability present challenges. Given VPI's potential in this area, its use could sustainably enhance clean water access and facilitate broader industry adoption. However, not much work has been done to assess VPI's utility in this space. We recommend further research on VPI's use for polymer membranes in order to fully explore the benefits it may offer for sustainable and cost-effective water treatment.

Development of a Novel Experimental Technique for Measuring Equilibrium Adsorption Isotherms Under Dynamic Conditions

Joseph Kuklewski, Armin Ebner, Charles E. Holland, James A. Ritter

A new technique has been developed for measuring equilibrium adsorption isotherms under dynamic conditions. This method involves pressurizing an adsorbent-filled column slowly (under a few sccm) and in such a way that the column is always instantaneously in equilibrium. The system works well for corrosive gasses such as sulfur dioxide, which is commonly found in flue gas and has been known to induce acid rain. In this project, a system was developed and improved to ensure the success of future experimentation. A packed bed of length 21.6 cm and diameter 2.1 cm was used to contain the adsorbent silica gel as well as two layers of glass beads. Experiments were conducted at three temperatures of 30, 50, and 75 °C. The system was fully regenerated at 200 °C for three hours between each run. SO_2 gas was flowed into the system at 5, 10, and 20 sccm to examine whether the resulting column loading plots (i.e., pressure vs column loading) are under isothermal equilibrium. Several plots showing pressure change over time were created and used to develop isotherms and demonstrate the repeatability of the experiments. The mass flow controllers for sulfur dioxide and helium were studied and calibrated for ease of use, and the volume of each system component was calculated. Efficient methods of regeneration for the adsorbent were also examined.

Simulations of Cloud and Aerosol-Phase Sulfate Formation Using GAMMA 6.0

Sandhya Sethuraman, Zifeng Tang, Kedong Gong, Vicki Grassian, V. Faye McNeill

Analysis of haze events under highly polluted conditions has indicated a gap in our understanding of the formation of particulate sulfate. In particular, air quality modeling studies inferred the importance of aerosol-phase S(IV) oxidation, which had previously received less attention than cloud water processes. Given the multitude of S(IV) oxidation pathways and their sensitivity to parameters such as ionic strength and pH, more data were needed before these processes could be incorporated into models. In the past decade laboratory studies have filled in this gap, with a proliferation of proposed mechanisms for sulfate formation in the aerosol phase and at the gas-aerosol interface.

We have updated the Gas Aerosol Model for Mechanism Analysis (GAMMA) to include updated S(IV) oxidation chemistry for cloudwater conditions following current recommendations (Tilgner et al. 2021), in addition to pH tracking and transition metal ion cycling. For aerosol conditions, we have implemented recently recommended mechanisms from the literature for S(IV) oxidation via HOOH, ROOH, NO2, TMI catalyzed processes and uncatalyzed oxidation, as well as updated organosulfate formation chemistry. Results will be presented for sensitivity studies performed under Beijing winter conditions, urban US, and rural conditions, and trends will be discussed.

Arrested Methanogenesis of Organic Waste: A Sustainable Source of Short-Chain Volatile Fatty Acids

Megan Harmon, Haoran Wu, Adrian Lomax, Meltem Urgun-Demirtas

Volatile Fatty Acids (VFAs) are high-value short-chain organic acids with applications in biofuels, industrial chemicals, and materials. However, commercial VFAs are traditionally produced from fossil fuels, contributing to greenhouse gas emissions and climate change.[1] In a circular economy, using waste streams to produce valuable fatty acids allows for a sustainable and cost-effective source of these products. This project aims to produce FAs through arrested methanogenesis of organic waste at pilot-scale digesters (50 liter). Volatile fatty acids are an intermediate product in anaerobic digestion, a natural biological process in which microorganisms break down organic matter in the absence of oxygen. The typical products of anaerobic digestion are methane and carbon dioxide. However, if the last step of anaerobic digestion, methanogenesis, is stopped, then VFAs can be obtained as final products. This is favorable over methane as VFAs are more versatile and have a higher market value.[2] In this project, cheese whey, a by-product of cheese production, was used as the feedstock. Methanogenesis was arrested by controlling experimental conditions including pH (~6), organic loading rate (OLR), and hydraulic retention time (HRT). Various operating conditions, including OLR and HRT, were evaluated to assess their impact on carbon conversion, VFA yield, productivity, and titer, and the concentration of different VFA products during the process scale up .

Over three different operation periods, the OLR was increased while keeping the HRT constant (4.8 days). The first condition had the lowest OLR (13.3 gCOD/L·day) and produced the lowest concentration of VFAs (29.20 g/L) and had the lowest productivity (6.08 g/L_{liq}·day) but had the highest conversion (0.97 $gCOD_{digested}/gCOD_{fed}$) and yield (0.76 $gCOD_{acid}/gCOD_{digested}$). The second condition had an intermediate OLR (16.7 gCOD/ L·day) produced an intermediate VFA concentration (31 g/L) and had an intermediate productivity (6.52 g/Lliq·day), but had the lowest conversion (0.94 $gCOD_{digested}/gCOD_{fed}$) and yield (0.71 $gCOD_{acid}/gCOD_{digested}$). The third condition had the highest OLR (20.1 gCOD/L·day) and produced the highest VFA concentration (34.6 g/L) as well as the highest productivity (7.24 g/L_{liq}·day), but had a slightly lower conversion than condition 1 (0.96 $gCOD_{digested}/gCOD_{fed}$) and slightly higher yield than condition 2 (0.72 $gCOD_{acid}/gCOD_{digested}$). In addition, higher OLR resulted in significantly longer system adjustment time (28 days more) to reach steady state than other lower OLR conditions, and had a higher amount (21.55 wt%) of lactic acid present than condition 1 (2.09 wt%) or condition 2 (6.51 wt%). Overall, this project was able to successfully arrest methanogenesis, shown by a lack of methane produced in the bioreactor, and produce the desired product, VFAs.

References:

[1] Holtzapple, M. T., Wu, H., Weimer, P. J., Dalke, R., Granda, C. B., Mai, J., & Urgun-Demirtas, M. (2022). Microbial communities for valorizing biomass using the carboxylate platform to produce volatile fatty acids: A Review. *Bioresource Technology*, *344*, 126253. https://doi.org/10.1016/j.biortech.2021.126253

[2] Wu, H., Scheve, T., Dalke, R., Holtzapple, M., & Urgun-Demirtas, M. (2023). Scaling up carboxylic acid production from cheese whey and brewery wastewater via methane-arrested anaerobic digestion. *Chemical Engineering Journal*, *459*, 140080. https://doi.org/10.1016/j.cej.2022.140080

Repurposing Commercial Nonwovens into Direct Air Capture Materials

Lucinda Nugent, Hannah Holmes, Matthew Realff, Ryan Lively

Global temperatures are currently rising due to increasing atmospheric carbon dioxide (CO_2) levels. Anticipated environmental and societal consequences include negative impacts on our climate, biodiversity, human health, food and water supply, and the economy. Direct air capture (DAC) technologies can mitigate these consequences by removing carbon dioxide from the atmosphere. A key component in DAC technologies is the adsorbent material used to separate CO_2 from the air. Solid adsorbents offer advantages such as low regeneration energy requirements, modularity, alleviation of corrosion and evaporation challenges, and rapid mass transfer to active sites. Amine-based adsorbents are frequently used due to their high CO_2 affinity, which proves essential at the low CO_2 partial pressures relevant to direct air capture. However, these amines are often added to expensive support materials for immobilization, which can comprise a major portion of the overall direct air capture cost.

Here, we evaluate the use of repurposed commercial nonwovens for carbon dioxide removal from the atmosphere. Wipes and pads found in many hygiene, cosmetic, and medical products are comprised of non-biodegradable polymeric nonwoven materials. Repurposing these inexpensive materials could not only reduce the cost of carbon capture but additionally address a second global challenge of polymer waste from the commercial use of nonwovens in relevant industries. We infused polyamines into several types of nonwoven materials and determined that the weight fraction of polyamine in a methanol solution could be used to tune the resultant polyamine amount on the immobilizing material. We then analyzed the CO_2 adsorption performance of these amine-loaded nonwovens and compared the kinetics and equilibrium uptake amounts. The relationship between a material's water contact angle and CO_2 uptake was subsequently assessed. We then estimated the direct air capture cost as a function of contactor cost, ultimately showing the potential economic benefits of repurposing inexpensive commercial nonwovens to form direct air capture materials.

The Effect of Ammonium Sulfate on Succinic Acid

Julia Silvia, Ryan Snyder

The Effect of Ammonium Sulfate on Succinic Acid

Julia Silvia

Under Advisement of Professor Ryan Snyder

Bucknell University

In our atmosphere, there are aerosol particles that affect global cooling and warming. The more understanding we can have of these particles, the more we can understand the climate change we face today. The formation and behavior of ideal (single substance) aerosol particles in laboratory conditions are in many cases well understood. However, some systems have behaviors that make them the exception to general trends. One of these systems is the Succinic Acid and Ammonium Sulfate system. Previous studies have shown that when various organic and an inorganic substance mixtures form aerosols, they typically form a one-peak distribution for particle diameter. However, Succinic Acid and Ammonium Sulfate mixtures do not follow this behavior which can influence their abilities to form cloud-producing nuclei. Particle sizes are determined using a Scanning Mobility Particle Sizer (SMPS). In this work, I will describe our quantification of this behavior, as well as present a potential explanation for it. I will describe the impact of solute ratio, as well as a post-particle formation temperature equilibration on the particle sizes.

Biosynthesis and Incorporation of Aromatic Non-Standard Amino Acids from Readily Accessible Precursors

Abigail Spangler, Shelby Anderson, Aditya Kunjapur

The genetic code consists of twenty amino acids, however, with today's knowledge of biology, chemistry, and engineering, scientists are able to create and incorporate nonstandard amino acids, or nsAAs, into proteins for specialized purposes for pharmaceutical and industrial applications. Synthetically produced nsAAs are often not enantiomerically pure and may be costly. When supplementing nsAAs to cells, often high concentrations are required in order to achieve the desired incorporation. Coupling nsAA biosynthesis with an orthogonal translation system (OTS) all within a single cell allows for a cheaper and stereoselective path for nsAA production and incorporation. So far, the production of aromatic nsAAs with a β-hydroxy group has been established using L-threonine transaldolase enzymes from diverse aldehyde substrates and L-threonine co-substrate. Unfortunately, the β-hydroxy group prevents previously engineered OTSs from incorporating the nsAA into proteins to our knowledge. We have designed an enzyme cascade which can remove the β-hydroxy group to form an aromatic nsAA ready for incorporation. Through in vitro and in vivo testing in E. coli, the pathway enzymes were characterized and conditions optimized for production and incorporation of nsAAs into protein using genetic code expansion techniques with an engineered aminoacyl-tRNA synthetase and tRNA pair and a green fluorescent protein (GFP) reporter. The biosynthetic pathway enzymes show broad specificity on aromatic compounds and can further extend to include a carboxylic acid reductase in order to initialize from carboxylic acids, which may be less expensive, less toxic, or more soluble than the aldehyde analogs. Using high-performance liquid chromatography (HPLC), we can quantify the yield of the nsAA biosynthesis. HPLC yields for a variety of nsAAs in the in vitro cascade are above 95% with enantioselectivity at greater than 99% e.e.. In a live cell context for coupled biosynthesis and incorporation, the GFP reporter can be purified and mass spectrometry used to quantify the accuracy of the desired incorporation of nsAA. This enzymatic pathway may open a world of new chemistries which can be incorporated into proteins for various functionalities.

Computationally Motivated Mutagenesis for Intelligent Incorporation of Noncanonical Amino Acids

Nathan Phan, Roman Adomanis, Blaise Kimmel

Recombinantly expressed variable heavy chain regions of heavy chain antibodies, otherwise referred to as nanobodies, condense the highly selective characteristics of the larger antibody into a smaller package. In turn, these proteins maintain the ability to serve as a vehicle for intelligent drug design, something commonly found in cancer and other rare disease therapeutics. Despite nanobody's major presence in immunoengineering, there lacks a distinct set of guidelines to rationally select conjugation sites on these nanobodies, resulting in effectively blind labeling and loading of drugs. Noncannonical amino acids (ncAAs) extend the preexisting 20 canonically encoded amino acids, introducing novel interactions, specifically the capacity to perform click chemistry, a common method in drug loading. Improper incorporation of these noncannonicals can yield detrimental results, reducing or even preventing binding. Due to the lack of guidelines, many incorporation attempts result in trivial results, due to nonbinding or inactivity. By providing intuition of incorporation sites, ncAAs become more accessible as a technology platform, and thus synthetic biology as a whole can be expanded to solve relevant issues beyond immunoengineering, such as biocatalysis. Additionally, intelligent selection of conjugation sites may elevate the time and financial expense of blind mutagenesis.

We seek to define these guidelines, creating a workflow *in silico,* in which conjugation sites can be rationally selected based off a variety of parameters. By this introduction of computational methods, we quickly screen and select promising conjugation site candidates for more informed mutagenesis rounds. The selection explores (1) the surface exposure of the various residues on the nanobody, (2) the distance between conjugation site and binding regions, and (3) the relative hydrophobicity of the side chains. We create various scripts interfacing PyMol and Python to extract and screen the nanobody. Each trait was screened and mutated independently to one another, with combination of screens tested afterwards. After selection *in silico*, we transition to *in vitro*, experimentally validating our findings through site directed mutagenesis, and expression of our mutated nanobodies. We then characterize the function of resultant nanobodies through examination of cell viability and target receptor binding affinity *in* vitro using relevant disease model cells. Upon collection of data, we plan to revisit the original computational model, leveraging statistical analysis to identify key features that have the highest correlation to both the binding interaction and intercellular activity, with the end goal of creating a predictive model accounting for necessary parameters.

Enhancing Cyanobacterial L-Alanine Production Via Nutritional and Process Optimization

Collin Travis, Cody Kamoku, Christopher M. Jones, David Nielsen

The cyanobacterium *Synechococcus sp. PCC 7002* (PCC 7002) represents a promising photosynthetic chassis for the sustainable production of biochemicals, including L-alanine, an essential nutritional supplement. The photoautotrophic nature of PCC 7002 enables efficient utilization of light and CO2, making it a strong candidate for large-scale outdoor cultivation, particularly in nutrient-rich wastewater environments. While genetic engineering can enhance production metrics, further optimization of nutritional and process conditions is crucial. This research focuses on the influence of various nitrogen sources on L-alanine production, including urea—an abundant component of wastewater—along with traditional sources such as sodium nitrate and ammonium sulfate. The impact of light and CO2 availability on L-alanine production has also been assessed. By leveraging the robust growth and environmental stress tolerance of PCC 7002, this research contributes to more sustainable methods of amino acid production, potentially reducing production costs in the food and pharmaceutical industries.

Exploring CXCL10-Derived Antimicrobial Peptides (AMPs): The Role of Polymer Conjugation and Cysteine Oxidation on Properties and Performance

Elliot Brna, Zixian Cui, Puthayalai Treerat, Rachel Letteri, Matthew Crawford, Molly Hughes

The rise of drug-resistant bacteria presents a serious global health threat, driving the need for novel antimicrobial agents. Two antimicrobial peptides (AMPs) derived from the human chemokine protein CXCL10: stapled-P9, a 22-amino-acid sequence from the C-terminus with an alpha-helical structure, and D8, an unstructured 8-amino-acid sequence, from the N-terminus. To improve the effectiveness of stapled-P9, we developed various peptide-polymer conjugates aiming to reduce toxicity to mammalian cells, increase resistance to enzymatic degradation, and enhance antimicrobial activity. Three different conjugate architectures—linear, star, and comb—were compared using alamarBlue assays to assess their bactericidal activity. Our results indicate that the comb-like conjugates exhibited higher bactericidal activity compared to their linear and star-shaped counterparts, all while maintaining low levels of toxicity to human red blood cells. Peptide D8 ($_{D}$-RTVRCTCI), despite lacking a defined structure (unusual for an AMP) also showed broad-spectrum bactericidal activity. To better understand which amino acids influence its function, we exchanged each with $_{D}$-alanine, noting decreased activity upon replacing arginine (R), valine (V), isoleucine (I), and cysteine (C). While the reduction in activity from replacing hydrophobic residues (V and I) and cationic residues (R) was expected, given their roles in promoting interaction with hydrophobic and anionic bacterial membranes, we are interested in exploring the role the two cysteines have in D8's activity. To investigate this, we synthesized three D8-derived peptides with protective caps on one or both cysteines confirming these modifications via MALDI and NMR analysis. Since capping either or both cysteines reduced antimicrobial activity, we examined the relative reactivity of each cysteine to determine if differences in oxidation rates influence D8's structure and solution behavior. We hypothesized that the cysteine next to the arginine would be more reactive due to the arginine's positive charge stabilizing the negative charge of the thiolate intermediate. Using Ellman's reagent, we monitored free (unoxidized) thiol concentrations over time and compared the behavior of D8 with its 5th cysteine capped versus 7th cysteine capped. Understanding the role of cysteines and their oxidation in the solution behavior and antimicrobial performance will advance the molecular design of cysteine-containing peptides as novel antimicrobial agents.

Designing a Peptide Ligand for a Novel Cancer Therapeutic

Nada Mohamed, Vivek Kumar, Nellone Reid, Joseph Dodd-o

Half a million women die every year with breast cancer, of which 150,000 cases are estimated to be Triple Negative Breast Cancer (TNBC). TNBC is an aggressive type of breast cancer that lacks the Estrogen (ER), Progesterone (PR) and Human Epidermal Growth Factor (HER2) receptors. There is a lack of targeted therapies available for TNBC, as current endocrine and drug therapies that target ER, PR, or HER2 are ineffective. Sigma 2 receptors are overexpressed in proliferative cells and tumors, especially breast cancer cells. Some of the ligands that bind to the receptors are known to have cytotoxic effects. The aim of this study is to computationally randomize amino acid sequences that bind to the sigma 2 receptor to make a cytotoxic drug. By utilizing the Rosetta Scripts software, an XML-like language used for parsing functions within the Rosetta suite, 4,000 different amino acid sequences that bind to this receptor were generated. The top ten unique sequences with the lowest binding energy were then used for further analysis. Using the FlexPepDock docking protocol, PackRotamersMover, and relaxation algorithms, one thousand different amino acid sequences were generated. The top 10 most energetically favorable structures were further analyzed by their Root Mean Square Deviation (RMSD) relative to the top scoring pose to evaluate the potential entropic penalty of adopting the energetically favorable pose. By analyzing the RMSD values and corresponding energy scores, the top three peptides were then chosen to synthesize and characterize to further test for cytotoxicity and affinity.

Using Supercharged Dark GFP As a Tool to Engineer Protein Translation Droplets

Clara Victorio, Jane Liao, Allie Obermeyer

Cells, as the most basic units of life, have evolved to be self-sustaining systems composed of thousands of biomolecules and numerous biochemical reactions. The depth of our understanding of these living systems is reflected in our ability to engineer similar life-like properties in synthetic cells. An important challenge of engineering artificial cells lies in replicating the complex spatiotemporal organization and compartmentalization of cells while integrating key functions. Much of the research in this area has focused on using membrane-bound microcompartments such as liposomes, polymersomes, and proteinosomes. However, this fails to emulate the crowded, molecularly dense milieu of cells, which is essential for many biological functions. To address this, we explored complex coacervation as an alternative, membrane-free compartmentalization approach to mimic the crowded, viscoelastic, and highly charged nature of cellular cytoplasm, using bacteria as an example. Previously, our group overexpressed a panel of globular cationic proteins in *E. coli* cells and found that the protein net charge was a key determinant of phase separation. Here we report the use of *E. coli* lysate as a model *in vitro* system. Using this cell free model, we have characterized the phase separation of endogenous biomolecules with engineered cationic scaffold proteins. Specifically, starting from superfolder GFP(+12) as a model scaffold protein, we mutated the chromophore (T65G, Y66G) to engineer a "DarkGFP" that exhibited the same structure and phase separation behavior but lacked fluorescence. As expected, in subsequent phase separation assays DarkGFP formed coacervates, but also reduced the spectral overlap between the scaffold protein and other fluorophores in the system, including an *mScarlet*-tagged ribosomal protein in the cell lysate. Overall, both in-cell and *in vitro* experiments showed that phase separation is primarily driven by heterotypic interactions between anionic RNA and cationic scaffold proteins. Notably, the cationic scaffold protein significantly concentrated macromolecules necessary for gene expression such as RNA and the large ribosomal subunit protein, bL9. Future work will focus on replicating life-like biochemical processes such as cell-free gene expression and understanding phase separation dynamics in response to internal and external perturbations.

Engineering Beta Sheet Peptide Structures or Lack Thereof through Computational Design and Experimental Testing

Justin Le

Recent advancements in peptide engineering focuses on creating and designing beta sheet structures for nanomaterials and biomedical applications. I have explored anti-parallel beta sheets through computational design, emphasizing the challenges in distinguishing between different beta sheet conformations. I further built on this by designing parallel β-sheet nanofibrils using Monte-Carlo simulations, demonstrating that computational methods can differentiate between parallel and anti-parallel structures.

Furthermore, I was able to expand this research to include bioactive domains in β-sheet peptides for hydrogel applications in wound healing and other biomedical uses. They identified three potential outcomes for peptide self-assembly mentioned below.

1: Formation of nanofibers with beta sheets covering bioactive domains, allowing receptor binding.

2: Beta sheets blocking bioactive domains, preventing receptor interaction.

3: Bioactive domains interfering with self-assembly, resulting in no nanofibers.

This research integrates computational design with experimental validation to engineer peptides with desired structural and functional properties. I focused on Transmission Electron Microscopy (TEM) which is crucial for revealing atomic details, bridging the gap between molecular design and practical applications in nanomaterials and biomedicine.

Designing Peptides for Multiphase Complex Coacervation

Ethan Rivers, Arvind Sathyavageeswaran, Isaac Ramírez Marrero, Sarah Perry

Complex coacervation is a liquid-liquid phase separation driven by the electrostatic and entropic complexation of oppositely charged polyelectrolytes. One can create distinct complex coacervates from combinations of polymers with different chemistries, such that if the two sets of droplets were to mix, they would remain distinct, as multiphase coacervates. Currently, scientific knowledge is limited with respect to the formation and phase behavior of multiphase coacervates. As a result, we are proposing the creation of multiphase coacervates using peptides in order to better understand their physical characteristics. Our research is centered on forming coacervates using both poly(α-lysine) and poly(ε-lysine) in complex with poly(glutamate). α-lysine and ε-lysine are isomers of each other, but create polymers with different charge spacing, solubility, etc. Our goal is to determine whether this difference in the electrostatic interactions represents a large enough difference in properties to facilitate multiphase coacervation.

The peptides were prepared using different Fmoc/Boc protecting groups during solid-phase peptide synthesis to obtain the two different isomers of poly(lysine). The poly(lysine) was mixed with poly(glutamate), and the resulting coacervates were characterized using a well plate reader and optical microscopy.

As we move forward, we hope to understand the parameters that allow for the creation of multiphase coacervates. Ultimately, our goal is to devise strategies for understanding the phase behavior of coacervation and learn how to modulate the physical properties of coacervates for applications in protein encapsulation.

Computer-Aided Ligand Design for the Sigma-2 Receptor to Increase Anti-Neuropathic Pain Activity

Hatice Aygun, Nellone Reid, Vivek Kumar, Joseph Dodd-o

Neuropathic pain has a massive impact on a patient's quality of life, typically occurring in the aftermath of diabetes, chemotherapy, aging, or obesity. Despite currently affecting between 7-10% of the U.S. population, established treatments directed towards neuropathic pain are limited. The treatments currently in use for neuropathic pain are tricyclic antidepressants or non-steroidal anti-inflammatory drugs (NSAIDs). These medications are not made to address the biological mechanisms of neuropathic pain, instead focusing on mitigating its symptoms. Therefore, there is an unmet need for safe and effective therapies that address the underlying molecular mechanisms of neuropathic chronic pain. The sigma-2 receptor (S2R) is abundant among the peripheral nervous system (PNS) and has a proven capacity to produce anti-allodynic effects. This study will utilize computational methods to de novo design therapeutic peptides that selectively block S2R binding of its cognate ligand to alleviate S2R-associated neuropathic pain. Using the Rosetta suite of macromolecular modeling functions parsed with RosettaScripts, five thousand mutagenesis runs were completed to produce 405 unique peptide sequences. The top ten sequences with the lowest binding energies were tested further using conformer RMSD analysis. The ligands will be then synthesized via solid-phase peptide synthesis and tested in vitro to validate in silico peptide design. Subsequent in vivo studies will assess the anti-neuropathic pain effects of the ligands generated.

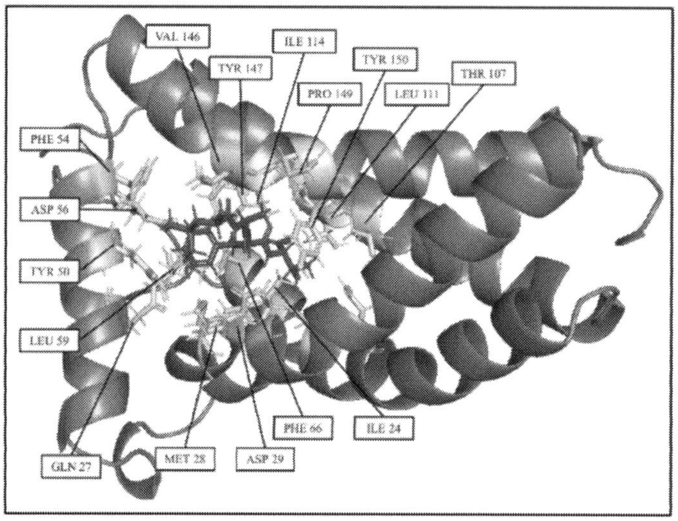

Figure 1: 3-D Model of the generated ligand, in red, bound to the S2R, in green. The interacting amino acid residues between the ligand and S2R are represented in yellow. Screen capture taken from PyMol.

Purification and Self-Assembly of Outer Membrane Protein F

Kartik Kher, Hyeonji Oh, Manish Kumar

Membrane proteins are very promising targets for designing selective ion and water purification processes because of their high selectivity. In this work, we describe the purification and self assembly of wildtype Outer Membrane Protein F (WT OmpF) and one of its variants designated as PARA OmpF. WT OmpF has a pore size of 1.08 nm2 while PARA OmpF has been engineered to have a pore size of 0.3 nm2 by the insertion of bulky residues on the pore wall. Identical purification methods consisting of cell lysis, multiple ultracentrifugation and solubilization steps, and two rounds of chromatography were used to purify both variants of OmpF. Post-purification analysis using an SDS-Page Gel and the Bradford Assay revealed concentration and relative purity differences between WT and PARA OmpF despite the use of the same purification methods. Both purified and detergent solubilized OmpF variants were mixed with poly(butadiene-b-ethylene oxide) (PB-PEO) in an attempt to induce self-assembly into planar 2D nanosheets using detergent dialysis. The resulting materials were analyzed with transmission electron microscopy which revealed a high and ordered packing of WT OmpF, indicating its potential for rapid and selective transport while the PARA OmpF had formed vesicle structures. We hypothesize that lower polymer to protein ratios may induce successful 2D nanosheet formation in the future. The successfully created WT OmpF 2D nanosheets can then be deposited onto a polymer support with any gaps being filled in by an amphiphilic lipid to create a biomimetic membrane similar in structure to human skin. This skinlike biomimetic membrane displays incredible vapor transport properties compared to other breathable fabrics despite having a pore size that is multiple magnitudes of order lower. Ultimately, the purification and self assembly of WT OmpF and its variants is an important first step in creating novel materials with unique transport properties.

Development of a Surrogate Virus Neutralization Test for Middle East Respiratory Syndrome Coronavirus

Bingcheng Huang, Jeong Soo (Jennifer) Lee, Augustine Duffy, Ravindra Kane

Middle East Respiratory Syndrome Coronavirus (MERS-CoV) is the causative agent of a highly infectious viral respiratory illness with a high mortality rate of over 30%, and the current medical system has no licensed vaccines or specific antiviral treatments for the disease, making it especially threatening to the public health. Therefore, to support the MERS-CoV vaccine and therapeutics development, we are developing a surrogate virus neutralization test (sVNT) in the format of a competitive enzyme-linked immunosorbent assay (ELISA) to detect the immunodominant neutralizing antibodies in the humoral serum samples. To date, we have expressed Human Dipeptidyl Peptidase 4 (hDPP4), the receptor for MERS-CoV, and MERS-CoV Spike protein (MERS-CoV S) in mammalian cells and purified these proteins. The proteins were quantified with bicinchoninic acid protein assay (BCA), and the purities of each protein was evaluated with sodium dodecyl sulfate polyacrylamide gel electrophoresis (SDS-PAGE). Subsequently, the binding of hDPP4 and MERS-CoV S was verified using dynamic light scattering (DLS) which showed an increased hydrodynamic radius of the mixture of hDPP4 and MERS-CoV S compared to the measurements of the individual proteins in the control samples. Additionally, a direct ELISA indicated binding of hDPP4 to immobilized MERS-CoV S on a polystyrene 96-well microplate. We will discuss ongoing work, guided by these preliminary results, to develop an sVNT assay, which could be used to detect MERS-CoV RBD-targeting neutralizing antibodies in sera of immunized animals as well as in patient serum samples.

Steam Distillation of Biomass for Insect Repellents

Gloria Hernandez, Eric Palmer

This study investigated the mosquito repellency of essential oils from hedge apples extracted by steam distillation. Hedge apples (Maclura pomifera) are also known as "Osage oranges" and can be found growing in the south-central U.S. Hedge apple trees are commonly planted as hedges due to the hardness and rot-resistance of the wood. The fruit can be eaten but usually is not because of the bitter taste. This study was inspired by the traditional practice of placing hedge apple fruits in outdoor areas to repel mosquitos.

Building on previous methods used for extraction of essential oils from guayule and hemp biomass, steam distillation was selected for extraction of essential oils from hedge apples. Chemical characterization will be used to identify compounds that have been found to be effective in mosquito repellency from previous experiments and related literature. To test the efficacy of the extracts against yellow fever mosquitoes (Aedes aegypti), an arm-in-cage assay will be conducted to observe the mosquitos' behavior in the presence of these oils in an isolated environment. Some factors that need to be considered for the steam distillation extraction process include the parts of the fruit to be used, fruit ripeness, and changes during storage. This research will contribute to the understanding of the chemical composition of effective mosquito repellents, identification of alternative sources for these compounds, and development of pest management solutions.

Development of an Efficient Biosensitized Solar Cell Using Bacteriorhodopsin Producing Haloarchaeon *Halobacterium Noricense*

Ruchi Mathad, Vivek Chakraborty, Mehwish Ali, Sagar Kanekar, Anand Kulkarni

The escalating global demand for energy, coupled with the adverse environmental impacts of conventional energy sources, necessitates a paradigm shift towards cleaner and greener alternatives. Fossil fuel dependence and high carbon footprints from contemporary energy sources contribute significantly to global warming and climate change. Solar cells offer a promising avenue for sustainable energy production. Unlike traditional cells, solar cells leverage the photovoltaic effect in a P-N junction diode, converting light energy into electrical energy without relying on chemical reactions or fuel. The amalgamation of individual solar cells forms solar panels, which can efficiently harness the abundant and economical energy provided by the sun.

Biosensitized Solar Cells, as an alternative to traditional solar cells, employ Titanium Dioxide (TiO_2) electrodes and bacteriorhodopsin as a light-absorbing material. These cells can be seen as an enhancement of Dye-Sensitized Solar Cells, utilizing bacteriorhodopsin in place of traditional Ruthenium dyes. Bacteriorhodopsin is derived from Haloarchaea, specifically from hypersaline environments, where unique conditions foster the growth of these microorganisms. Haloarchaea, thriving in hypersaline environments, are known to produce bacteriorhodopsin, a light sensitive protein that converts light energy into electrochemical energy. Notably, *Halovivax asiaticus, Halobacterium noricense* and *Halostagnicola larsenii* are some haloarchaea which have been shown to yield substantial amounts of bacteriorhodopsin, offering a potential sustainable source for biosensitized solar cells.

Experiments involving bacteriorhodopsin harvested from cultures of *Halostagnicola larsenii* RG2.14 have demonstrated promising results. When immobilized on TiO_2 nanocrystalline films, the biosensitized solar cells exhibited an open-circuit voltage (OCV) of 0.52 V, photocurrent density of 120 $\mu A/cm^2$, and an overall efficiency of 0.11%. These findings underscore the viability of harnessing bacteriorhodopsin for solar energy conversion. The aim is to develop a lower cost biosensitized solar cell with minimal environmental impact while ensuring comparable efficiencies to traditional solar cells. In conclusion, this research offers a comprehensive exploration of biosensitized solar cells utilizing bacteriorhodopsin from haloarchaeon, Halobacterium noricense presenting a promising avenue for sustainable energy production. The outcomes of this study contribute to the ongoing global efforts towards cleaner and more efficient energy solutions. This paper explores the potential of biosensitized solar cells as a cost-effective and environmentally friendly solution to meet the rising energy demands.

A Kinetic Model for Single-Yeast Sour Beer Fermentation

Sofi Jeffrey, James E. Maneval

During the fermentation process in which glucose and or other sugars are metabolized by yeast to form ethanol, certain strains of yeast, called 'sour yeasts', will first metabolize sugar to produce lactic acid, after which, the mechanism switches to produce only ethanol. We sought to adapt an existing mathematical model for fermentation to describe this product-switch phenomenon.

The math model of González-Hernández, et al. [1], was adapted as the starting point of our model. The key feature of the present model is a switch function that controls the change in metabolism from lactic acid to ethanol during a fermentation. Utilizing the existing rate equations and stoichiometric coefficients within the model, we were able to construct a rate expression for lactic acid and integrate it into the model proposed [1,2] and extended the model to describe the use of maltose and maltotriose as primary substrates [2].

We conducted four trials, fermenting Philly Sour yeast with increasing initial glucose concentrations. We calibrated our model to specific gravity and pH data, and then data from one trial was used to estimate model parameters using optimization. Results were then applied to the other three trials to assess the capability of the model. We accomplished this through the switch function model approach [1] using parameters to control each unique component, including lactic acid production, and used optimisation algorithms to best match our modeled behavior to data.

Our model coincides well with experimental data when comparing our modeled specific gravity and pH to measurements taken over time. The switch function captures essential features of the specific gravity and pH data. Our model also provides plausible component curves which capture key features of how sour yeasts metabolize sugars in this process. For example, our model demonstrates lactic acid production coinciding with cell mass growth causing an early pH decrease to about 3.2, after which a change in metabolism converts the remaining sugar into ethanol. Predictions comparing our model with fitted parameters to non-calibration trials was encouraging.

The switch function model is effective in modeling complex changes in metabolism in this setting. Our results show that this model captured the change in metabolism from acid production to ethanol production. We recommend that more detailed experiments measuring composition of individual components should be used to improve the performance of the model and the accuracy of its parameters.

[1] *González-Hernández, Y., Michiels, E., Perré, P. A Comprehensive Mechanistic Yeast Model Able to Switch Metabolism According to Growth Conditions. Fermentation 2022 (8), 710. doi.org/10.3390/fermentation8120710.*
[2] *Gee DA, Ramirez WF. Optimal temperature control for batch beer fermentation. Biotechnol Bioeng. 1988 31(3),224-34. doi: 10.1002/bit.260310308.*

The Effectiveness of Solid-Phase Antimicrobials Against Gram-Positive Bacteria in Food Processing

Ian Brown, Ryan Summers, Stephen M.C. Ritchie

The dairy industry is searching for additional methods to eliminate or mitigate bacterial growth, most prominently that of *Listeria monocytogenes*. *Listeria monocytogenes* (LM) is a ubiquitous gram-positive bacterium first identified as a food born pathogen in 1983. Listeriosis, caused by LM, possesses an alarmingly high mortality rate between 20-30% for infected individuals. While pasteurization effectively eliminates LM during milk processing, Listeriosis outbreaks remain prevalent in the dairy industry due to post pasteurization contamination. Consequently, technologies that kill or mitigate growth of LM have exceptional value.

Solid-phase antimicrobial surfaces represent one such method of targeting and destroying these bacteria. Membranes are utilized throughout food processing, especially within the dairy industry. Various polymeric membranes for microfiltration, ultrafiltration, nanofiltration, or reverse osmosis applications are packaged in spiral wound membrane elements. The polypropylene (PP) feed spacer mesh within the spiral wound membrane element is an ideal surface for functionalization with an antimicrobial agent. The feed spacer mesh is in intimate contact with the process stream and endures frequent cleaning. The cleaning in place (CIP) procedure consists of various rounds of acid, caustic, and enzyme treatments and is typically applied every 24 hours.

Quaternary ammonium compounds (QACs) are frequently incorporated into polymers and applied to surfaces to combat pathogenic bacteria. In the food and dairy industry, solid-phase covalent attachment of QACs ensures no leaching of chemical compounds into the final product. The QAC functionalized feed spacer combats bacterial growth through adsorption. The negatively charged phosphate head of the lipid bilayer of the bacteria cell wall is electrostatically attracted to the positively charged nitrogen head of the QAC. Once adsorbed, the lengthy hydrophobic tail of the QAC intercalates into the cell wall disrupting membrane integrity to induce cell lysis and death. Adsorbed bacteria are also killed when CIP is applied to the feed spacer mesh.

In this work, we functionalized PP by covalently grafting polymer chains with quaternary ammonium groups to the mesh surface through photoinitiated free radical reactions. The resultant quaternary ammonium poly(vinylbenzyl chloride) functionalized polypropylene mesh (QA-PVBC-PP) was structurally verified through Raman spectroscopy and chemical dyes. The QA-PVBC-PP was also found to be non-leaching. The QA-PVBC-PP was tested against *Lactobacillus plantarum* and *Listeria innocua* in solution. Gram positive *L. plantarum* and *L. innocua* were chosen due to similar antimicrobial resistances as other pathogenic bacteria, including *L. monocytogenes*, making them effective and safe surrogates. The QA-PVBC-PP demonstrated significant antimicrobial activity and bacterial adsorption over 3 hours compared to non-functionalized PP. Furthermore, the QA-PVBC-PP retained bacterial adsorption capabilities following CIP procedure demonstrating chemical stability and potential for repeated use.

SH3 Binding Titrations with Scp(12)-Modified Gold Surface

Maylee Valentin, Sarah Bramlitt-Harris, Zahraa Albeshir, Carolyn Curley, Eleenah Sanders, Rose M. Balog, Jeffrey Halpern

Biosensors are defined as having a biorecognition element, which serves to target a specific analyte of interest, and a transduction element, which converts the surface binding response into a readable signal.[1] Progress has been made in the development of an SH3 electrochemical biosensor, which will function as a model system for other future analyte targets. Two modification procedures were pursued to investigate the recognition element Scp(12). The first sensor was initially modified with 10 mM 3-Mercaptopropionic Acid (3-MPA) in solvent anhydrous MeOH. The second sensor was initially modified with a 1:1000 of 12-Malondialdehyde (12-MDA) and 1-Dodecanethiol (1-DT) in solvent anhydrous MeOH. Both sensors were subsequently modified with polyhistidine- tagged Scp(12), following several identical modification steps: (1) NHS and EDC; (2) ANTA; (3) $NiSO_4$; (4) polyhistidine-tagged Scp(12). Both modifications were evaluated using Electrochemical Impedance Spectroscopy (EIS) and were confirmed successful.

The interaction between surface-modified recognition element, Scp(12), and target analyte, SH3, were studied using EIS. EIS settings were set to have an initial frequency of 100000 Hz, a final frequency of 0.2 Hz, 10 points/decade, an AC Voltage of 10 mV rms, and a DC Voltage of 0.160 V vs Ag/AgCl. All EIS data was collected in a 10 mM FFCN in PBS solution. Using a small volume cell setup that initially held 210 μL of the FFCN solution, concentrations of SH3 ranging from 0.190 μM to 190 μM were titrated into the base solution via seven concentration additions. Initial findings suggest a response from the scp(12)/SH3 or recognition element/analyte interaction as there is a substantial difference between SH3 data and the buffer data.

Acknowledgements: Support was provided by BIO-SENS, National Science Foundation EPSCoR award (#2119237).

References:

1. Ren, T.; Bramlitt, S. E.; Joelle M.J. LaFreniere; Seitz, W. Rudolph.; Jeffrey Mark Halpern. Conformation-Based Stimuli-Response Sensors: Strategies for Optimizing Electrochemical and FRET Transduction. *Sensors and Actuators Reports* 2021, 3, 100066–100066. https://doi.org/10.1016/j.snr.2021.100066.

Riboswitch Engineering : Tracking Levofloxacin Accumulation and Oxidative Damage in Single Cells Using Biosensors

Nathaniel Rodney, Juliet González, Patricia Hare, Wendy W.K. Mok

Background & Significance: Antibiotic treatment failure occurs when bacterial cells survive treatment, which can lead to bacterial infection regrowth after the antibiotics are removed and the surviving cells begin to propagate. This phenomenon can be detrimental for a patient, as they may improve during treatment only for infections to worsen again as these surviving persisters regrow. Persisters are different from resistant cells. Persisters are genetically identical to other cells in the population that die during antibiotic treatment, but they can change their phenotypic responses to tolerate drugs. By comparison, resistant cells are driven by evolution and caused by genetic mutations that allow them to fight against the antibiotic. Single-cell techniques are needed to identify what is happening to these persister cells during antibiotic treatment.

Methods: In this project, we are engineering two biosensors, one for checking oxidative damage on 8-oxoguanine (8-oxo-G), and another to check for accumulation of Levofloxacin (Levo), our antibiotic of interest. We are generating different designs of the biosensors by site-directed mutagenesis (SDM) and molecular cloning. We are assessing the activity and ligand-sensing ability of these sensors by fluorescence spectroscopy and flow cytometry.

Results: I generated some variants of the original biosensors by SDM. Many of the variants can generate a fluorescent signal, but they can also do so in the absence of the ligands. I am optimizing the sensors to ensure that they are only fluorescent when the ligands are present.

Conclusions: Using these two biosensors in bacteria will enable us to determine if bacterial persisters are experiencing damage from the Levo antibiotic, and to see if Levo is accumulating in the persisters and cells that die.

Clinical Potential of EPR Spectroscopy in Assessing Radiation Resistance in Pathogens and Tumors

Jared Sealy, Robert Volpe, Andres Londono-Barbaran, Michael Daly, John Dumler, Jennifer Weiser, Brian Hoffman, Ajay Sharma, Venkatesan Kathiresan, Michael R. McDevitt, Ekaterina Revskaya

Clinical Potential of EPR Spectroscopy in Assessing Radiation Resistance in Pathogens and Tumors

Electron paramagnetic resonance (EPR) spectroscopy has emerged as a valuable tool for characterizing the resistance of pathogens and cancer cells to ionizing radiation. Gamma radiation generates reactive oxygen species (ROS) through radiolysis, which in turn induces oxidative stress in organisms. These ROS damage proteins, the primary indicator of cellular survivability, through oxidation, leading to reduced or lost functionality[1,2]. The primary cause of oxidative protein damage during irradiation is superoxide, formed when oxygen gains an extra electron. Superoxide is effectively neutralized by manganous (Mn^{2+}) antioxidants (H-Mn), which accumulate in various cell types, compared to large manganous accumulating proteins (L-Mn). EPR spectroscopy has identified Mn antioxidant levels in prokaryotes, fungi, cultured cancer cells, and whole organisms like nematodes as the strongest biological markers of ionizing radiation resistance in living cells[3].

This study highlights the effectiveness of EPR in assessing radioresistance in *Borrelia burgdorferi*, an Mn-dependent bacterium that causes Lyme Disease, as part of an investigation into the potential development of an irradiated Lyme Disease vaccine[4]. Cells were cultured in BSK-H media and grown to a concentration of approximately 10^6 cells/mL. Samples were irradiated to determine the dose required to achieve 90% population reduction (D10). These irradiated samples were compared to a set of non-irradiated samples for EPR analysis, which provides the percentages of HMn and LMn (fH). Despite high Mn/Fe ratio, polyploid B. burgdorferi is radiosensitive, with a survival-limit of less than 1 kGy, indicating potential for irradiated vaccine development.

Additionally, we explore the potential of EPR in assessing the radiation resistance of tumors. We propose that EPR analysis of a biopsy sample could provide insights into a tumor's radiation resistance, offering valuable information for customizing a targeted radiation or chemotherapy treatment plan.

Bibliography

1. Daly MJ. A new perspective on radiation resistance based on Deinococcus radiodurans. Nat Rev Microbiol. 2009 Mar;7(3):237-45. doi: 10.1038/nrmicro2073. Epub 2009 Jan 27. PMID: 19172147.
2. Daly MJ, Gaidamakova EK, Matrosova VY, Vasilenko A, Zhai M, et al. 2007. Protein oxidation implicated as the primary determinant of bacterial radioresistance. PLoS Biol 5:e92
3. Sharma A, Gaidamakova EK, Grichenko O, Matrosova VY, Hoeke V, et al. 2017. Across the tree of life, radiation resistance is governed by antioxidant Mn(2+), gauged by paramagnetic resonance. Proc Natl Acad Sci U S A 114:E9253-E60
4. Tobin, Gregory J et al. "A novel gamma radiation-inactivated sabin-based polio vaccine." PloS one vol. 15,1 e0228006. 30 Jan. 2020

DNA Based Biosensors for Horizontal Gene Transfer Detection

Arthur Chen, Heather Curtsinger, Allison Lopatkin

Horizontal gene transfer (HGT), one of the mechanisms used by bacteria to transfer genetic information between themselves, occurs in all microbial environments, and is one of the primary causes of the dissemination of antibiotic resistance among bacterial populations. Typical methods of tracking HGT typically involve the introduction of antibiotic-resistant genes and genes that code for fluorescent proteins. Although feasible in a controlled laboratory environment, these methods are time-consuming and cumbersome to perform. Fluorescently-tagged aptamers offer an alternative method of quantifying HGT without the need for the introduction of new plasmids. The development of other methods that could quantify bacteria in-situ and allow for high-throughput screening of bacteria would be greatly beneficial for the study of horizontal gene transfer in bacteria.

DNA aptamers are oligomers composed of nucleic acids that have been shown to be able to selectively bind to certain targets. Via a method known as SELEX (systematic evolution of ligands by exponential enrichment), selectively binding aptamers can be produced via an iterative process. The ability of aptamers to bind selectively to a wide range of targets makes them a possible candidate for detecting horizontal gene transfer. The successful development of aptamers would allow for the observation of horizontal gene transfer in bacterial populations, and allow quantification via high-throughput techniques such as flow cytometry.

Replication of literature results for aptamer binding to bacteria has shown that aptamer binding is highly sensitive to the conditions they were involved in. To find aptamers that are suitable for use in screening for horizontal gene transfer may necessitate the need to modify the SELEX procedure to encourage the enrichment of DNA aptamer candidates that are not as sensitive to environmental conditions. More iterations of the SELEX procedure will need to be performed in order to properly evaluate the viability of aptamers as candidates for detecting horizontal gene transfer.

A Microbial Platform for Recyclable Plastics with Customizable Properties

Weixi Hu, Weixi Hu, Seokjung Cheong, Jay Keasling

Most plastics today are single-use and rarely recycled, resulting in substantial resource and energy waste. This has led to the accumulation of over 6 billion metric tons of plastic waste, causing significant environmental stress. Our study introduces a new biological method for producing specific lactones, crucial for making a new type of recyclable plastic called vitrimers. Vitrimers combine the easy processing of regular plastics with the durability of more robust materials. Unlike traditional durable plastics, vitrimers can be broken down chemically into simpler substances such as lactones and diacids, enhancing their sustainability. Historically, the production of chemicals using microbes has faced limitations due to the narrow range of molecules produced by natural or engineered enzymes. Our technology, which utilizes recombined Type I polyketide synthases (PKSs) and other enzymes, offers a promising method to create a diverse array of lactones, enabling the development of vitrimers with customizable properties, promoting more sustainable material cycles.

Development of Paper-Based Point-of-Care Diagnostics for Detection of Human Respiratory Pathogens Using Loop-Mediated Isothermal Amplification (LAMP)

Shamalaa Sathiasealan, Josiah Davidson, Ryan Relich, Mohit Verma

Paper-based microfluidic analytical devices (µPADs) are ideal for pathogen detection at the point-of-care (POC) due to their ease of production, user-friendliness, and scalability. This project focuses on developing a spatially multiplexed µPAD for detecting respiratory pathogens using Loop-mediated Isothermal Amplification (LAMP). This is a robust method suited for clinical settings with the goal to enhance patient treatment and enable the detection of emerging pathogens, particularly during public health emergencies. It will provide a diagnosis that is typically only available through costly laboratory-based testing.

LAMP amplifies nucleic acids in the presence of a specific pathogen, with six primers, binding to eight distinct regions of the target genome to drive the specificity of the reaction. Nucleotide incorporation during phosphodiester bond polymerization releases protons, causing a red-to-yellow colour change with the pH indicator phenol red. The Verma lab previously designed a µPAD which incorporates pre-dried reagents on chromatography paper reaction pads. These 5 x 6 mm pads are arranged in a rectangular array, annealed to a Melinex backing with a neutral adhesive, and separated by polystyrene spacers to prevent crosstalk. Nasal swabs or saliva samples from patients were diluted in water and added to rehydrate reagents in the pads. The device is then sealed and heated at 65°C for 60 minutes in a water bath, and the results are visually read by the user.

The device was previously tested using saliva samples, and the Verma Lab found the limit of detection (LOD) was 200 copies per µL with 98% accuracy using digital image analysis (97% analytical sensitivity, 100% specificity) and 91% accounting for colour perception using user surveys (71% analytical sensitivity, 100% specificity; n=4 users). The devices included a no-primer control to ensure accuracy and prevent false positives.

We are taking this design and furthering its diagnostic capabilities to target Influenza A/B, Respiratory Syncytial Virus (RSV), and further validating its use for merging strains such as SARS-CoV-2. We intend to perform this validation by spiking viruses cultured in vitro and determine the diagnostic performance of the device with primer sets targeting the HA gene of Influenza A/B, the junction region between the G and F genes of RSV, and the continued targeting of the junction region between ORF-7A and ORF-7B regions of SARS-CoV-2 previously investigated by the Verma lab.

By changing a single component on our diagnostic platform, the µPADs can be adapted to detect emerging pathogens, making them a valuable tool for sentinel surveillance and public health monitoring. The µPADs are expected to perform similarly in other respiratory pathogens of public health interests, including Influenza A/B and RSV in clinical and contrived samples, as demonstrated using the SARS-CoV-2 virus. These devices offer an inexpensive and accessible option for diagnosing acute respiratory illnesses at the POC. This targeted approach allows for pathogen-specific treatment, improving patient outcomes compared to the current symptom-based management. Furthermore, these µPADs offer an invaluable tool for sentinel surveillance which can serve as an early warning system for pathogen outbreaks and support efforts to track the evolution of respiratory diseases.

Exploring Irreversible Protease Inhibition: Using Yeast Surface Display and Genetic Code Expansion to Engineer Covalent MMP Inhibitors

Ivy Le, Sean Williams, Rebecca Hershman, James Van Deventer

A challenge in drug discovery is the need to develop irreversible enzyme inhibitors with high specificity; genetic code expansion through the incorporation of noncanonical amino acids (ncAAs) into antibodies is a potential solution. ncAAs have more chemically diverse functional groups than canonical amino acids (cAAs), enabling the incorporation of chemical reactivities not found in the standard genetic code. This project aims to discover antibodies capable of forming photo-induced covalent (photo-crosslinking) interactions with an MMP, which belongs to a class of enzymes with diverse roles in physiological functions, using yeast-displayed antibodies containing ncAAs. Stop codon suppression is a tool used to introduce noncanonical amino acids (ncAAs) into antibodies in place of a TAG codon. This desired interaction is expected to be achieved by substituting a canonical amino acid (cAA) on an antibody with p-azido-L-phenylalanine (AzF), an ncAA, that becomes a highly reactive species when irradiated with light and crosslinks with other chemical groups.

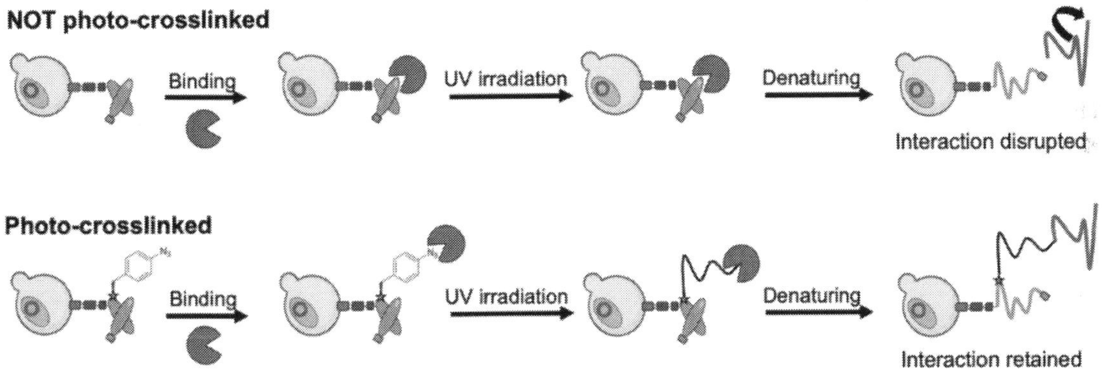

Figure 1. Illustration of photo-crosslinking interactions mediated by AzF. AzF-incorporated antibody constructs are expected to form crosslinking interactions with MMP under the presence of UV light, unlike WT constructs. Retained interactions are verified by denaturing the samples since crosslinking interactions are not disrupted by this process. These crosslinking interactions between the construct and the target enzyme are hypothesized to result in the irreversible inhibition or enhanced inhibition of the enzyme.

Phenotypic Characterization of Halotolerant Yeasts for the Valorization of Seaweed Biomass

Malik Resheidat, Sergio A. García-Echauri, Jae Won Oh, Paulina Romero, José Avalos

Seaweed biomass has enormous potential as a sustainable feedstock for the production of fuels, chemicals, and other valuable products. Using seawater, as opposed to freshwater, for the fermentation of seaweed sugars greatly increases the economic and environmental sustainability of these processes. Therefore, developing novel halotolerant yeasts for industrial applications in seawater offers great opportunities for these alternative feedstocks. Here, we present the results of screening over 40 different species of halotolerant yeasts for their ability to grow at the salinity of ocean water or higher. Based on the highest growth rates at increasing salt concentrations, a narrow selection of yeasts were further characterized for their ability to utilize different carbon sources and produce chemicals of interest. This led to the identification of some yeasts species with growth rates and ethanol yields that compare favorably to those of *S. cerevisiae* in fresh water glucose media. This work establishes the potential of previously uncharacterized halotolerant yeasts for further development into industrial hosts to valorize seaweed biomass, using seawater, for the sustainable production of fuels and chemicals.

Engineering Binding Affinity of Yth to m6A-RNA Leveraging Yeast Surface Display and Next-Generation Sequencing for Comprehensive Mutant Library Analysis

Katelynn Horvath, Yongku Cho

Engineered RNA-binding proteins (RBPs) show promise for diverse applications in research and therapeutic development. RBPs can be used to track and alter RNA localization in the cell. Harnessing the capabilities of RNA-binding proteins (RBPs) can enhance the analysis and characterization of RNA modification profiles. Moreover, RBPs can be used to regulate translation through influencing alternative splicing and inhibiting RNA function [1]. Modulating the interaction between YTH-domain containing proteins and methylated RNA ligands could advance the development of research tools in biotechnology. The YTH domain specifically binds methylated RNA ligands and is conserved across the five N6-methyladenosine (m6A) reader proteins in the human proteome. Both holo and apo structures of YTH have been solved through x-ray crystallography. Notably, tryptophan residues contained in the hydrophobic binding pocket have been shown to be important in binding m6A via an "aromatic cage," quantitatively evidenced by alanine scanning [2]. N6-methyladenosine (m6A) is the most prevalent internal modification in RNA across eukaryotic organisms. Consequently, the modification plays a significant role in various disease pathologies, particularly in cancer. Increased m6A modifications can promote the stability of oncogenic mRNA, driving tumor progression. Accordingly, to illustrate a possible application of engineered RBPs, characterizing m6A modifications in the transcriptome could serve as a biomarker for tracking cancer progression. Prior studies have determined that YTH has a 10-fold higher affinity to m6A modified RNA ($K_D \approx 0.18 \mu M$) as compared to unmodified RNA ($K_D \approx 2.8 \mu M$) [3]. Further affinity maturation of the YTH-m6A interaction through directed evolution of the YTH structure could produce a novel protein with nanomolar affinity better suited for the development of biological detection tools. Here, yeast surface display is utilized to characterize the binding affinity and specificity of the YTH domain to m6A modified RNA. Subsequently, fluorescence-activated cell sorting will allow for selection of high affinity, highly specific binders from mutant libraries. With this, multiple rounds of directed evolution can be performed. Mutant libraries generated through error-prone PCR are to be sequenced using nanopore sequencing to robustly analyze mutations tested.

References: **[1]** Chen, Y., & Varani, G. (2013). Engineering RNA-binding proteins for biology. *The FEBS Journal*, 280(16), 3734–3754. https://doi.org/10.1111/febs.12375 **[2]** Miller, L., et al. (2023). Characterization of epitranscriptome reader proteins experimentally and in silico: Current knowledge and future perspectives beyond the YTH domain. *Computational and Structural Biotechnology Journal*, 21, 3541–3556. https://doi.org/10.1016/j.csbj.2023.06.018 **[3]** Wang, X., et al. (2014). m6A-dependent regulation of messenger RNA stability. *Nature*, 505(7481), 117–120. https://doi.org/10.1038/nature12730

Heterologous Biosynthesis of Curcuminoids in Mammalian Cells

Remy Baskind, Leah Schrass, Daniel W. Pack

The plant molecule curcumin possesses useful medicinal properties. The abilities to scavenge free radicals, activate antioxidant response genes, and downregulate inflammatory pathways make curcumin a potent anti-inflammatory molecule. However, therapeutic use of curcumin is drastically limited by its poor bioavailability, low aqueous solubility, and rapid metabolism. To overcome these limitations, mammalian cells were engineered to express a set of plant derived enzymes catalyzing the biosynthesis of curcumin from tyrosine. Use of CRISPR/Cas9 gene integration can allow for stable curcumin production in-vitro to reduce cellular levels of reactive oxygen species. To facilitate CRISPR delivery, polymer-based vectors are preferable due to their high carrying capacity. Polyethylenimine (PEI) is considered the "gold standard" of polymer-based gene delivery. Despite this, PEI demonstrates low transfection efficiency in serum. Previous work in this lab determined that modifying PEI through succinylation improves gene delivery. In this study, the transfection efficiencies of various percent modified succinylated polyethylenimine (zPEI) types were evaluated using pGL3 firefly luciferase to determine optimal polyplex composition. From this analysis, zPEI-2% demonstrated the best transfection efficiency at the desired low polymer:DNA weight ratios. To evaluate CRISPR introduction HEK cells were transfected with optimized polyplexes containing plasmids for green fluorescent protein (GFP) and Cas9. Stability of CRISPR integration was analyzed by assessing GFP fluorescence over multiple passages. GFP assays and fluorescence imaging proved cells treated with the Cas9 enzyme obtained lasting gene expression. Additionally, cells treated with zPEI-2% polyplexes outperformed those treated with uPEI polyplexes. Finally, optimized polyplexes facilitated CRISPR introduction of the TAL/4CL1/CUS plasmid encoding curcumin producing enzymes. Fluorescence analysis of curcumin production produced inconsistent results, indicating the need to improve this method.

Multi-Pronged Skeletal Muscle Fiber Type Characterization through Immunostaining, Contraction Analysis, and qPCR

Pavankumar Umashankar, Tamara Rossy, Ritu Raman

Skeletal muscles mediate all voluntary movement and are composed of different fiber types that enable a wide range of functionality by tuning contractile strength and metabolic needs. Muscle fiber types change with physiological and pathophysiological stimuli. To study fiber type transformation *in vitro*, we need a way to reliably characterize and monitor fiber types. In this study, we optimized a three-pronged strategy to map fiber types in engineered muscle tissues differentiated from murine C2C12 myoblasts: immunostaining, quantitative PCR (qPCR), and analysis of contraction dynamics in response to electrical stimulation. These studies enabled us to understand how varying fiber morphology and gene expression impacts function.

After an initial screening to identify optimal dilutions and combinations of antibodies against specific myosin heavy chain isoforms representative of slow- and fast-twitch fibers, we investigated the changes in fiber type composition throughout the muscle maturation process. We analyzed muscle fibers on day 6, 10, and 14 of differentiation through stimulated contraction and calcium imaging assays, and performed qPCR and immunostaining analysis at the end of the experiment. Overall, the optimization of this three-pronged strategy has enabled better investigation of fiber type changes, with applications ranging from medicine to biohybrid machines.

Airway *Prevotella* Promote Inhibition of *Streptococcus Pneumoniae* Infection through Chemokine and Neutrophil Modulation

Arianna McCarty, Sara Stoner, Sam Fulte, Ana Fairbanks-Mahnke, Sarah Clark

The human body is host to many naturally occurring (commensal) bacteria as part of individualized microbiota varying in composition by anatomical location and throughout one's life. These bacterial communities play a crucial role in host-pathogen interactions and the immune response, both through direct harming of pathogens as well as regulation of immune signaling pathways. *Prevotella* species are highly prevalent components of these communities, including within the respiratory microbiota. *Streptococcus pneumoniae*, a notable respiratory pathogen, is the leading cause of community acquired pneumonia, and the largest infectious cause of death in children under 5 years of age. Furthermore, pneumococcal vaccines are more protective against invasive pneumococcal infections rather than lung infections, emphasizing the need to identify other therapeutics as well as risk factors for pneumococcal pneumonia. This study investigates how specific *Prevotella* species can regulate early immune responses to respiratory infection by *S. pneumoniae*, with the aim of utilizing this knowledge to generate novel immunotherapies.

Investigation of *Prevotella*'s effects on the presence of common chemokines such as CXCL2 and IL-8 revealed a marked increase of chemokine secretion after A549 and D562 (respiratory epithelial cells from the lung and pharynx respectively) were inoculated to *Prevotella melaninogenica.* Furthermore, flow cytometry studies of mouse lung and bronchoalveolar lavage (BAL) samples showed an increase in neutrophil percentage and total recruitment in response to *P. melaninogenica* inoculation. Lastly, adherence and infection assays demonstrated increased abundance of *P. melaninogenica* correlates with reduced infection by *S. pneumoniae*, whereas increased abundance of *P. intermedia* correlates with increased infection by *S. pneumoniae.* Together, these findings suggest innate immune priming behaviors of the respiratory microbiota by *P. melaninogenica* through chemokine and neutrophil modulation help to reduce *S. pneumoniae* infection. Furthermore, future investigations into *P. intermedia*'s behavior are warranted to better understand what signaling causes upregulation of *S. pneumoniae* infection. With a heightened knowledge of these commensal mechanisms, we seek to harness these regulatory pathways to allow the creation of translational therapies aiming to mitigate and prevent pneumonia worldwide.

Exploration of Tryptophan Modifying Enzymes: An Expansion of Nonnatural Psilocybin Derivatives

Felicia Guagliardo, Jessica Flower, William Gibbons, Andrew Jones

Current medical treatments for mental illnesses such as anxiety, depression, and post-traumatic stress disorder are often limited and can cause many unwanted side effects. The lack of accessibility to effective mental health treatment across the globe calls for new and creative solutions. A promising alternative has been found in psychedelic treatment, mainly psilocybin. Psilocybin is naturally found in over 200 species of 'magic' mushrooms, and has recently been the subject of a growing number of clinical studies testing its ability to treat mental health disorders including depression, anxiety, and PTSD.

In *E. coli*, the biosynthesis of psilocybin starts with the formation of 4-hydroxytryptophan (4-HTP) by tryptophan synthase from a 4-hydroxyindole supplement. 4-HTP is then decarboxylated, phosphorylated, and methylated to form the final psilocybin product. The incorporation of new tryptophan modifying enzymes into the base psilocybin biosynthetic pathway has the potential to create new biosynthetic psychedelic derivatives to increase the drug candidate pool for development of next-generation mental illness treatment. We have evaluated the activity of halogenases, which catalyze the incorporation of chlorine or bromine atoms onto their substrates, as well as prenyltransferases, which add a dimethylallyl group onto tryptophan and indole derivatives. Both of these directions have the ability to result in novel, new-to-nature compounds which could be promising drug candidates. We have successfully shown both halogenase and prenyltransferase activity on 4-HTP, and more importantly have created chlorinated psilocybin compounds through the use of an E. coli co-culture. This observation is the first example of de novo production of chloropsilocybin in a recombinant host and is a step forward in the creation of bioproduction approaches for pharmaceutical drug candidates for mental health treatment.

Directed Evolution of Oxidized Bicyclic Monoterpene Biosensors

Victoria Sagne

Terpenes are the largest and most diverse class of naturally occurring molecules with various biological activities, making them valuable for pharmaceuticals and industrial products. Despite their diverse applications, sourcing terpenes is challenging because most are extracted from their native producers, and several are minor products of chemical reactions. The oxidized bicyclic monoterpenes myrtenol and verbenol are extensively used as fragrances and flavorings and possess several medicinal properties, making them promising candidates for therapeutic agents and the treatment of certain diseases. However, their natural abundance is less than 0.70% and constitutes less than 40% of the native oxidation products of α-pinene. The optimization of the myrtenol and verbenol biosynthetic pathways may be achieved via a directed evolution approach. However, this approach accumulates a large number of enzyme variants that are difficult to efficiently screen using existing technology. We propose using transcription factor-based genetically encoded biosensors for this purpose. By engineering transcription factor proteins through multiple rounds of directed evolution, biosensors can be tailored for a specific molecule. Here, we report the initial screening and tailoring of different biosensor constructs with verbenol and myrtenol.

Examining the Effects of Common Vape Additives on the Interfacial Properties of Model Lung Surfactant Monolayers

Madelyn Atkins, Estephanie Nottar Escobar, Emmanuelle Ong, Prajnaparamita Dhar

The number of adolescents who report vaping has drastically increased in recent years which is alarming due to the rise in E-cigarette or Vaping Use-Associated Lung Injury (EVALI) which has symptoms including difficulty breathing, shortness of breath, and chest pain. Studies on the bronchoalveolar lavages of EVALI patients have found vape additives within the sample of fluid extracted from the lungs which were not found in healthy patients. One possible mechanism for EVALI is that the vape additives are interacting with the lung surfactants thus causing a change in the mechanical properties of the surfactant. The lung surfactant forms a monolayer film at the air-liquid interface of the alveoli which, consists primarily of surfactants such as Dipalmitoylphosphatidylcholine (DPPC) and 1-palmitoyl-2-oleoyl-phosphatidylglycerol (POPG), and proteins such as SP-B and SP-C, reduces the work of breathing by reducing the surface tension in the alveoli. Conversely, damage to the film caused by foreign matter can lead to increased surface tension and difficulty in breathing. Thus, our work hypothesizes, using a model lung surfactant composed of DPPC and POPG, that common additives to vape liquids, such as cannabidiol (CBD) and vitamin E, damage the film via the displacement of lung surfactants, resulting in increased surface tension in the alveoli of the lungs. While the surface activity of lung surfactant mixture containing vitamin E have been previously studied, the effects of CBD and CBD mixed with vitamin E is an area in need of more research. In this work, we utilize a Langmuir-Pockels Ribbon Trough to emulate the mechanical dilatation of the lungs during breathing. Briefly, our experiments consist of spreading a monolayer of surfactant to an air-liquid interface and applying compression and expansion cycles while monitoring the film's surface pressure, which is used to evaluate the effects of common vape additives on the film behavior. Additionally, dilatational rheology studies can be used to provide understanding into EVALI progression and therapeutic intervention, by examining changes in the viscoelastic properties of the film upon the addition of vape additives. The results with the model lung surfactant and CBD showed an earlier sharp increase in surface pressure upon the compression. This corresponds to a shift in the film state transition from expanded to condensed liquid, which suggests tighter interfacial packing. In addition, the maximum surface pressure upon compression decreased with increased number of cycles indicating a damaged ability of the film to return to its unstressed state. Preliminary results of the dilatational rheology did not show a significant impact of CBD addition over the measured interfacial viscoelastic moduli despite the changes in surface pressure. However, studies indicate that a concern with CBD in vaping liquids is the decomposition of CBD into cannabidiolquinone (QCBD) which could impact both the surface pressure and viscoelastic properties of the lung surfactant. Further studies should include the effects of heat and light degradation on surface activity of model lung surfactants added by vape additives such as vitamin E and CBD.

Investigating the Effects of HIV Nef Sequence Variants on Pulmonary Endothelial Cell Function

Eli Heath, Minh Nguyen, Sharilyn Almodovar

While significant progress has been made to improve the quality, length of life, and health of people living with human immunodeficiency virus (HIV), this patient group remains at an increased risk of developing chronic vascular complications and comorbidities. The rising incidence rate of HIV-associated pulmonary vascular disease (HIV-PVD) is a major cause of concern in HIV research, yet the molecular mechanisms for HIV-PVD remain largely unknown. Here, we explored a potential mechanism of HIV-PVD by examining the effects of HIV Nef on human pulmonary endothelial cell function *in vitro*. Endothelial cells line the interior of all human vasculature and play a critical role in blood vessel constriction/dilation, platelet direction, immune cell recruitment, and angiogenesis. Nef is an accessory protein of HIV and a primary determinant of viral pathogenesis. This project investigated the effects of HIV Nef on pulmonary endothelial cell function to determine its potential role in causing pulmonary hypertension and PVD.

Here, we utilized banked Nef molecular constructs previously generated from patients with known cases of HIV-associated pulmonary hypertension. Briefly, the plasmid DNA encoding these nefs were expanded using Top10 competent cells spread on agar plates treated with ampicillin, kanamycin, and G418. Colonies were picked, inoculated in Luria-Bertani broth, and pelleted. Plasmids were purified using QIAprep Miniprep Kit (Qiagen), and plasmid DNA was confirmed on previously sequenced plasmids with restriction endonucleases and gel electrophoresis. The unique nef sequence variants were then purified as endotoxin-free plasmids using the Endofree Prep Maxi Kit (Qiagen) and transfected into human pulmonary arterial endothelial cell (HPAEC) cultures. Transfected cultures were incubated at 37°C, 5% CO_2 overnight, and work is in progress to test three endothelial cell functional endpoints: intracellular nitric oxide (NO) production (by live cell imaging), angiogenesis biomarkers, and cell adhesion biomarkers (using cell supernatants). We expect to identify the mutational profiles of HIV Nef that contributes to PVD in our *in vitro* models.

This work was supported by an NIH/NHLBI Grant R01 HL172709 to SA and funding from Texas Tech University Honors College to EH and MN.

Regulating Peptide Amphiphile Micelle Properties Via Lipid Content for Drug Delivery Applications

Emma McDougal, Agustin Barcellona, Megan Schulte, Bret Ulery

Current cancer treatments are well known for their considerable cytoxicity and undesirable off-target effects. To address these significant shortcomings, new classes of therapeutics need to be developed to improve treatment efficacy and patient comfort. Peptide-based therapeutics are a promising alternative to traditional chemotherapies, but *in vivo* delivery of peptides is challenging due to their low membrane permeability, loss of secondary structure, and low local therapeutic concentration. Conjugating hydrophilic peptides comprised of therapeutics and bioactivity modulating components (*e.g.*, cell penetrating peptides – CPPs) with hydrophobic lipid tails yields modular peptide amphiphiles (PAs). These PAs self-assemble in water to form micelle nanoparticles (PAMs) when at high enough concentrations by hiding their hydrophobic lipid components in their cores and displaying their hydrophilic peptide components on their surface. Excitingly, micelle-mediated delivery of peptide therapeutics can address many, if not all, of the primary concerns associated with free peptide delivery. While promising, most work to date with PAs focuses on the use of only a small number of lipid moieties, primarily focusing on using one or more palmitic acids. To expand the number of PAs available, this work focuses on exploring the impact that altering lipid chain length, number of desaturation events, and equivalents of lipids per peptide has on PAM size, shape, and stability. As a test bed for this effort, novel PAs were explored for their capacity to alter the bioactivity of a therapeutic peptide that has shown broad-spectrum anti-neoplastic effects in hematological cancer. Specifically, PAM physical and biological characteristics were studied *in vitro* employing a human B cell leukemia to determine the impact lipid tail size has on desirable PAM cancer cell cytotoxicity. Based on this work, lipids can be selectively chosen to optimize material characteristics for drug delivery applications of PAMs not only for cancer therapy but also a wide variety of other drug delivery applications.

Development and Characterization of Quercetin-Loaded Microparticles Via Spray Drying for Pulmonary Drug Delivery Applications

Daniel Perez Torres, Camila Cersosimo, Amanda Pepler, Samantha Meenach

Pulmonary inflammation is associated with respiratory disorders such as pulmonary arterial hypertension and chronic obstructive pulmonary disease. A hallmark of these diseases is an increase in reactive oxygen species in the body, leading to further inflammation and acerbation of these diseases. Fortunately, a class of natural products termed plant-derived flavonoids have been found to alleviate inflammation and oxidative stress in lung tissue, making them promising therapeutic agents for the treatment of pulmonary diseases. Flavonoids are accessible and cost-effective, with quercetin being one of the most commonly consumed flavonoids in the human diet. Despite its therapeutic potential, the efficacy of quercetin is limited due to its poor water solubility, thereby limiting its oral bioavailability. This can be overcome through the encapsulation of quercetin in polymeric-based nanoparticles (NP) and microparticles (MP) for localized delivery to the lungs. Both NP and MP can increase the bioavailability of quercetin while lowering its dosage amount and frequency. Quercetin-loaded nanoparticles (Q-NP) were formulated via nanoprecipitation and tested for their size, homogeneity, and drug loading. The particles were <200 nm in diameter, with a low polydispersity index, and high drug loading. An antioxidant assay was used to confirm that Q-NP retained high antioxidant properties of quercetin. Quercetin-loaded nanocomposite microparticles (Q-nCmP) were successfully prepared by spray-drying Q-NP to formulate a dry powder aerosol. The Q-nCmP demonstrated desirable morphology and aerosol dispersion properties, high drug loading, and were successfully redispersed back into parent Q-NP formulation. Our ongoing research aims to further the development of effective and safe therapies for lung inflammation and chronic respiratory disorders.

Advancing Silk-Elastin-like Protein Polymers: Biomolecular Coatings with Fatty Acids and Nanoparticles for Drug Delivery

Jed Quiaoit

Current drug delivery systems account for the biomolecules involved (e.g., small molecules, peptides, antibodies, nucleic acids, live cells, etc.), the bioavailability of the drug molecule, environmental factors (e.g., pH, permeability, intracellular interactions, etc.), and devices compatible with said therapeutic molecules (e.g., microneedle patches, micro/nanoparticles, polymer films, implants, etc.). A major challenge, however, in drug delivery centers on overcoming physiological, chemical, and biological barriers when entering the human body. Silk-elastin-like protein polymers (SELPs) offer a promising solution due to their chemical stability, easily tunable physicochemical structures, and biocompatibility. Leveraging silk's abundance in nature and uniform properties, SELPs serve as a promising sustainable drug delivery platform.

This study focuses on developing the SELP system and demonstrating its versatility with various biomolecules, which includes coating lipid membranes and silk protein-based nanoparticles (SNPs) with SELPs. The SELP platform, augmented with fatty acid groups, has been successfully proven to be biocompatible with live human cells and capable of covalently binding with silk-based nanoparticles. Upon refining these models, SELPs can be engineered with cell-targeting oligopeptides such as GE11 and LT6 to target specific cell types like EGFR high-expressing cancer cells. Future iterations of the SELP-fatty acid conjugate platform can be improved by adjusting parameters such as temperature, incubation time, and fatty acid composition to enhance cellular binding. Additionally, SELP-SNP results suggest that an alternative coating method (e.g., layer-by-layer assembly of oppositely charged particles) may offer better optimization for SELPs and SNPs.

[Attached: Figure 1: Simplified schematic of SELP platform with fatty acid conjugate]

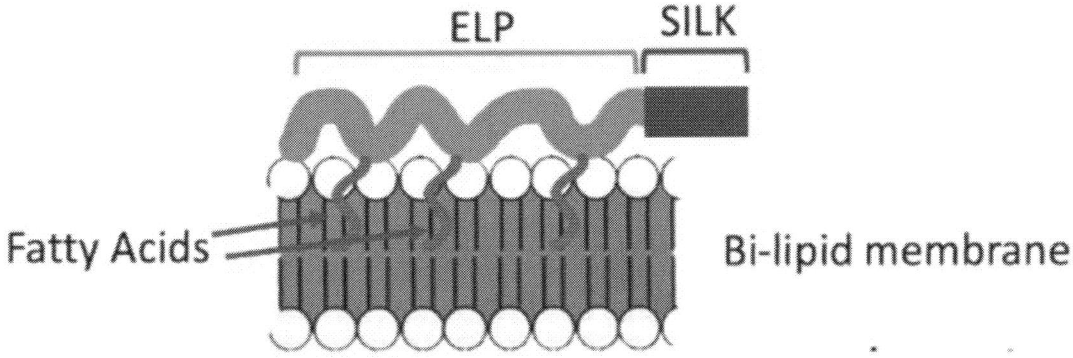

Cationic Carbon Nanodots for the Delivery of Nucleic Acids

Robert Russum Jr., Jesus Galeana, Gözde Demirer

The changes in climate due to the substantial increase of anthropogenic carbon dioxide and other greenhouse gasses have attracted widespread public attention in recent years due to their impact on global surface and atmospheric temperatures. Consequently, these changes are causing environmental stresses to occur far more often and severely; which is predicted to decrease crop yields by 25% by 2050. It is clear that novel solutions are needed to improve crop yields by making them more resilient to changing climate conditions. In working towards a solution to these issues, CRISPR/Cas9 genetic engineering has immense potential to introduce genes that allow for greater crop productivity and the improvement of their resistances to both biotic and abiotic stressors. However, for this, tools to effectively deliver nucleic acids into plant cells need to be developed. To tackle this problem, we are exploring nanomaterials, specifically carbon dots (CDs), for their simple synthesis, high water solubility, biocompatibility, low toxicity, nano size, and fluorescent properties, which make them highly suitable candidates for cargo delivery for gene editing applications. Herein, we have successfully functionalized CDs with polyethyleneimine (PEI), a cationic polymer, and loaded DNA on them via electrostatic attraction. Ongoing work is optimizing this platform to deliver genes into plant leaf cells for gene editing. Our results lay a basic foundation for future work in creating universal plant gene delivery systems.

Assembling Neural Endosome Targeted Nanoparticles for Pain Drug Delivery Via Flash Nanoprecipitation

Xiuxian Li, Rachel Pollard, Ashley Han, Parker Lewis, Nigel Bunnett, Nathalie M. Pinkerton

Chronic pain, a neurological condition affecting one-third of the adult population, is inadequately treated with existing medications. Recent pharmacological research suggests that pain receptors within neuronal endosomes mediate pain. However, current medications only block cell surface protein receptors, while endocytosed receptors continue to release pain signals. Drug-loaded nanoparticles offer a potential solution by delivering drugs directly to endosomes.

In this work, we are designing and optimizing nanoparticle vehicles to deliver pain therapeutics by studying nano-neuro interactions and pain behavior in sensory neurons. We have assembled a tunable core-shell nanoparticle library composed of three polymeric surfaces: polyethylene glycol (PEG), poly(ethylene oxide-co-glycidyl methyl ether) (rPEG), and poly-2-ethyl-2-oxazoline (P2OX), using Flash NanoPrecipitation (FNP). Two fluorophores, Rubrene and Dioctadecyloxacarbocyanine (DiO), were encapsulated in the core to visualize nano-neuro interactions via confocal microscopy. We observed colocalization between endosomes and nanoparticles by transfecting cells with Rab5a-GFP. The association rates of the nanoparticles in immortalized dorsal root ganglion neurons (F11) were characterized using flow cytometry.

Importantly, we present the first use of the F11 cell line with nanoparticle delivery and support its validation as a sensory neuron model for pain research. FNP is validated as a successful approach for synthesizing tunable and modular fluorescent nanoparticles. Our findings show that PEG, rPEG, and P2OX each exhibit distinct uptake rates and trafficking patterns in F11 cells. Endosome colocalization signals indicate a decreasing level of overlap from PEG to rPEG to P2OX.

Our results suggest that rPEG and P2OX may serve as alternative nanoparticle surfaces to PEG. Nanoparticles observed in endosomes along axon regions indicate that the vehicles have entered cells at nerve endings and can potentially interfere with pain signaling. Evidence of retrograde transport of nanoparticles highlights their potential clinical efficacy. Future work will involve using endocytosis and inhibitors to study endosomal pathways and further substantiate nano-neuro interactions in vivo. With these findings, we are closer to achieving more efficient drug delivery for chronic pain and other neurological disorders.

Local Delivery of an Adenosine Signaling Agonist Via Polymeric Microparticles for Suppression of Inflammation in Allergic Contact Dermatitis

Stacia Subick, Elizabeth Bentley, Steven R. Little

Vascular composite allotransplantation (VCA) involves transplantation of multiple tissue types, including skin, making these transplants highly prone to rejection [1, 2]. One potential way of developing new strategies for immunosuppression in VCA is to study a disease with similar characteristics to skin transplant rejection, such as allergic contact dermatitis (ACD) [3]. In ACD, activated immune cells release extracellular ATP (eATP), which promotes inflammatory signaling. eATP is naturally converted to adenosine (Ado), which promotes anti-inflammatory signaling [4]. Therefore, it is possible that blocking eATP signaling and/or amplifying Ado signaling may lead to immunosuppression in ACD.

This study aimed to determine if polymeric microparticles could be used to encapsulate and locally deliver three eATP or Ado signaling modifiers *in vivo* to reduce the innate immune response associated with ACD. The three drugs of interest for this study, chosen for their success in clinical trials, mechanism of action, and hydrophobicity, were IBMECA (adenosine receptor agonist), Regadenoson (adenosine receptor agonist), and JNJ-55308942 (eATP receptor antagonist). In this study, the aforementioned drugs were encapsulated in poly(lactic-co-glycolic-acid) (PLGA) microparticles (MPs) using a single emulsion solvent evaporation method [5]. Microparticles were characterized for release kinetics and size distribution. To study the effects of drug-loaded MPs in ACD, a murine contact hypersensitivity (CHS) model was used [6]. In this model, ear thickness measurements are taken as a method to quantify inflammation (indicative of swelling) [6]. To induce ACD-like symptoms, mice were sensitized topically with 1-fluoro-2,4-dinitrobenzene (DNFB), a known chemical allergen. Prior to sensitization, ear thickness was measured, and drug-loaded-MPs or Blank-MPs (containing no drug) were injected subcutaneously at the base of the ear. 24 hours after sensitization, ear thickness was re-measured, and mice were sacrificed to harvest cervical draining lymph nodes for flow cytometry.

It was found that IBMECA-MPs significantly reduced the change in ear thickness when compared to Blank-MPs. Regadenoson-MPs and JNJ-MPs did not reduce change in ear thickness when compared to Blank-MPs. Flow cytometry analysis of dendritic cell populations showed that IBMECA-MPs altered expression of several surface markers related to inflammatory responses. More specifically, IBMECA-MPs significantly reduced CD40 expression across multiple dendritic cell subtypes. Together, these results suggest that treatment with IBMECA-MPs could have a therapeutic effect in ACD.

References:

[1] Uluer, Mehmet C., et al. "Vascularized composite allotransplantation: Medical complications." *Current Transplantation Reports*, vol. 3, no. 4, 15 Aug. 2016, pp. 395–403. doi: 10.1007/s40472-016-0113-x.

[2] Abdallah, Florence, et al. "Skin immune landscape: Inside and outside the organism." *Mediators of Inflammation*, vol. 2017, 2017, pp. 1–17. doi: 10.1155/2017/5095293.

[3] Starzl, Ravi, et al. "Review of the early diagnoses and assessment of rejection in vascularized composite allotransplantation." *Clinical and Developmental Immunology*, vol. 2013, 2013, pp. 1–9.

doi: 10.1155/2013/402980.

[4] Faas, M.M., et al. "Extracellular ATP and adenosine: The Yin and Yang in immune responses?" *Molecular Aspects of Medicine*, vol. 55, June 2017, pp. 9–19. doi: 10.1016/j.mam.2017.01.002.

[5] Jhunjhunwala, Siddharth, et al. "Controlled release formulations of IL-2, TGF-β1 and rapamycin for the induction of regulatory T cells." *Journal of Controlled Release*, vol. 159, no. 1, Apr. 2012, pp. 78–84. doi: 10.1016/j.jconrel.2012.01.013.

[6] Balmert, Stephen C., et al. "In vivo induction of regulatory T cells promotes allergen tolerance and suppresses allergic contact dermatitis." *Journal of Controlled Release*, vol. 261, Sept. 2017, pp. 223–233. doi: 10.1016/j.jconrel.2017.07.006.

Multi-Compartment Pharmacokinetic-Pharmacodynamic Modeling of Drug Delivery from Bi-Layered Polymeric Microspheres for Wet AMD

Sarita Das, Koki Kanehira, Yaman Oklla, Eduardo A. Chacin Ruiz, Mohammad Aminul Islam, Katelyn E. Swindle-Reilly, Ashlee N. Ford Versypt

Wet (neovascular) age-related macular degeneration (nAMD) is an aggressive eye condition caused by the formation of abnormal leaky blood vessels located in the central area (macula) of the retina. These leaky blood vessels are associated with increased vascular endothelial growth factor (VEGF) levels. Current treatment options for nAMD involve repeated intravitreal injections into the eye with drugs including abicipar, ranibizumab, and bevacizumab, which all play a part in suppressing VEGF levels in the eye [1-3]. However, repeated injection is not an ideal treatment plan and can cause pain, anxiety, and disturbance for patients. Drug delivery systems (DDS) offer a solution to provide extended delivery of drugs.

Here, we connected pharmacokinetic/pharmacodynamic models of intravitreal injection of abicipar [1], ranibizumab [2], and bevacizumab [3] to the mathematical modeling of drug release from bi-layered polymeric microsphere DDS [4] to investigate drug release dynamics and VEGF suppression in the eye. The DDS consists of a drug-loaded core made of chitosan and an outer polycaprolactone layer. The inner layer controls the extended drug release after an initial burst release, whereas the outer layer protects the core from degradation. Our aim is to quantify the drug release dynamics and VEGF suppression from the DDS compared to existing regular injection models and optimize the DDS parameters, i.e., drug loading amount and radii of the inner and outer layer of the microsphere.

Our first model for abicipar mapped the resulting drug concentration and half-maximal inhibitory concentration (IC50) in eye compartments. This model helped us conclude that the DDS was more effective in holding the drug concentration above the IC50 value for an extended period compared to regular injection cycles. Our models for ranibizumab and bevacizumab also determined that the DDS could provide increased VEGF suppression over time. Additionally, we simulated the effect of variation in the initial drug loading amount and variations of the inner and outer radius of the DDS. All models found that decreasing the drug loading amount and the thickness of the sphere's inner layer decreases the duration of drug release and VEGF suppression.

Our simulated results demonstrate the potential of DDS in nAMD treatment to decrease injection frequency, thereby reducing patient discomfort while still allowing extended drug release for longer VEGF suppression. Future goals of this research are to optimize the DDS model for the different drugs and to use the results in mathematical models for patient outcomes, including their visual acuity and retinal thickness.

Lipid Nanoparticle Modifications for Improved mRNA Delivery through Microneedle Patches

Gianna Fiduccia, Sophia Sakers, Elisa Schrader Echeverri, James Dahlman, Mark Prausnitz

mRNA is a powerful technology developed for a diverse set of uses, such as vaccination, cancer treatment, and protein therapeutics.[1] The development of lipid nanoparticles (LNPs) has aided in the delivery of mRNA into cells, but intramuscular injection of liquid LNP formulations poses a challenge, as they are not stable at low temperatures and require extensive resources for distribution and administration.[2] Microneedle patches (MNPs) are arrays of micron-scale needles made from dissolvable polymers used to delivery therapeutics transdermally, and they offer a viable solution.[3] MNPs are less painful, are easier to transport, store, and administer, and may increase thermostability compared to intramuscular injection.[3] However, research pertaining to drying LNP formulations for MNP use is limited, and initial tests determined that LNPs are destabilized during MNP manufacturing. Optimizing LNP design may improve the stability and functionality of LNPs for mRNA delivery in an MNP.

Nanoluciferase encoding mRNA was formulated into LNPs using microfluidic mixing. LNP composition included ionizable lipid SM-102, cholesterol, DMG-PEG2000, and distearoylphosphatidylcholine at various molar ratios, with 50:38.5:1.5:10 used as a positive control to mimic Moderna's SpikeVax COVID vaccine LNP formulation. LNPs were then mixed into a solution of 6.25% polyvinyl alcohol and 10% sucrose, dried under a vacuum in PDMS molds, and backed with epoxy to make MNPs. The dried MNPs were reconstituted in phosphate-buffered saline and evaluated for size using dynamic light scattering, encapsulation efficiency (EE) using Ribogreen assay, and in-vitro cellular expression in a RAW 264.7 cell line.

A variety of lipid molar ratios were screened to determine their effect on LNP stability during the MNP fabrication process. Increasing SM-102 had a stabilizing effect on MNPs, indicated by small LNP size, high EE, and high cellular expression post-drying. Increasing distearoylphosphatidylcholine content had a destabilizing effect. Increasing cholesterol content or DMG-PEG2000 content did not have a significant effect on markers of stability. The mass ratio of lipid: mRNA was also modified to improve the function and biophysical characterization of dried LNPs ranging from 20:1 to 100:1. Increasing mass ratios to 50:1 and 75:1 generated high cellular expression in particles but had minimal effect on EE and size. Large mass ratios decrease the loading capacity of mRNA in MNPs, which is undesirable in a low-dose MNP. Finally, phospholipids were screened, and initial data suggest that positively charged DOTMA increases *in vitro* expression before and after MNP manufacturing.

Modifications to LNP molar and mass ratios suggest an improvement in MNP drying stability. Additional formulation optimization can improve the function of LNPs post-drying. Future directions of this work include investigating the effect of ionizable lipid identity on stability and further investigations of alternate phospholipids.

References:

1. Schlake, T., Thess, A., Fotin-Mleczek, M., & Kallen, K.-J. (2012). Developing mrna-vaccine technologies. *RNA Biology*, 9(11), 1319–1330. https://doi.org/10.4161/rna.22269

2. Wilson, B., & Geetha, K. M. (2022). Lipid nanoparticles in the development of mrna vaccines for COVID-19. *Journal of Drug Delivery Science and Technology*, 74, 103553. https://doi.org/10.1016/j.jddst.2022.103553

3. Prausnitz, M. R. (2017). Engineering microneedle patches for vaccination and drug delivery to skin. *Annual Review of Chemical and Biomolecular Engineering*, 8(1), 177–200. https://doi.org/10.1146/annurev-chembioeng-060816-101514

Designing Diblock Polypeptides for Polyelectrolyte Complex Drug Delivery of Globular Proteins

Megan Jen, So Yeon Ahn, Allie Obermeyer

Protein therapeutics play a significant role in medical advancements, however, intracellular uptake of protein-based therapeutics remain a challenge. The majority of developed protein therapeutics do not readily cross cell membranes, but rather act primarily on extracellular or membrane-bound targets. Thus, intracellular protein delivery methods must be developed to realize the full therapeutic potential of exogenous protein therapeutics. To allow for cellular uptake, liquid peptide-based droplets can be designed by mimicking the physicochemical properties of endogenous membraneless organelles. Droplet formation occurs by complex coacervation, the liquid-liquid phase separation of oppositely charged polyelectrolytes, to achieve mesoscale self-assembly. We biosynthesized block copolypeptides with a net neutral block and a positively charged block, to assemble polyelectrolyte complex (PEC) micelles when combined with model anionic globular protein cargo. The two components spontaneously form uniform nano-sized micelles when mixed and are held together by non-covalent interactions that allow the condensation of the cargo protein while ensuring its solubility in the liquid phase. Here, we describe the design of biosynthetic PEC micelles that encapsulate anionic GFP as a model cargo protein by complexing with a net neutral fluorescent protein mScarlet-I fused to the disordered region of Histone H5, a highly charge dense cationic protein. Optimal parameters for stable micelle formation were systematically studied using genetic modification to vary protein net charge and charge density. The engineered block copolypeptides were mixed with anionic cargo protein at varying mixing ratios to promote assembly. The phase behavior of engineered proteins were evaluated via microscopy, dynamic light scattering, and turbidity assays to assess stable micelle formation.

Hydrogels with Microsphere Drug Delivery Technology to Promote Healing Post-Pelvic Radiotherapy and Prevent Vaginal Stenosis

Isha Dave, Emma Cohen, Johanna Hannan, Deborah Marshall, Woojin Han, Jennifer Weiser

Women's healthcare is often underfunded and overlooked, contributing to suboptimal outcomes in treating conditions such as gynecological cancers. In 2023 alone, there were 114,810 new cases[1] of gynecological cancer and about 60% received radiotherapy as treatment[2]. As a side effect of radiotherapy, approximately half of these patients[3] will experience vaginal adhesions and stenosis, which is the formation of scar tissue, causing patient discomfort with exams, poor vaginal health, and limiting sexual function. Use of current solutions, like vaginal dilators or topical treatments, are limited in the immediate post-radiation timeframe when the primary injury response occurs, because of patient discomfort. To address this issue, an injectable bioadhesive hydrogel was created to prevent post-radiation damage in the vaginal canal by keeping the injured tissue apart. This bioadhesive, based on crosslinked polyethylene glycol, offers a protective, physical barrier to reduce vaginal adhesions and stenosis. Additionally, the hydrogel contains polylactic-co-glycolic acid microspheres (PLGA-MS) loaded with estradiol (E2) to help promote the healing of the tissue and reepithelization of the vaginal mucosa.

The polymers chosen in this study provide properties that allow the gel to adhere to tissue and have tunable gelation kinetics for easy injectability. To optimize the formulation of the gels for the treatment of vaginal toxicity, the gels must be mechanically characterized, easily injectable, bioadhesive, and compatible with a controlled drug delivery system.

To meet these criteria, tests were performed regarding gelation kinetics of the PEG-based bioadhesive, the size of the PLGA-MS, and loading efficiency of the PLGA-MS. Rheometric analysis was done using ASTM D2556 to determine gelation kinetics. The bioadhesive solutions (n=4 per formulation) were casted onto the rheometer to undergo polymerization while the test was being run at 37 °C. Preliminary bioadhesive testing was done using mice skin tissues. The hydrogels were cast on the tissue after being soaked in 1X PBS and visual inspection confirmed strong bioadhesion. To assess the feasibility of incorporating PLGA-MS into the hydrogels, the spheres were prepared using a single emulsion technique and imaged under a microscope. ImageJ was used to analyze the size and distribution of the spheres. This resulted in an average particle size of 45-50 microns, suitable for embedding in the bioadhesives during polymerization. High-performance liquid chromatography of the loaded E2 PLGA-MS was done using an Avantor C18 column. Standard curves of E2 in acetonitrile were created using a mobile phase of water/acetonitrile 40:60. After placing the microspheres in acetonitrile (varying concentrations of E2 mg/mL), the loading efficiency was determined to be 87.2%. Further studies aim to optimize the gel formulation and study the controlled release of E2-loaded PLGA-MS embedded in the hydrogels.

References

1. *Gynecological cancers - Foundation for Women's Cancer.* Foundation For Women's Cancer - Eradicate or lessen the impact of gynecologic cancer. (2024, August 21). https://foundationforwomenscancer.org/gynecological-cancers/
2. Karlsson, J. A., & Andersen, B. L. (1986). *Radiation therapy and psychological distress in gynecologic oncology patients: Outcomes and recommendations for enhancing adjustment.* Journal of psychosomatic obstetrics and gynaecology. https://www.ncbi.nlm.nih.gov/pmc/articles/PMC2763444/#:~:text=Of%20the%20approximately%2078%2C000%20women,surgery%20or%20chemotherapy%20%5B1%5D
3. Varytė, G., & Bartkevičienė, D. (2021, April 1). *Pelvic radiation therapy induced vaginal stenosis: A review of current modalities and recent treatment advances.* Medicina (Kaunas, Lithuania). https://www.ncbi.nlm.nih.gov/pmc/articles/PMC8066324/

Evaluating Gene Integration Efficiency in Ial-PiD2, a *Plodia Interpunctella* Cell Line

Cecilia Rodriguez, Bryce Shirk, Jasmine McTyer, Marisa O. Pacheco, Liam Rodgers, Paul Shirk, Whitney Stoppel

Insect cell lines have demonstrated exceptional potential within applications in vaccine technology, therapeutics, and recombinant protein production. Through genetic modification, insect cells have been engineered to replicate the protein glycosylation patterns of mammalian cells, offering a potentially safer and more cost-effective alternative for cell lines used in the production of human therapeutics requiring these patterns.[1] One such example is the Sf9 cell line derived from the pupa ovarian tissue of the fall armyworm, which has been utilized in the baculovirus expression system.[2] However, there are over 100 insect cell lines commercially available, yet few have been assessed for applications in the production of recombinant proteins.[1] IAL-PiD2 cells derived from the wing imaginal disks of *Plodia interpunctella*, could provide a valuable platform for recombinant technology. Targeted gene editing using CRISPR/Cas9, and similar technologies can allow for precise and stable genomic integrations, improving the potential for some future applications compared to random genomic integrations seen with piggyBac-based methods. Moreover, the non-viral transfection platform reduces risk to human health should the proteins produced by the cell line have future human applications.[1] However, the use of IAL-PID2 cells for protein production has not been evaluated, and research in their genetic manipulation is lacking. Therefore, we aimed to determine the feasibility and efficiency of modifying IAL-PiD2 using tools like CRISPR/Cas9.

Herein, we designed constitutive promoter-driven constructs with the mNeonGreen protein insert for our DNA template, hypothesizing that these constitutive promotor-driven sequences would lead to sustained expression of an easily characterizable phenotype (green fluorescence). IAL-PiD2 cells were seeded in triplicate within a 24 well plate at 100,000 cells per well and reached roughly 65% confluency after 48 hours before transfection with the CRISPR/Cas9 complex with the template. This confluence level was chosen to represent a point where the cells were rapidly dividing while allowing space for continued cell growth during the 7 day period we monitor gene expression intensity.

We first sought to evaluate the efficiency of commercially available transfection reagents within the IAL-PiD2 cell line. Three transfection reagents from Mirus Bio® were used to deliver the constructs into the cell line. We analyzed their respective efficiencies in the expression of the mNeonGreen phenotype and evaluated the number of transfected cells and fluorescence intensity over time using fluorescence microscopy. To quantify cell expression and fluorescence intensity across many images, we developed a CellProfiler® pipeline. This software allowed us to count the total cells in each well and quantify the cells expressing the desired phenotype by setting desired threshold limits for fluorescence intensity. Results show increased transfection efficiency over time with substantial expression beginning on Day 5 after transfection. Based on Day 7 data, *Trans*IT®-Insect, the transfection reagent synthesized for insect cell lines, was the most efficient (**Figure 1**). Future work aims to compare transfection efficiency in SF9 cells with the results obtained in IAL-PiD2 cells.

1. Yee CM, Zak AJ, Hill BD, Wen F. The coming age of insect cells for manufacturing and development of protein therapeutics. Industrial & engineering chemistry research. 2018;57(31):10061-70.
2. Hong Q, Liu J, Wei Y, Wei X. Application of Baculovirus expression vector system (BEVS) in vaccine development. Vaccines. 2023;11(7):1218.

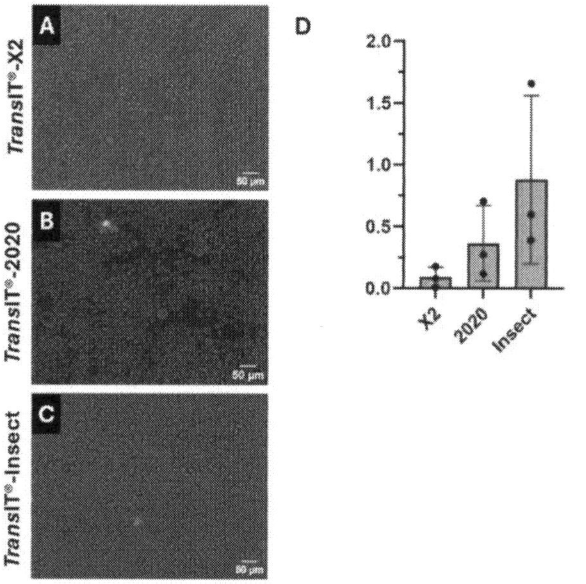

Figure 1: Transfection Efficiency and Expression Intensity on Day 7. A-C) Representative images of fluorescent gene expression using **A)** *TransIT*®-X2 transfection reagent **B)** *TransIT*®-2020 transfection reagent, and **C)** *TransIT*®-Insect transfection agent in IAL-PID2 cell culture, 7 days after transfection. **D)** Transfection efficiency for each transfection reagent, determined by comparing the number of fluorescent cells to total cells within a sample.

Single Emulsion Manufacturing Method for Encapsulation of Low Molecular Weight Chemokines

Michael Pezzillo, Julie Kobyra, Stephen C. Balmert, Steven R. Little

For decades, poly(lactic-co-glycolic acid) (PLGA) has been a widely utilized FDA-approved polymer due to its biodegradability and biocompatibility. PLGA is often utilized in the development of a variety of different drug delivery systems, such as microparticles. These PLGA microparticles (MPs) are often used to encapsulate hydrophobic small molecule and peptide-based drugs to enable sustained drug release and are manufactured using a single emulsion solvent evaporation method. More recently, there is growing interest in expanding the use of PLGA to encapsulate biologic drugs, such as proteins, to increase their bioavailability and protect them from degradation. It has previously been demonstrated that local, sustained delivery of CCL22, a chemokine known to recruit regulatory T cells, promotes immune homeostasis in various inflammatory disease models. However, encapsulating biologics typically requires a more complex double emulsion technique, which limits scalability and clinical translation. This study explored the feasibility of using a single emulsion technique to encapsulate CCL22, aiming to develop a more streamlined process for large-scale production.

PLGA with varying end caps was used in a single emulsion water-in-oil evaporation method to encapsulate CCL22 into MPs. CCL22 was dissolved in DMSO and combined with PLGA in dichloromethane to create a discrete phase, which was then added to an aqueous continuous phase containing PVA and homogenized to form an emulsion. After solvent evaporation through continuous stirring, the MPs were washed, flash-frozen, and lyophilized. The MPs were characterized for size, morphology, and release kinetics using Coulter Counter measurements, scanning electron microscopy (SEM), and *in vitro* release assays. Encapsulation with 10 kDa acid-terminated PLGA was successful but resulted in a significant lag phase in drug release. Changing to 10kDa hydroxyl-terminated PLGA improved the magnitude of the initial burst release; however, following the burst a lag phase was still seen. These findings suggest that low molecular weight chemokines, like CCL22, can be effectively encapsulated using a single emulsion technique. To further enable translational potential, CCL22 MPs were also manufactured using a continuous microfluidic process. Taken together, this study suggests that a single emulsion encapsulation technique, through homogenization or microfluidics, may be a viable option for low molecular weight biologics, thus expanding the therapeutic potential of PLGA delivery systems.

Exploring Liposome-Chitosan and Liposome-Xanthan Gum Interactions

Charles White

My work focuses on studying the interactions between liposomes and two different polyelectrolytes (chitosan, a polycation, and xanthan gum, a polyanion). We are specifically interested in understanding liposome stability upon interaction with these polyelectrolytes. To assess liposome stability in the presence of the polyelectrolytes, I utilize dynamic light scattering (DLS), zeta-potential measurements, and a calcein leakage assay via fluorescence spectroscopy. When combining chitosan with a 0.25 mg/mL solution of liposomes, a substantial increase in measured effective diameter was observed from 1 ppm to 10 ppm chitosan. Alternatively, xanthan gum did not seem to change effective diameter over the same incorporation range. The calcein leakage assay provided additional confirmation for the interaction of liposomes and chitosan and lack thereof for liposomes and xanthan gum. I concluded that this is due to the tendency of the polar head group of a liposomes bilayer to bind more effectively to positive charges due to a negative zeta-potential being present around a bare liposome, which in this case allows for chitosan to bind more effectively than the xanthan gum. Additionally, calculation of the number of charged groups per mass of each polyelectrolyte shows that xanthan gum has 2.5 times more charge density than chitosan, further amplifying its charge repulsion with liposomes.

How Many Passes Does It Take? An Investigation to Determine the Optimal Number of Liposome Extrusion Cycles

Kasey Piper, Kenneth Mineart

Liposomes are lipid-based nanoparticles commonly known for their drug carrying capacity and value in membrane and nanoreactor research. One commonly used method of fabricating liposomes of particular size is membrane extrusion where liposomes are pushed through a porous membrane and form smaller liposomes. However, the details pertaining to liposome extrusion, such as the number of passes, membrane pore sizes, and pre-extrusion sample preparation, vary between labs. The goal of this project is to establish a relationship between the number of passes through a track-etched, polycarbonate membrane, and the average diameter and lamellarity of extruded liposomes. To investigate, we extrude samples to different amounts varying from 1 to 1,000 passes and measure the average hydrodynamic diameter, diameter distribution, and radius of gyration of each using dynamic and static light scattering (DLS/SLS). We measure lamellarity with small angle x-ray scattering (SAXS) and cryogenic transmission electron microscopy (cryoTEM). When liposome size distribution and lamellarity data start to flatten out, it is suggested that extrusion has been completed, and further extruding would not change the liposome size and lamellarity. These results provide researchers with the optimal number of passes for liposome extrusion that will allow them to form ideal distributions in the shortest amount of time.

Characterization of Catechol O-Methyltransferase Activity and Promiscuity in *E. coli*

Nolan Brown, Abhishek Sen, Andrew Jones

Heterologous enzyme expression in *E. coli* often experiences nuanced activity and substrate preferences. Our aim was to study 14 catechol *O*-methyltransferases (COMT) individually and in combination with a tryptophan decarboxylase (PsiD) to determine the preferred metabolic route towards the synthesis of the naturally occurring hallucinogen, 5-MeO-DMT. An intermediate in the pathway, 5-hydroxytryptophan (5-HTP), can either be decarboxylated first by PsiD and methylated second by COMT (Route 1), or methylated first and decarboxylated second - both routes leading to the production of 5-MeO-tryptamine. To determine the preferred metabolic route in *E. coli* BL21star^{TM}(DE3), we cloned a 4 member mutant T7 promoter library for each of the 14 COMTs individually and in combination with PsiD (polycistronic configuration). Individual colonies of BL21 star^{TM}(DE3) expressing the promoter library (xx^4) were grown in a semi-defined medium (AMM) and supplemented with pertinent intermediates. The fermentation-based screening was carried out in 48-well plates over 48h at 37°C. The samples were centrifuged and supernatants were analyzed by HPLC-MS. The data suggests that COMTs prefer 5-HTP over serotonin as a substrate while PsiD is able to decarboxylate 5-HTP and 5-MeO-tryptophan equally well. Pairing this information in the context of the biosynthetic routes available leads to a clearly preferred metabolic route which will guide the use of COMTs and PsiD in the development of extended pathways towards various tryptamine natural products.

Production and Characterization of Autumn Berry Wine

Camden Jackson

Elaeagnus umbellate Thunb., commonly known in Kentucky as the "Autumn Berry' or "Autumn Olive", is an invasive nitrogen-fixing shrub that was initially introduced to control erosion in eastern Kentucky. The plant is drought tolerant and can survive in a variety of climates and soil conditions. While autumn berries should not be purposely cultivated, they are common in Kentucky and harvesting the berries can prevent further spread in the environment. Autumn berries are also a rich source of lycopene antioxidant, a valuable food ingredient. Therefore, the goal of this project was to look into the feasibility and viability of fermenting autumn berry juice to create alcoholic beverages (red and white wine). The berries have a large pit, resulting in an inefficient juice output of 0.30 milliliters per gram of berry but sugar content measurements by HPLC indicated promise for use in yeast fermentations. Lab-scale fermentations were performed and pH, sugars, esters, titratable acidity, and alcohols were measured before, during, and after fermentation to characterize the wine produced.

Expansion of Non-Natural Psilocybin Derivative Synthesis in *E. coli*

Grace E. Kemmerly, J. Andrew Jones, William Gibbons Jr.

Psychedelic drugs have garnered much clinical attention for their promising therapeutic potential in the treatment of mental health conditions. However, the presence of hallucinogenic side effects remains an issue that greatly limits their clinical use and legalization. It is suspected that derivatives of psychedelics may elicit the observed positive mental health effects while minimizing the negative hallucinogenic side effects. Production of a strong drug candidate pool is fundamental to the testing and discovery of such drugs. Recent derivatization of psilocybin already expanded this candidate pool to non-natural psychedelics; however, derivative synthesis was limited by the promiscuity of the traditional enzymatic pathway. Synthetic bottlenecks have been observed at tryptophan synthase (TrpB) and L-tryptophan decarboxylase (PsiD), thus alternative enzymes were explored. Incorporating TrpB variants Q90, Tri, and Azul and PsiD variant RgnTDC has enabled the expansion of derivative production from monosubstituted indoles beyond what was previously observed. Both tri-culture and co-culture designs were employed using optimized psilocybin production strains in conjunction with the new alternate enzymes in either wild-type *E. coli* or *E. coli* containing genomic integration of the traditional pathway enzymes. These systems were found to be successful in expanding non-natural derivative synthesis. They enabled production of seven new tryptophans and four new tryptamines and greatly enhanced the production of 18 more. The tri-culture setup successfully produced three new psilocybin derivatives, 7-aldehyde-psilocybin, 7-cyano-psilocybin, and 7-isopropyl-psilocybin, and the genome-integrated co-culture has produced 5-chloro-psilocybin. This testing primarily focuses on the use of Q90 and RgnTDC and suggests that use of these alternative enzymes allows for synthesis of bulkier indole derivatives than observed with traditional pathway enzymes. Further testing of monosubstituted indoles is required to realize the complete potential of the integrated co-culture setup.

Quantifying Coffee Freshness for Sustaining the Specialty Coffee Industry

Maya A. Fetzer, Katsuyuki Wakabayashi

The modern coffee industry places emphasis on the freshness of beans and grounds, which is not necessarily predictable or easily measurable. The current industrial method is to use customer feedback and the production date to measure product freshness, which lacks objectivity in the actual amount the freshness has decayed. This could result in coffee getting discarded unnecessarily. Because coffee is a natural, resource intensive product, it is important to reduce production waste by understanding freshness degradation. This research aims to quantify freshness in coffee beans by investigating how roast level affects the rate of degradation in roasted coffee beans. The findings of this research will contribute to more sustainable practices in the coffee industry by optimizing the use of coffee beans and reducing the environmental impact associated with improper handling of coffee beans.

Green coffee beans were roasted in house to two different roast levels, measured in terms of percent weight loss and color scale. The roasted beans were stored in ambient, airtight containers to simulate consumer/home coffee storage settings. To measure freshness, chemistry-based measurements, such as gas chromatography/ mass spectroscopy (GC/MS) and thermogravimetric analysis (TGA), were employed. The chemical properties were used to characterize freshness of the beans as a function of storage time for the two roast levels, which in turn was correlated to other physical properties (e.g. density) and brew qualities (e.g. pH).

Optimizing Polyhydroxyalkanoate Production: Computational Design and Engineering of Phac for Enhanced Biodegradable Plastics

Snigdha Jagarlapudi, Kyle Sander, Adam P. Arkin

PhaC is a critical enzyme for synthesizing polyhydroxyalkanoates (PHAs), biodegradable polymers from renewable resources that offer an eco-friendly alternative to petroleum-based plastics. PhaC has been effectively utilized in microbial systems to produce PHAs, with *Cupriavidus necator* being a particularly promising host for such applications. A novel potential application for PHAs is in crewed Mars missions, where they could provide regenerative bioplastics for material manufacturing *in situ*. However, to generate durable, high-performance plastics, it is necessary to optimize substrate-binding and increase substrate breadth of PhaC. This study advanced this goal by developing a PhaC mutant library through epPCR, coevolutionary analysis of protein sequences, and MutComput, a neural network algorithm. This effort led to the identification of 124 promising sites and the creation of a mutant library of several hundred thousand members. Molecular visualization tools revealed that these sites lie within critical protein domains, such as substrate-binding domains and plastic egress channels.

Development of a Tunable Collagen/Heparin Matrix for Investigating Mechanical Cytoskeletal Crosstalk

Laurel Fishburn, William Weeks, Leigh Hardin, Omayma Al Azzam, Jorge Almodovar, Dana Reinemann

The cytoskeleton plays a crucial role in cellular architecture, directing both cellular shape and a variety of mechanically influenced tasks, such as cell division, sensing and adaptation, and motility. The cytoskeleton can dynamically alter its length and stiffness through sensing its environment, yielding specific cellular responses. Mechanical changes can also occur from the bottom up through changes in cytoskeletal crosslinking and motor protein activity, altering molecular-level stiffness and thus further complicating mechanochemical feedback loops. The relative influences of "top down" versus "bottom up" mechanics on cellular activity and to what extent they are integrated within cytoskeletal systems remain unclear. This inherent dynamic nature proves difficult to mimic in *in vitro* studies designed to probe protein interaction and function of the cytoskeleton. To understand how the local mechanical environment plays a role in cytoskeletal crosstalk, we integrated a mechanically tunable collagen/heparin matrix constructed with a layer-by-layer technique combined with actomyosin ensembles. The layers are characterized using a quartz crystal microbalance with dissipation monitoring (QCM-D), which can detect nanoscale changes in mass and viscoelastic properties. Using the QCM-D, we were able to show increases in thickness and viscoelasticity of the matrix with an increasing number of collagen/heparin bilayers. We have also demonstrated that actin will stably bind the matrix at multiple layer depths. The presence of these layers will mimic different cytoskeletal environment stiffnesses and be coupled with actomyosin ensembles to understand how changing the system-level mechanical environment will alter force generation dynamics of myosin II using optical trapping. Preliminary bundle experiments suggest that the presence of the collagen/heparin bilayers improves the stability of actin filaments and the rate of photobleaching seen on the microscope. This approach allows us to investigate the biophysical crosstalk mechanisms of actomyosin ensembles and elucidate the sensory force feedback loops that ultimately scale and influence large-scale cellular tasks.

Effects of Varying Rapamycin Concentrations and Timing on Cellular Functions in Yeast Surface Display

Riley Basinger, Luke Abbott, Balaji Rao, Albert J. Keung

Rapamycin is a commonly used reagent in cell and molecular biology. One use is chemically inducible dimerization. In FKBP-FRM dimerization, rapamycin binds to FKBP (FK506 Binding Protein 12) which it has a high affinity for. It will bind FKBP to the FRB (FKBP Rapamycin Binding) domain of mTOR and control their interactions. These two domains can then be fused to proteins of interest whose interaction the researcher might want to control. However, rapamycin also has direct effects on the cell. It inhibits the mTOR pathway, slowing cellular metabolism. This inhibition of the mTOR pathway is useful in controlling protein synthesis, cell growth and potentially aging, but in the context of controlling synthetic biology, components would yield off-target pleiotropic effects in engineering applications. Therefore, our goal is to determine what concentrations of rapamycin can be used without inhibiting cell growth, gene expression, and other potentially useful functions such as the display of proteins on the surface of yeast.

The experimental platform we used was a yeast surface display system. In our previously developed system (Waldman et al., 2021), we utilize a yeast surface display system that uses a bidirectional galactose promoter. This allows for well-regulated exogenous gene expression of the desired proteins. Specifically, we are transcribing p300 (with a V5 tag) and histone proteins, which include the N-terminus histone tail. P300 is a histone acetyltransferase enzyme that acetylates the lysine residue of histone tails. Aga 2 binds the desired proteins to the surface of our yeast. Through this system, we can detect the display of histone tails, p300 (through a V5 tag), and the occurrence of acetylation between the enzyme and substrate. This detection will be done through the use of two antibodies. The first antibody binds to the desired protein. The second antibody will attach to the first antibody and utilize fluorescence to be detected through flow cytometry. We ran two experiments that tested the effect of rapamycin on this synthetic biology system.

The first experiment tested rapamycin's effect on cell growth in yeast. We tested at different concentrations of rapamycin (0 ng/mL-5000 ng/mL). By adding rapamycin during the growth phase, substrate and enzyme expression were affected, but acetylation (enzymatic function) was not. This also shows that the galactose promoter was inhibited. The second experiment tested rapamycin at different concentrations over different time intervals throughout rapamycin's induction phase (where gene expression of the plasmid occurs). When rapamycin was added at the same time as the galactose induction, rapamycin greatly attenuated the amount of displayed substrate and enzyme produced. If added 3 hours (or more) after induction, it had almost no effect on substrate and enzyme production. These results could show that all of the displayed substrate is coming from the first 3 hours of induction due to the complete binding of Aga 2. Ongoing analysis will deepen our understanding as to why substrate and enzyme expression is reduced when rapamycin is added at the same time as the galactose promoter was induced.

Integration of Degron Tags in Plasmid Cloning for Controlled Protein Translation

Hyunjun Ryh

Synthetic biology facilitates the engineering of biological parts within organisms to unlock new functions and purposes. One way to accomplish this is by introducing proteins not native to the organism via protein translation—a vital tool in biomanufacturing. However, expressing these proteins often proves challenging due to toxicity, which induces metabolic stress to cells and can cause cell death. Even proteins native to the organism can sometimes be toxic when overexpressed, posing significant challenges to the advancement of protein synthesis.

Addressing these issues, this study examines the integration of degradation tags to genes of interest for plasmid cloning. When proteins are expressed in *E. coli*, Lon protease enzymes, native to the cell, recognize and degrade the tagged proteins; this allows the cell to grow before succumbing to toxicity. As the strain grows and amplifies the plasmid number, these DNA copies can be collected for employing a cell-free system (CFS) to translate the protein of interest. CFS is an in vitro tool that utilizes transcription-translation (TXTL) machinery that enables the rapid production of functional proteins from cell extracts. The strain used in CFS is Lon protease deficient; thus, most tagged proteins remain intact.

To validate this approach, researchers selected three degradation tags shown in the literature with the highest degradation rate and energy efficiency with Lon enzymes. These tags were then attached to the end of sfGFP biosensor protein from different promoter sites: pJ23119 (strong constitutive), pJ23105 (weak constitutive), and pT7 (strong constitutive) via site-directed mutagenesis PCR cloning method. The plasmid was then transformed into DH10β (strain used for cloning), and the plate reader was used to measure the protein concentration and cell growth after 24 hours. Preliminary data indicate significant degradation of sfGFP with tags compared to untagged ones at different promoter sites, for which 43.68-53.80% degradation in pJ23105, 73.11-76.70% degradation in pADS4, and 22.83-32.80% degradation in pJ23119 was observed. Results indicated that the overall degradation between the tags did not differ significantly, but the degradation rates could vary. Therefore, a growth curve will be measured to determine the most energy-efficient tags out of the three. Furthermore, with CFS, the protein expression for tagged and untagged plasmids has all shown high sfGFP expression (~25,000 RFU). This signifies that the protein degradation seen in DH10β was indeed by Lon protease enzymes, not due to other factors like misfolding that could have played a role. The following steps are to replicate data in a controlled setting, find ways to have even more degradation in vivo, and apply the novel mechanism to express other toxic proteins. The success of this research holds immense potential in the realm of biotechnology. Such development offers numerous real-world applications, including the ability to express various biosensor proteins that cannot be translated with the current methodology.

A Bioinformatic Pipeline for Identifying Differentially Abundant Biosynthetic Gene Clusters in IBD Patients

Amanda Cai

Bacterial secondary metabolites play crucial roles in environmental adaptation, microbial defense, and interspecies communication. Inflammatory Bowel Disease (IBD) is a complex disorder with a significant microbial component, yet the role of secondary metabolites in disease progression remains poorly understood. This study investigates the potential involvement of ribosomally synthesized and post-translationally modified peptides (RiPPs), a major class of secondary metabolites, in IBD. We focused on two important classes of RiPPs: lanthipeptides and lasso peptides. We developed a novel computational pipeline using hidden Markov model (HMM) profiles to identify RiPP biosynthetic gene clusters (BGCs) in genomic sequences, overcoming limitations of current machine learning-based methods, which are often trained on limited datasets. We applied this tool to analyze gut microbiome metagenomes from a cohort of IBD patients and healthy controls. The abundance of each identified lanthipeptide and lasso peptide BGC was quantified for each individual, followed by statistical analyses to determine significant differences between groups. Our analysis revealed that four RiPP BGCs were underrepresented in IBD patients compared to healthy individuals, suggesting a protective role for these secondary metabolites in maintaining gut health. These findings provide new insights into microbial contributions to IBD and identify potential biomarkers for disease diagnosis and progression. This work establishes a foundation for future studies on the mechanisms of RiPP-mediated effects in IBD and opens avenues for developing novel therapeutic strategies targeting the gut microbiome. Future work will focus on experimentally characterizing the identified lanthipeptides and lassopeptides to determine their structures and bioactivities, as well as expanding our analysis to other RiPP classes and microbiome-associated diseases.

Dissecting the Role of Mutant RAS Signaling to Drive High-Yield Cellular Reprogramming

Jane Atkinson, Brittany Lende-Dorn, Kate E Galloway

The oncogene HRAS regulates cell proliferation, differentiation, growth, and cell death. However, its role in direct conversion remains poorly understood. Direct conversion converts common, easily accessible cell types into rare, difficult to isolate cell types. While proliferation increases conversion rates, the mechanistic contribution of HRAS and signaling networks in these cell fate transitions remains poorly characterized. Addition of two oncogenic mutants, *HRAS G12V* and *p53DD,* to a transcription-factor based reprogramming 'cocktail' results in a 500-fold increase in the direct conversion of mouse embryonic fibroblasts (MEFs) to induced motor neurons (iMNs). Here, we identify the biphasic effect of MAPK signaling on reprogramming and identify an optimal RAS expression for conversion. Our findings indicate that genetic encoding through multicistronic cassette design impacts reprogramming via differential *HRAS G12V* expression levels. We hypothesize that differences in gene order, linkers (*IRES* vs *P2A*), and cloning 'seams' lead to different levels of expression of *HRAS G12V* and *p53DD,* which in turn drive different reprogramming efficiencies. Our results revealed that cassettes with either too-low or too-high *HRAS G12V* levels resulted in lower iMN yields, while cassettes with moderate *HRAS G12V* expression produced higher yields. Since excessive RAS levels reduced yield, we tested to see if inhibiting MAPK signaling downstream of RAS with a small molecule MEK inhibitor could rescue reprogramming. We found that slight inhibition of the MAPK pathway increased reprogramming yield for cassettes with high *HRAS G12V* expression levels. Unlike MAPK inhibition, inhibition of the AKT pathway did not rescue reprogramming. This result indicates that it is excessive MAPK signaling, but not AKT signaling, that reduces reprogramming at high *HRAS G12V* expression levels. These findings suggest a 'sweet spot' of MAPK activation for optimal proliferation and reprogramming. Small molecule inhibition of downstream RAS targets offers an additional level of control over RAS signaling levels. In the future, high-yield reprogramming has tremendous potential for use in disease modeling, therapeutic screening, and cellular therapies. Understanding and engineering signaling pathways during reprogramming are important steps towards this goal.

Enhancement of Photosynthetic Sorbitol Production in the Cyanobacterium *Synechococcus* sp. PCC 7002

Pranav Bhavaraju, Cody Kamoku, Luis Taquillo, David Nielsen

Sorbitol is a sugar alcohol with broad uses in the food, cosmetic, and pharmaceutical industries. It is also considered a 'top 12' biobased chemical by the U.S. Department of Energy due to its utility as a high value building block chemical. As a potential and more sustainable production route, sorbitol biosynthesis has been engineered in *Synechococcus* sp. PCC 7002, a photoautotrophic microorganism that utilizes CO_2 and light for growth. While initial strains successfully produced sorbitol, final titers were low, and growth was impaired due to the promiscuous activity of the final pathway enzyme, sugar alcohol phosphatase. To address this, a homologous enzyme with high specificity towards sorbitol-6-phosphate was identified. Next, to further improve sorbitol production, additional Calvin cycle enzymes were over-expressed to overcome inherent kinetic bottlenecks. Lastly, the growth media was optimized to overcome nutrient depletion, thereby further enhancing sorbitol production.

Purification and Functional Studies of Surface Protein from Segmented Filamentous Bacteria from Commercial Turkey

Julius Yoh, Radmila Janjusevic

Segmented filamentous bacteria (SFB) are gram-positive, spore-forming bacteria that colonize the gut. SFBs are naturally found in the small intestine of vertebrates, including humans, mice, rats, chickens, and turkeys, during early periods of life. In the host, SFB induces maturation of immune system partially through development of Th-17 cells, antigen producing B-cells and other immune cells. The Th-17 cells produce cytokines that promote epithelial cell proliferation. Additionally, Th-17 cells upregulate antimicrobial peptides, which provide a defense against pathogens invading the epithelial cells. Significant investigations have been conducted on SFB strains isolated from turkey due to their unique metabolic characteristics and potential pathogen-protective role for these commercially important animals. However, the function of the proteins found on the surface of SFB have not been extensively explored. We are interested in adhesins- bacterial surface proteins important for attachment to the small intestine epithelial cells and potential role in priming immune cells. One such protein is SFB-NYU-003340 (P3340m) isolated from mice-specific SFB. The homolog of this protein in SFB isolated from humans was identified in a protein P00774. In this work, we have identified potential homolog from turkey SFB, P3340T. However, the role of these three homologs from different SFB for bacteria themself is not understood fully. In this study we have cloned a synthetic DNA for P3340t for expression in *E. coli* and developed a protocol for purification of P3340t. Furthermore, we studied the ability of P3340t to bind the most abundant antimicrobial peptide in the small intestine of commercial turkey in the period when SFB potentially attach to epithelial cells and survive in ileum.

Targeted AAV Gene Therapy for Triple-Negative Breast Cancer: Anti-EGFR Monoclonal Antibody-Conjugated Exosomes

Alyssa Fairgrieve, Tanvi Varadkar, Kai Chen, Yingnan Si, X. Margaret Liu

Triple-negative breast cancer (TNBC) is an aggressive subtype with poor prognosis, high recurrence, and lack of HER2, progesterone, and estrogen receptors, making therapeutic targeting challenging. This study explores a novel gene therapy targeting mitochondria, which disrupts the inner mitochondrial membrane (IMM) potential, inducing cell death and suppressing tumor growth. While adeno-associated virus (AAV) is an effective, well-characterized gene delivery vector, immunogenicity limits its safety and efficacy. This study evaluates AAV-mediated delivery of the mLumiOpto gene therapy, utilizing nanoparticles and monoclonal antibodies for precision targeting of TNBC cells. mLumiOpto was delivered via anti-EGFR monoclonal antibody-tagged exosome-associated AAV (mAb-Exo-AAV). In vitro and in vivo experiments confirmed the expected size, morphology, specific binding to TNBC cells, cytotoxicity, and no off-target effects. These results suggest mAb-Exo-AAV delivery of mLumiOpto as a promising targeted TNBC therapeutic.

Engineering Nanobody-Drug Conjugates for Cancer Immunotherapy

Jonah Finkelstein, Neil Chada, John T. Wilson

The stimulator of interferon genes (STING) pathway is an emerging and clinically relevant target for cancer immunotherapy. The innate pathway functions to detect cytosolic DNA through the activation of cyclic GMP-AMP Synthase (cGAS). Activated cGAS generates cGAMP, which then activates ER-resident protein STING, leading to the downstream production of immunostimulatory molecules. Natural cyclic dinucleotide agonists suffer from rapid clearance and poor stability, whereas powerful synthetic diamidobenzimidizole (diABZI) agonists can cause immune-related adverse effects when administered systemically. To combat these delivery problems, we developed a STING-activating nanobody conjugate targeting Integrin Alpha M, also known as CD11b, a cell surface receptor present on myeloid cells. The STING pathway is active in myeloid cells so targeting CD11b allows for the specific delivery of drugs to this cell population.

To test our hypothesis, we generated a CD11b-targeting nanobody by transforming a plasmid vector containing the nanobody DNA sequence, an enzymatic conjugation tag, and a 6xHistidine affinity tag into *E. coli*. We grew and induced large bacterial cultures, then lysed the cells by sonication. The lysate was centrifuged, then purified by nickel affinity chromatography and an endotoxin-removal column. Endotoxin-free protein was then purified using size exclusion chromatography. Initial production revealed the sequence, derived from literature, did not express in our production and purification system. Site-directed mutagenesis was performed to eliminate histidine residues in the sequence, resulting in strong expression. Computational modeling provided estimates of the effects of each mutation on folding. Binding was confirmed by flow cytometry with RAW macrophages. An azide-containing linker was then conjugated to the nanobody using an engineered sortase enzyme. A DBCO-functionalized diABZI STING agonist or Cy5 dye was then covalently conjugated to the nanobody through click chemistry. The binding affinity was determined by biolayer interferometry. Future work involves performing dose-response studies with the agonist-conjugated nanobody and *in vivo* biodistribution and pharmacokinetic studies using mice.

As CD11b has been shown to internalize slowly, a cleavable linker is necessary o ensure internalization of the drug. A urokinase-cleavable linker was added to the nanobody conjugates after a literature search revealed high levels of urokinase expression in breast tumors and glioblastoma. Cleavage was confirmed *in vitro* using purified enzyme. Future work involves dose response studies *in vitro* compared to the previously developed conjugates and characterizing the biodistribution and pharmacokinetics.

Effect of Lactoferrin on *Staphylococcus Epidermidis* biofilm Development

Nicole Calandra, Sydney Packard, Elizabeth Stewart

Lactoferrin was found to significantly reduce *Staphylococcus epidermidis* biofilm growth in a concentration dependent manner. *S. epidermidis* is one of the most common causes of nosocomial (hospital-acquired) infections in the US. The primary virulence factor of *S. epidermidis* is biofilm formation. Bacterial biofilms are surface-adherent communities of bacterial cells encapsulated by a self-produced matrix. Lactoferrin, a protein prevalent throughout the human body, including in blood, human milk, and the tear film, is known to have an antimicrobial effect on planktonic, free-floating *S. epidermidis* cells by iron chelation and cell wall disruption. However, the antimicrobial effect of lactoferrin on *S. epidermidis* biofilms is not fully understood. Quantitative image analysis of confocal laser scanning microscopy images was utilized to evaluate the effect of lactoferrin on biofilm development, bacterial cell wall integrity, and biofilm mechanics. Biofilm development was assessed by evaluating biofilm biovolume, height, and porosity. Cell wall integrity was assessed through analysis of the prevalence of cells labeled with propidium iodide. Biofilm mechanics were evaluated using microrheology measurements. Lactoferrin influences *S. epidermidis* biofilm development in a concentration dependent manner where treatment with higher concentrations of lactoferrin results in shorter, more porous biofilms, an increased prevalence of cells with compromised cell membranes, and biofilms with softer mechanics. Understanding the effect of lactoferrin on *S. epidermidis* biofilm development is important for developing antibiofilm therapeutics and ultimately reducing the prevalence of nosocomial infections.

Chemomagnetic Differentiation of Neuron-like Cells Using the Neurotransmitter Capsaicin

Pearl Martisek, Gabriela Romero Uribe

Neurodegenerative diseases often lead to significant loss of neural function and tissue integrity. Recent studies suggest that using capsaicin as a neurotransmitter may enhance neuronal growth and differentiation. This work investigates the effects of capsaicin on the differentiation of ND 7/23 cells, a hybrid neuronal cell line. By employing MNPs to facilitate the controlled release of capsaicin, we aim to determine its potential to induce neuronal differentiation. Capsaicin will be encapsulated within a polymer coating grown from magnetic nanoparticles (MNPs) surface. These MNPs will be applied to the ND 7/23 cells, and capsaicin will be released from the polymer coating upon applying alternating magnetic fields (AMFs). The purpose of using AMFs is to control the release rate of the neurotransmitter onto neurons through magnetic heating of MNPs and temperature response of the polymer coating. The cell morphology of ND 7/23 cells was evaluated to assess the effects of capsaicin on cell differentiation. ND 7/23 cells were classified into three stages: stage 0-1 (rounded cell with lamellipodia and filopodia), stage 2 (cells exhibiting multiple neurites), and stage 3 (cells demonstrating longer axonal propagations). Preliminary data indicates that a seeding density of 10,000 cells showed the most differentiation after 7 days of exposure to 0.003 mg of capsaicin, combined with growth factors (NGF + cAMP). The goal of this research is to investigate the potential effects of the localized release of capsaicin on nerve differentiation and regeneration. Future studies will explore various concentrations of capsaicin to further evaluate its effect on neuronal differentiation.

Optimization of Hyaluronic Acid-Pnipaam Co-Polymers for Mitochondrial Transplantation

Savanna Sheffield, Davis Anum, Jamie Ahmed, Samir P. Patel, Thomas Dziubla, Jason E. DeRouchey

Organelle transplantation is a biomedical technique that involves the transfer of healthy organelles, such as mitochondria, from donor cells to damaged or dysfunctional recipient cells. This approach aims to restore cellular function by replacing impaired organelles with fully functional ones. Mitochondrial transplantation is an emerging treatment to serve as a therapy for mitochondrial dysfunction diseases such as secondary spinal cord injury. One of the key limitations of this therapy is the lack of localization of exogenous mitochondria to the intended site of action. This study investigates the synthesis and optimization of poly(N-isopropylacrylamide) grafted hyaluronic acid (HA-pNiPAAm) hydrogels for mitochondrial delivery as a therapeutic treatment for secondary spinal cord injury. The primary objective is to optimize and enhance reaction conditions and yields while reducing synthesis times. Reaction kinetics and dialysis purification were evaluated using amine detection assays. Along with this, a library of HA-pNiPAAm hydrogels were synthesized, with varying HA molecular weights, pNiPAAm molecular weights, and grafting percentages, using the refined protocol. The hydrogels were characterized for their lower critical solution temperature (LCST), viscosity, and gelling properties using ultraviolet-visible spectroscopy and viscometry. This comprehensive characterization aims at identifying the optimal hydrogel formulation for efficient mitochondrial delivery. The results provide valuable insights into the hydrogel properties that most improve mitochondrial delivery, laying the groundwork for future in vivo studies and clinical applications.

Enhancement of Aerosol Pulmonary Drug By Employing Surfactants

Minkyoung Kim

More than 34 million Americans suffer from chronic lung diseases. However, current treatment of pulmonary drug delivery heavily relies on aerodynamics. This is far from an ideal treatment, because inhaled aerosols follow the path of least resistance through the airways, and these paths are subject to change by flow direction. Thus, there is an increase in aerosol deposition in proximal airways, leaving a substantial portion of downstream lung regions untreated. These untreated areas lead to persistence of bacterial infections as well as antibiotic resistance. Hence, there is an urgent need to address the issue of ineffective treatment. The aim of this project is to find corroborating data that can confirm that surfactants enhance the spread of a drug in the lungs. Results from initial experiments have suggested that surfactants can enhance the spread of the drug. What is more remarkable is that the surfactant-containing drops do not just merge upon contact but go through a series of deformation patterns that are consistent with drop size. It has been observed that ridges form even before the gap between the drops close and that the bigger drops get deformed more. In addition, talc particles were used to study the field flow of the spreading of the drops. An image from the experimental data is attached to this abstract. Investigating the interaction of multiple surfactant-containing drops is important for it could pave ways for an enhanced spreading of the drug and thus an advanced treatment.

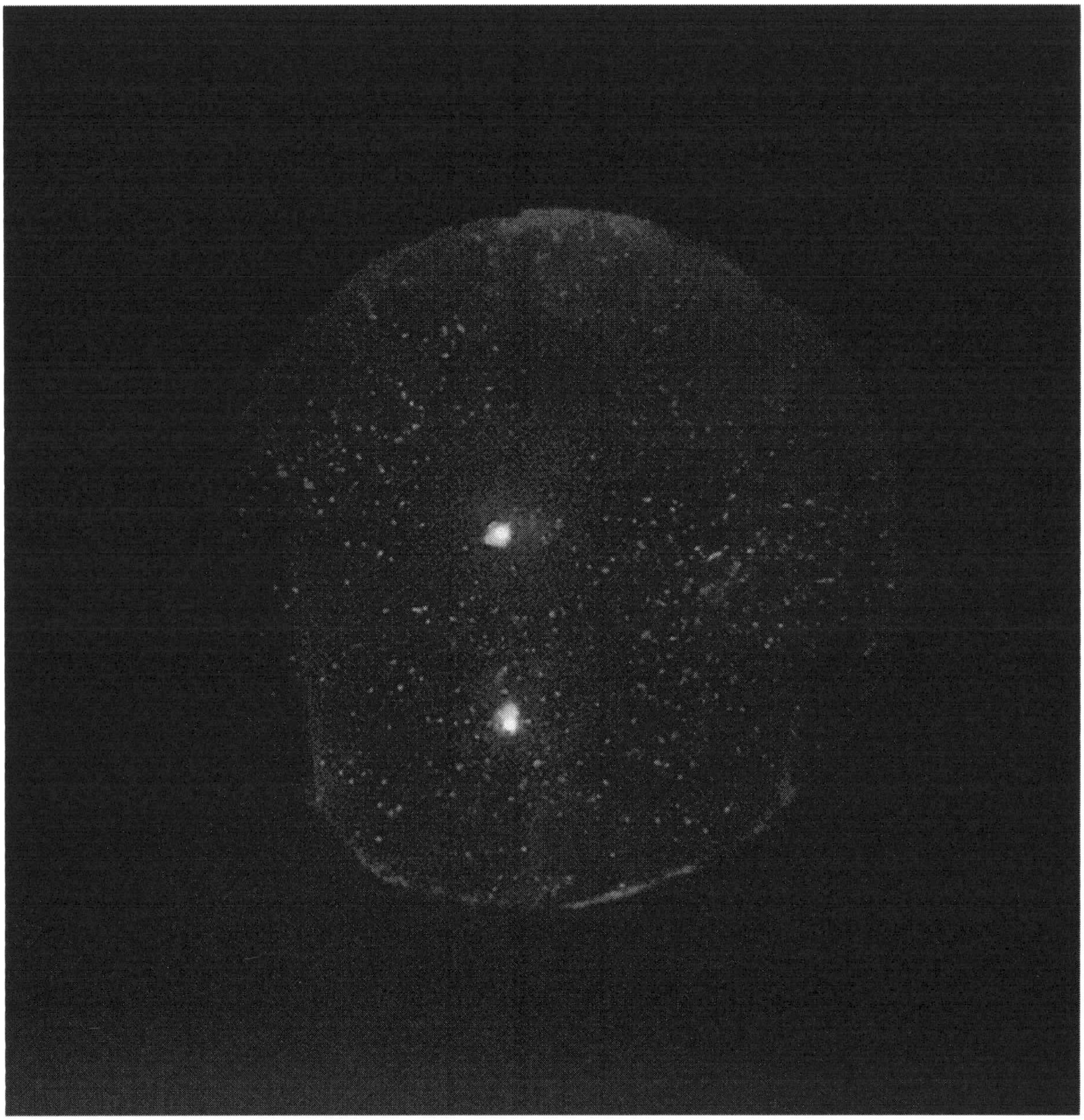

Modifying Poly(beta-amino ester) Hydrogel Formulations to Achieve Sustained Release of Bupivacaine Anesthetic

Allison Raley, Claire Rowlands, Pranto Paul, Nikita Gupta, J. Zach Hilt, Brittany Givens Rassoolkhani

The current standard of care to treating post-operative pain is opioids. Opioids are powerful pain blockers but are a highly addictive substance responsible for many overdose deaths. Nerve blocks can reduce opioid use but typically last no more than 4 hours with an anesthetic, Bupivacaine, injection. Another option is the sustained-release nerve block, Exparel®, which only provides pain management for 72 hours (3 days). The challenge with Exparel® is that facial surgery requires at least 7 days of post-operation pain relief, which means that the highly addictive opioids are still prescribed. The aim of this work is to improve upon the current sustained release of nerve-blocking anesthetics to reach the needs of more patients. In this research, a variety of formulations and structures of Poly(beta-amino ester) (PBAE), a biodegradable hydrogel, have been tested to achieve a more sustained release of Bupivacaine, which is an anesthetic with a long half-life. The PBAE used was formed from diethylene glycol diacrylate (DEGDA) or Poly(ethylene glycol) diacrylate (PEGDA) and isobutylamine (IBA). All formulations kept a molar ratio of 1.2:1 diacrylate to amine, but the molar ratios of the diacrylates were differentiated: molar ratio 3:1 DEGDA to PEGDA and a homopolymer of DEGDA. The purpose of having a formulation with just DEGDA as the diacrylate is to increase crosslinking, due to its shorter chain length within the PBAE, to have a more sustained release over time. The structures of each formula were tested in a particle format and a variety of disc thicknesses. Results from release studies showed that utilizing the homopolymer formulation had a more sustained release of Bupivacaine over a 7-day period in comparison to the copolymer formulation. However, this release profile was only seen in disc form, and did not extend to microparticles. Overall, this suggests that future developments will need to keep the integrity of the hydrogel structure to achieve a sustained release with this formulation.

Engineering a More Efficient De Novo Protein for Quantum Dot Synthesis

Ashley Kang, Leah Spangler

Quantum dots (QDs) are semiconductor nanocrystals with optical properties similar to fluorescent dyes but offer longer fluorescent durations and higher photon emission. These properties make QDs suitable for bioimaging and cancer cell-tagging. However, common QDs like cadmium sulfide (CdS) are potentially toxic for in vivo applications. Non-toxic alternatives, such as biomineralized $CuInS_2$ quantum dots, are being explored. Traditionally, the protein cystathionine γ-lyase (CSE) synthesizes QDs by breaking down cysteine into H_2S, pyruvate, and ammonia, which react with metals to form nanocrystals. Recently, de novo protein Construct K (ConK), designed through computational models, was discovered to catalyze QD synthesis in a similar manner. Although ConK can produce QDs, its reaction time is significantly longer (10+ hours) compared to CSE (30 minutes).

To address this, I investigated mutations in ConK to accelerate the synthesis process. I focused on mutating aromatic amino acids near the active site, specifically histidine residues at positions 43 and 46, to alanine or lysine. Site-directed mutagenesis was performed on ConK plasmids, followed by transformations into DH5α E. coli cells for plasmid amplification. Sequencing confirmed successful mutations in H43K and H46A. These plasmids are now being transformed into BL21 E. coli cells for protein overexpression.

The next phase involves purifying the mutated ConK proteins and testing their ability to synthesize non-toxic QDs. The ultimate goal is to enhance the reaction rate and scale up QD production for industrial and medical applications, making cancer diagnostics and treatments more accessible.

Processing and Characterization of Alginate Tubes for Use in Vascular Tissue Grafting

Logan Merriam, Laura A.E. Brunmaier, Kara L. Huse, Abigail Flanegan, Arushi Thatola, Isabel Grinager, Travis Walker

Alginate is a naturally occurring polysaccharide derived from brown algae, which is normally extracted into a sodium alginate (NaAlg) form. NaAlg will crosslink when reacted with a bivalent cation. This reaction gives NaAlg several distinct features allowing for use in biomedical applications such as wound dressings, drug delivery, and tissue engineering. To this end, shear rheology was used to characterize the viscosity of NaAlg to assess the processability of NaAlg into tubes for use in cardiovascular grafting. A co-flow setup was utilized to fabricate NaAlg tubes by pumping NaAlg around an 8-gauge flow of aqueous CaCl2 into a CaCl2 bath for complete crosslinking. The data was used to construct a phase-shift diagram documenting several hundred extrusion points, outlining the boundaries for successful extrusion of NaAlg tubes. The tubes were characterized by burst pressure and tensile strength tests to measure the mechanical properties of the material. This work aims to establish relationships between rheological properties of different concentration alginate solutions, extrusion limits of NaAlg tubes, and their subsequent mechanical properties, thereby defining process-structure-property-performance relationships for NaAlg.

Protein 14-3-3 Expression on Yeast Surface Display

Aislin Robb, Yongku Cho

The protein 14-3-3 has seven different isoforms and is known to cause many diseases including neurological diseases, neuropsychiatric diseases, and cancers. 14-3-3 is known to have a weak binding affinity to several other proteins which could explain why it is implicated in a wide variety of disorders. 14-3-3ζ is known to be linked to neurological diseases such as Parkinson's Disease and Alzheimer's Disease. 14-3-3η is linked to dementia and Parkinson's Disease [2]. The goal of this research is to gain a deeper understanding of the 14-3-3 protein and express 14-3-3 using yeast surface display to detect weak protein binding to its target. Yeast surface display is commonly used in protein engineering to express recombinant proteins which are linked to the surface of yeast cells in the extracellular space [1]. Expressing 14-3-3 will lead to insights about how this protein interacts and binds to other proteins in a living organism. Yeast surface display has been used by our lab previously to screen libraries of nanobodies, antibodies, and single chain variable fragments (scFv) against novel binders to a variety of targets. By understanding the avidity or the strength of multiple affinities, more can be learned about 14-3-3's selectivity. With past data supporting yeast surface display to detect weak protein binding interactions, this technique will be employed in a similar way. Overall, 14-3-3 will be expressed on yeast surface display.

References:

[1] Cherf GM, Cochran JR. Applications of Yeast Surface Display for Protein Engineering. Methods Mol Biol. 2015;1319:155-75. doi: 10.1007/978-1-4939-2748-7_8. PMID: 26060074; PMCID: PMC4544684.

[2] Foote M, Zhou Y. 14-3-3 proteins in neurological disorders. Int J Biochem Mol Biol. 2012;3(2):152-64. Epub 2012 May 18. PMID: 22773956; PMCID: PMC3388734.

Controlled Release Profiles of Resiquimod from a Novel Melanoma Vaccine Using in Vitro Subcutaneous Models

Conner McClelland

Metastatic melanoma, the deadliest form of skin cancer, is limited in the number of viable treatment methods. Early-stage melanoma can often be removed surgically, provided the tumor remains localized in the body or metastasizes in regions that permit surgical removal. Upon metastasis to parts of the body such as distant lymph nodes, other areas of the skin, or other organs, melanomas become harder to detect and less viable for surgical removal, thus creating the need for alternative treatments. The majority of melanoma cases are attributed to genetic mutations in the BRAF oncogene resulting in the malignant growth of melanocytes. Given the immunogenic nature of melanoma tumors, immunotherapy proves a promising treatment method. This project works to develop a safe and feasible vaccine as a late-stage treatment method for BRAF-mutant metastatic melanoma using a novel polycaprolactone multiblock copolymer containing orthoester linkages (CAP-OA1) as a drug delivery vessel. The polymer's liquid state provides suitable drug encapsulation that allows for administration via subcutaneous injection while the biocompatibility associated with polycaprolactone allows for safe degradation in the body. The vaccine consists of CAP-OA1 loaded with tumor-associated antigens and resiquimod, an imidazoquinolinamine derivative and Toll-like Receptor 7/8 agonist that has anti-tumor and immunostimulatory effects. Upon subcutaneous injection, the liquid polymer serves as a depot for the continuous release of the antigen and resiquimod. This study provides valuable insight into the transport mechanisms and release kinetic profiles of resiquimod-loaded CAP-OA1 through in vitro release studies that aim to mimic the release kinetics of the components in vivo. Release studies were performed using two different models of subcutaneous delivery: a bulk fluid model and a hydrogel model. The most common setup for these studies consists of the release of a drug from its polymeric encapsulation directly into an aqueous collecting solution that resembles the pH of the human body. This is referred to as the bulk fluid model. This method offers ease of setup but lacks the representation of drug release into a native tissue environment which may affect degradation mechanisms of the encapsulation and diffusion of the drug through native tissue. Agarose hydrogels have been shown to mimic the mechanical properties of native tissue, offering a more biomimetic drug-release environment. Resiquimod concentrations were quantified using UV-vis spectrophotometry. The models resulted in similar resiquimod release profiles, with the majority of resiquimod being released after nine days. This implies that the immune response to the vaccine will begin no sooner than nine days post-injection. Higher initial rates of release were observed in the bulk fluid model, indicating that bulk fluid release studies may be less accurate in predicting release rates for up to seven days after injection. Higher variability was observed in the bulk fluid model, pointing to the nonuniform nature of polymer degradation in a bulk fluid compared to in a hydrogel structure. These findings demonstrate the increased performance of hydrogel models over bulk fluid models; it is recommended that hydrogel structures be implemented in future release studies.

Protein Release Kinetics of Proinsulin Variants from HA-Collagen-Fibronectin Hydrogels

Maryann Melendrez Cuadros

Hydrogels, because of their adaptive properties and biocompatibility, have become promising platforms for controlled drug delivery. In this study, protein release kinetics from hyaluronic acid (HA)-based hydrogels incorporating fibronectin and collagen were examined to elucidate the influence of the hydrogel composition on the release profiles of proinsulin variants. HA- collagen-fibronectin hydrogels were synthesized according to a standardized protocol, with variations in collagen and fibronectin concentrations. Three proinsulin variants, F25D, F25D- K3, and F25D-K12, were used along with the hydrogels to conduct protein release assays over 24 hours using fluorescence endpoint measurements. Distinct release profiles were observed for each proinsulin variant. For F25D, hydrogels containing fibronectin at 0.125 mg/mL and collagen at 1.0 mg/mL exhibited the highest protein release, with a significant increase observed from 12 to 24 hours. Collagen, particularly at 1.0 mg/mL, showed the most considerable protein release across all variants, while fibronectin demonstrated comparatively lower release rates, especially at 0.25 mg/mL. The findings underscore the importance of hydrogel composition in modulating protein release kinetics. For instance, collagen emerged as a critical component in protein release, whereas fibronectin exhibited lower release rates. The results provide valuable insights for hydrogel-based drug delivery systems, emphasizing optimizing hydrogel compositions to achieve desired and realistic release profiles for real-life therapeutic applications. Further research might explore factors influencing protein release dynamics, facilitating more effective drug delivery strategies.

Long-Term Release of Large Biologics Via High-Affinity Supramolecular Polymeric Hydrogels

Rebecca Avrutin, Myriel Kim, Honggang Cui

Monoclonal antibodies (mAbs) have emerged as a major class of therapeutic agents due to their high specificity, sensibility, and much-reduced side effects. Despite these advantages, significant challenges remain in maintaining a therapeutic dose in the body for an extended period. Developing biocompatible materials that can release specific therapeutic antibodies at a controlled rate is highly sought after in drug delivery. Peptide-based hydrogels have garnered growing interest as effective drug delivery systems for their tunability, biodegradability, and biocompatibility. mAbs can be easily incorporated into the hydrogel matrix and released at a rate that depends on the hydrogel swelling (physical disintegration), chemical degradation, and their binding affinity to the hydrogel networks. Our lab has developed a class of supramolecular hydrogels using a protein A mimicking peptide that demonstrates a high binding affinity to mAbs. Previous work has shown that the antibody release from these hydrogels can be tuned based on the ligand density and the initial ligand loading ratio relative to the delivered proteins. My research focuses on hydrogel mesh size as another potential variable for further optimizing these hydrogels to model the release profile of current mAb-based therapeutics. The goal is to extend the duration of their therapeutic effect and reduce the frequency of injections. My recent findings suggest that the linker chemistry that bridges the mAb-binding ligand to the immunofiber has an important impact on the release profile of mAb, with longer linkers resulting in a more linear release profile. Given that the length of the linker also affects the hydrogel mesh size, we speculate the smaller mesh size as a result of employing a longer linker would have a smaller confinement ratio, thus limiting the mAb diffusion and extending their release period. By varying the mesh size and mAb diffusion mechanics, I envision that these hydrogels can be eventually optimized for controlled mAb release, enabling sustained therapeutic levels and enhancing the overall patient experience.

Assessing Release Profile of Electrosprayed Polymeric Coatings

Mirna Sofia Guerra-Torres, Hannah Schaeffer

Electrospraying is an efficient technique for producing polymeric coatings with smooth topographies. This technique stands out for its ability to create reproducible coatings in ambient conditions from a plume of charged droplets. These polymeric coatings can be uniformly fabricated to incorporate potential drug molecules for delivery. The primary goal of this study is to develop model drug-loaded films that exhibit controlled release into a medium, specifically Phosphate Buffered Saline (PBS). To simulate *in vivo* conditions, we employed a thermal shaker and replenished the PBS after each model drug concentration measurement. We determined the total cumulative percentage of drug release for model drug Rhodamin B (RhB) over time. The controlled release observed was a result of exploiting the change in the glass transition temperature (Tg) of the polymer, Polyvinyl Acetate (PVAc). Tailored drug release is highly sought after in various medical treatments, including cancer therapy. Moving forward, we intend to investigate controlled released coatings to generate tunable and distinctive release curves.

Hypoxia and Chemotherapy Promote Secretion of EMT-Driving Growth Factors By Fibroblasts in Pancreas Cancer

Jason Wieder, Karl Kowalewski, Matthew Lazzara

Pancreatic ductal adenocarcinoma (PDAC) has a five-year survival rate of just 12%. PDAC is characterized by a dense fibrotic stroma laden with cancer-associated fibroblasts (CAFs) and numerous hypoxic tissue subdomains due to hypovascularity. Interactions between CAFs and malignant PDAC cells remain poorly understood, but their elucidation will undoubtedly nominate new targets for precision medicine approaches to curb chemoresistance. We hypothesized that hypoxia and chemotherapy enhance CAF secretion of growth factors that promote malignant cell epithelial-mesenchymal transition (EMT), which leads to tumor chemoresistance and invasiveness. We first explored how chemotherapy alters paracrine signaling between CAFs and malignant cells by using publicly available single-cell RNA sequencing data from human PDAC tumors. CellChat analysis of the annotated data predicted that ligand-receptor interactions were generally increased between CAFs and malignant cells in response to chemotherapy. In particular, AREG-EGFR and HGF-MET signaling between CAFs and malignant cells were elevated. Importantly, EGFR and MET signaling cooperate to promote EMT. To investigate the possibility that malignant cells regulate growth factor secretion by CAFs, we next explored how juxtacrine interactions between CAFs and malignant cells impact CAF phenotypes. We focused on CAF phenotype regulation via NF-κB and histone 3 lysine 36 dimethylation (H3K36me2) because we recently showed that these pathways cooperate to promote CAF secretion of EMT-promoting growth factors. PDAC CAFs were cultured with paraformaldehyde-fixed cancer cells (to prevent paracrine interactions), and immunofluorescence imaging and quantitative image analysis were performed. CAFs in contact with malignant cells exhibited increased NF-κB nuclear localization and elevated H3K36me2. CAFs in contact with cancer cells while cultured in hypoxia displayed even greater NF-κB nuclear translocation, suggesting hypoxia and cancer cell-CAF interactions had cooperative effects that may be crucial for regulation of CAF subtype. Our findings highlight the ability of chemotherapy and tumor hypoxia to enhance CAF secretion of EMT-promoting growth factors and identify drug targets whose antagonism is predicted to sensitize PDAC to chemotherapy by impeding CAF-cancer cell crosstalk.

Screening Small Molecule Inhibitors of Prostate Cancer with an Adaptable Reporter Gene Assay

Talia Benducci, Mickey Huang, Thomas Jonsson, Ph.D, Hagai Ligumsky, MD, Ph.D, David B. Agus, MD, Katherin Patsch, Ph.D

The androgen receptor (AR) is the predominant therapeutic target for men with prostate cancer. Patients with androgen-dependent prostate cancer are often treated with surgical castration or androgen ablation therapy. Patients with castration-resistant prostate cancer (CRPC) are treated with small-molecule AR inhibitors or chemotherapy. However, patients treated with FDA-approved AR small molecule inhibitors (e.g. enzalutamide) often develop resistance (1). In order to leverage a model system that would allow the screening of novel drug candidates for enzalutamide resistance, we developed an AR-based cell line that facilitates high-throughput in vitro drug screening.

The PC3 prostate cancer cell line does not have endogenous AR. We ectopically expressed AR with a modified lentiviral luciferin linked to an androgen response element (2). To benchmark these cells, we screened a 12-point dilution of small molecule AR antagonists with R1881 (20 nM, synthetic testosterone) and a vehicle control. Our results indicate that this cell line can provide a quantitative readout of androgen receptor activity with R1881 within physiological levels of androgen-dependent cell lines (ENZ IC50: 358.28 nM ± 14.74 nM). This system can also be genetically modified with AR mutations that have been found in men with enzalutamide resistance. This would facilitate a readout of the continuum of prostate cancer, starting from androgen-dependent to CRPC-resistant disease in an isogenic model.

This cell line and screening can also be applied beyond small molecule inhibitors. Small interfering RNAs (siRNA) are able to silence gene expression by degrading mRNA. This model can detect a knockdown of AR activity by testing siRNAs that target AR against R1881, a scramble and transfection reagent alone. We have screened two different siRNAs that target a different binding site of AR and received 82% and 86% knockdown when compared against scramble with R1881. We are currently optimizing this model to test multiple dilutions of siRNA and produce a dose response curve.

This assay successfully characterized an AR-based prostate cancer cell line while leaving room for novel studies on enzalutamide resistance. This system is also not limited to drug screenings, but can also be used to detect knockdown of AR activity. Our preliminary data with siRNAs targeting AR achieved detectable levels of knockdown, and ongoing optimization of this model will enable more insightful dose-response studies. This assay is a versatile and robust tool for drug screening and the development of new therapeutics.

References:

1. P. A. Watson, V. K. Arora, and C. L. Sawyers, "Emerging mechanisms of resistance to androgen receptor inhibitors in prostate cancer," vol. 15, no. 12. Springer Science and Business Media LLC, p. 701, Feb. 29, 2016, doi: 10.1038/nrc4016.

2. K. Patsch *et al.*, "Paradoxical androgen receptor regulation by small molecule enantiomers," vol. 118, no. 12, p. e2100918118, 2021, doi: 10.1073/pnas.2100918118.

Investigating the Role of Net Charge and Charge Patterning in the Encapsulation of Green Fluorescent Protein in Complex Coacervates

Arjun G. Iyer, Xianci Zeng, Yi Zheng, Jeanne A. Hardy, Sarah L. Perry

Complex coacervation is the liquid-liquid phase separation of oppositely charged polyelectrolyte solutions resulting in a polymer-rich coacervate phase and a polymer-poor supernatant phase. Complex coacervation has numerous applications, ranging from the formation of membraneless organelles to drug delivery. Coacervates can also be formed with protein solutions in which the polyelectrolytes interact with the protein and incorporate it into the polymer-rich phase. Protein encapsulation is especially interesting in fields like food science and drug delivery, where the sequestration of proteins allows for a much greater concentration than in solution while still maintaining solubility and protein viability. Prior studies on protein encapsulation have found that the net charge of proteins plays a major role in how they interact with polyelectrolytes and how effectively they are sequestered into the coacervates. Existing research also shows that the charge patterning of surface amino acid residues affects protein encapsulation along with pH and salt concentration. We seek to explore the convolution of these factors on protein incorporation through the use of tailored supercharged green fluorescent proteins (GFPs). Supercharged GFPs can be engineered to have a specific net charge by altering specific amino acid residues, and this tailoring can be used to control the patterning of charge on the protein surface. In this study, we will explore the incorporation of GFP variants with different net charges and charge distributions into complex coacervates made with poly-L-lysine and poly-D,L-glutamate. The incorporation will be assessed by measuring the turbidity and fluorescence of coacervates and calculating the concentration of protein in the coacervate versus the supernatant. Lastly, thermal stability assays will be performed to assess the temperature at which the GFPs become denatured. These results will give insight into how different parameters of GFPs affect protein encapsulation. This insight can be used as a foundation in future applications of GFPs, but can also be extended to encapsulation studies of other biologics ranging from other proteins to viruses to bacteria.

Material Properties of Neutral Globular Protein-Based Condensates

Heather Binion, Rachel Fisher, Allie Obermeyer

In the world of cellular biology, the study of membrane-less organelles has recently gained significant traction, revealing a complex and dynamic world of cellular organization. These organelles are formed through phase separation, where biomolecules transition from an originally homogeneous solution to a solution containing a distinct dilute phase and a dense phase, termed a biomolecular condensate. Complex coacervation, involving the phase separation of oppositely charged polyelectrolytes, has been a focal point in exploring these phenomena, often highlighting the role of electrostatic interactions in protein aggregation and the formation of condensates.

We have previously concentrated on the interactions between positively charged green fluorescent proteins (GFP), and negatively charged biomolecules. While these studies provide valuable insights, they have predominantly emphasized electrostatic forces. However, it is becoming increasingly clear that non-electrostatic interactions, such as hydrophobicity, π-π stacking, and van der Waals forces, also play a crucial role in the formation and stabilization of biomolecular condensates.

In this study, we address this gap by engineering a library of eight neutral GFPs, each tagged with short peptides containing a few aromatic residues, to investigate the non-electrostatic mechanisms that drive phase separation in cellular contexts. By utilizing *Escherichia coli* as a model system, we can analyze how these neutral proteins phase separate to form condensates, focusing particularly on the influence of aromatic interactions. This approach allows us to dissect the contribution of specific amino acid properties and patterning to the overall material characteristics of the condensates formed.

Preliminary findings suggest that the pairing of amino acids like phenylalanine, tryptophan, and tyrosine enhances phase separation through increased hydrophobic interactions and π-π stacking, resulting in more stable condensate structures. However, it remains unclear whether the observed biomolecular condensates are formed through protein-RNA interactions or self-protein interactions known as simple coacervation. On-going work, both *in vivo* and *in vitro* works on probing the conditions under which these proteins condense and identifying the type of interactions leading to condensation. With this foundational understanding, we can begin to build biomolecular condensates with more complex, and biologically relevant, structures such as multiphase, multi-layered condensates.

Optimizing Hydrogel Microsphere Formation for Suspension Cell Culture

Gabriella Faircloth, Rachel Moen, Ethan Lippmann, Jamey Young

Extracellular vesicles (EVs) are lipid membrane-bound nanoparticles secreted by cells that contain nucleic acids, proteins, and metabolites derived from their cells of origin. While there is much to be learned about their structure and function, they are known to play key roles in intercellular communication and are involved in many physiological and pathological processes. Although EVs have wide therapeutic potential, current GMP that focus on the large-scale biomanufacturing of EVs is very limited. The use of hydrogel microspheres has been recognized as a potential way to scale up the culture of adherent cell lines, and thus the production of EVs. Gelatin methacryloyl (GelMA) is a hydrogel that is UV crosslinkable when mixed with the photoinitiator LAP upon exposure to UV light. Its crosslinked form exhibits robust mechanical properties that provides a 3D structure in which to culture cells at an increased seeding density and for extended culturing time, making it a good candidate for microsphere medium. One proposed method of encapsulating cells consists of a flow-focusing microfluidic chip that facilitates formation of GelMA microspheres through the union of an aqueous phase and an oil phase. Previous optimization efforts studied the ratio of the oil phase to the aqueous and determined that the ratio of the oil phase and gel phase remaining the same but at higher velocities does consistently produce the same-sized spheres but at an increased rate, but spheres are not stable in culture conditions. In an attempt to continue this work and gain further understanding of how to manufacture consistently sized, fully crosslinked, spherical microspheres using this method, parameters such as flow rates, the concentration of hydrogel used, the concentration of LAP photoinitiator added, and oil types were optimized. Microspheres formed under each condition were then characterized by their size, crosslinking efficiency, production, and structural integrity using fluorescent imaging coupled with image quantification and rheological analysis. Moving forward, additional microsphere optimization will be completed after the addition of cells. Ultimately, the goal of this project is to develop a scalable platform to produce EVs from tissue-specific cell types.

Sequential Photodegradation of Hydrogels to Spatiotemporally Control *in Vitro* Crypt Fission

Delaney McNally, Kaustav Bera, Peter Dempsey, Kristi Anseth

The inner lining of mammalian intestine is composed of an epithelial layer that is responsible for digestion, absorption of nutrients, and overall protection of the intestine. For its efficient functioning, the epithelial layer is organized in a crypt-villus structure. During its normal functioning, the epithelial cells experience multiple stresses which requires their frequent turnover. Intestinal stem cells (ISCs) in the base of the crypts continuously divide and differentiate to replenish the epithelium[1].

Crypt fission is the process of a crypt dividing into two daughter crypts which is required during development, homeostasis, and crypt regrowth after injury. During development, fission widens and elongates the intestinal tract[2]. Fission in adult crypts is more of a repair mechanism allowing for homeostasis. In healthy adults, intestinal crypts can be injured due to damaged tissue from inflammation, intestinal resection, or toxic injury. It has been shown that crypt fission is increased in patients who suffer with inflammatory bowel diseases (IBD) such as ulcerative colitis and Crohn's disease. In healthy tissue, crypt fission rates are low, and division is mainly symmetric; however, in diseased tissue, crypt fission occurs at a much higher rate with increased asymmetry. It is important to understand how crypt fission is mechanistically initiated and progresses during crypt regeneration under different physiological conditions.

The mechanisms of crypt fission and the roles of different epithelial cells in this process are not comprehensively understood due to a lack of *in vitro* models. The goal of this project is to create an *in vitro* model that simulates crypt fission that is comparable to that observed *in vivo* fission. This includes the mechanical and biochemical cues that drive crypt fission.

Using a Strain-Promoted Azide-Alkyne Cycloaddition (SPAAC) reaction to spontaneously form a polyethylene(glycol) (PEG) hydrogel, we encapsulate ISC derived organoids in 3D. Then, 405 nm light is utilized to degrade specific regions of the hydrogel, directing crypt formation and ideally, crypt fission. Preliminary data show bifurcation of crypts occurring at desired dimensions, and the system is being optimized to ensure optimal crypt length and width during the fission process. The system also allows differentiation of stem cells into secretory cells as observed *in vivo* crypts.

Ongoing studies indicate that the cell nuclei might play an active role in the process of crypt formation. Since variations in epithelium shape can dictate nuclear forces, I developed an analytic method to quantify cell shape, nuclear deformation, and protein expression which can identify pathways critical in crypt fission. Using immunostaining and integrin staining to identify specific biochemical pathways, we will observe extracellular matrix (ECM) deposition from the intestinal organoids as well as location of critical differentiated cell types, such as Paneth cells and Lgr5-GFP+ cells. Thus, this model provides a platform to study *in vitro* what factors regulate crypt fission and intestinal regeneration.

[1] Z. Zhao *et al.*, *Nature Reviews Methods Primers*, vol. 2, no. 1, Dec. 2022.

[2] G. M. Jowett and E. Gentleman, *Cell Stem Cell*, vol. 28, no. 9, pp. 1505–1506, Sep. 2021.

Cell Response to Stiffness Gradients in Alginate Hydrogels

Zoe Ostrowski, Azarnoosh Foroozandehfar, Fred Namanda, Ian Schneider

Directed cell migration is an important aspect in many physiological and pathological processes including cancer metastasis. Cells take in and process many signals including the mechanical signal of tissue stiffness. Durotaxis is the process in which cells migrate towards stiffer regions of the extracellular matrix. Alginate, which comes from brown algae, is commonly used hydrogels due to its biocompatibility and capability to retain large amounts of water. Using stencils to control diffusion, alginate hydrogels can be crosslinked with calcium to create 2D stiffness gradients imprinted into a 3D gel. These gels can be used to engineer an environment that mimics the extracellular matrix, providing insight into how cancer cells migrate and respond to mechanical signals. To make the gels cell adhesive, collagen was incorporated into the gel using two different functionalization methods, bulk functionalization and surface functionalization. Two different behaviors are observed under these conditions. Surface functionalization shows gradients in cell number and spread cells, which could be cause either by durotaxis or stiffness-dependent proliferation. Under bulk functionalization, we show that cells invade into the gel. By fixing and staining cells, invasion can be investigated using confocal microscopy to determine how far into the gels cells move. Future work is geared towards aligning fibers within the gel to create complex multi-cue environments to understand how cells respond when multiple cues are present.

Constructing Long-Acting Injectable Supramolecular Hydrogels Via Self-Assembly of Antiretroviral Drug Amphiphile

Jiarui Yang, Boran Sun, Han Wang, Maya Monroe, Honggang Cui

Despite significant progress in HIV treatment since the virus was discovered 40 years ago, around 38 million people worldwide still live with HIV, and the virus continues to cause nearly 700,000 deaths annually. A significant barrier to a cure is the presence of latent viral reservoirs in memory T cells, necessitating lifelong therapy. Current HIV treatments predominantly rely on nucleoside reverse transcriptase inhibitors like lamivudine (3TC), but these drugs have low molecular weight and are hydrophilic, leading to rapid systemic clearance and frequent dosing. When doses are missed or adherence decreases, the virus can mutate and progress to AIDS, where the immune system is severely compromised and life expectancy is significantly reduced.

The primary challenge in developing long-acting injectables for antiviral treatment is extending the drug release span. To address this, we exploit the reversible properties of peptide-based supramolecular polymers and their unique ability to self-assemble into filamentous nanostructures in aqueous environments via non-covalent interactions and form three-dimensional hydrogel networks under physiological conditions. Developing self-assembling drug amphiphiles (DAs), where therapeutic agents are chemically linked to peptide amphiphiles and incorporated into hydrogels, represents a novel approach to controlling drug release through controlling the kinetics of dissociation from the supramolecular state to the monomeric state and achieving sustained therapeutic effects.

We designed a series of ARV DAs based on lamivudine, a hydrophilic antiretroviral (ARV) drug and nucleoside reverse transcriptase inhibitor. These ARV DAs are constructed with alternating sequences of hydrophobic valine (V) and hydrophilic lysine residues (K) connected through a biodegradable succinic linker to lamivudine (K3TC), with varying numbers of negatively-charged glutamic acid (E) residues (1–3) at the C-terminus. When dissolved in deionized water, the ARV DAs spontaneously self-assembled into micrometer-length supramolecular filaments with varying degrees of lateral stacking: the increasing number of E residues provides enough electrostatic repulsion to maintain assembly integrity and enhances water-soluble ARV DAs. Adding 1× PBS caused rapid gelation in the ARV DAs that contained 2 or 3 E residues—in vitro studies revealed that dilution caused the dissociation of the filaments into monomers, enabling linear, sustained, long-term drug release. *In vivo* experiments further validated the injectability of these ARV DAs, their rapid *in situ* hydrogel formation upon injection, enhanced local retention at the injection site, and prolonged therapeutic release over a month. Pharmacokinetic data indicated that these supramolecular hydrogels maintained plasma levels of lamivudine above the IC50 for over 40 days in mice, with minimal inflammation and adverse immune response. These results provide important insights into the strategic design of long-acting injectable therapies through drug-based molecular assembly approaches, and the ARV supramolecular hydrogels demonstrated here show great potential for enhancing HIV treatment outcomes.

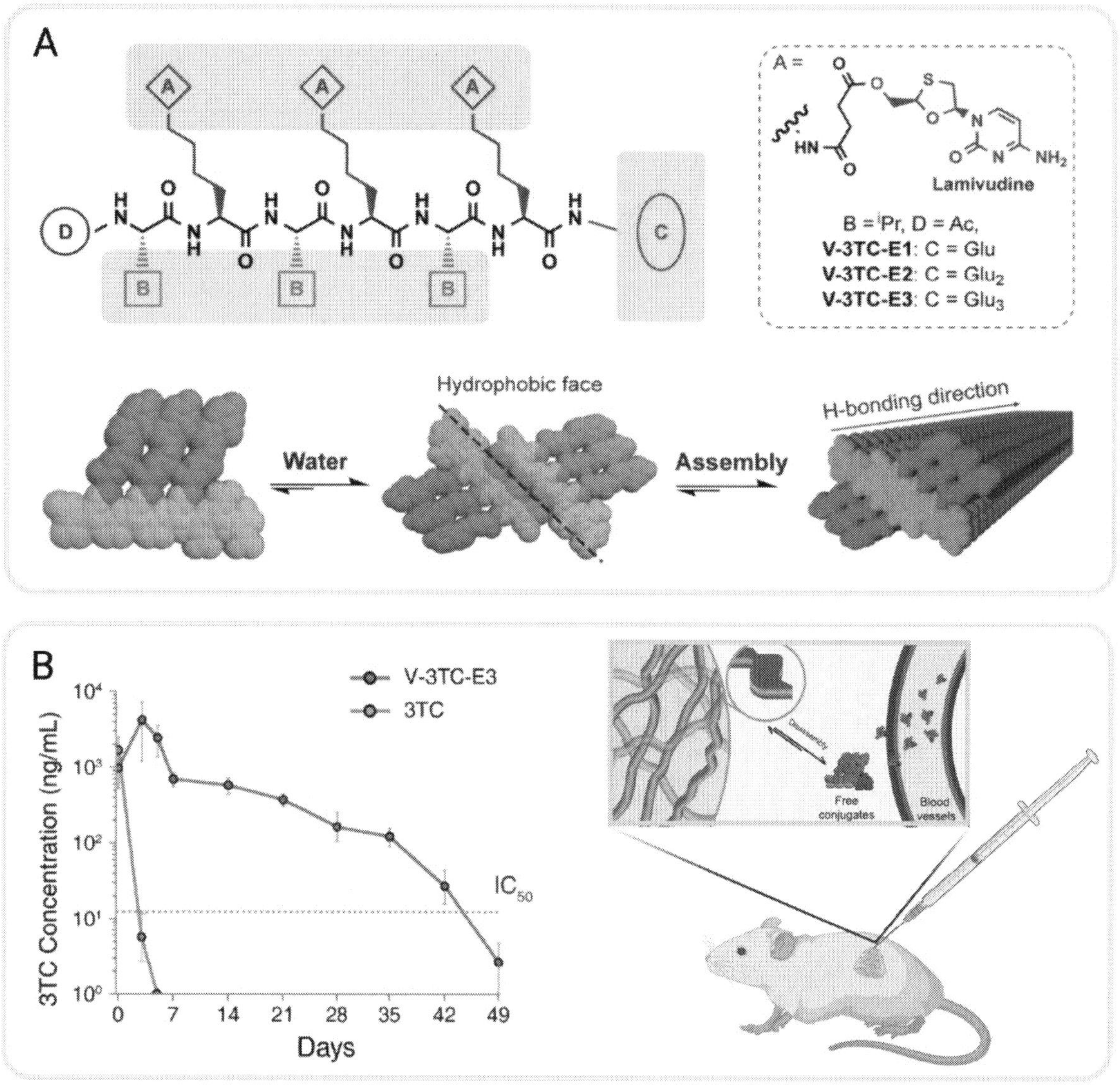
A
A =
Lamivudine
B = ^{i}Pr, D = Ac,
V-3TC-E1: C = Glu
V-3TC-E2: C = Glu_2
V-3TC-E3: C = Glu_3
Hydrophobic face
H-bonding direction
Water
Assembly
B
V-3TC-E3
3TC
3TC Concentration (ng/mL)
Days
IC_{50}
Free conjugates
Blood vessels

Alginate Hydrogel Formulations for Evaluating Staphylococcus Epidermidis Biofilm Growth in Soft Microenvironments

Erin McNeill, Sydney Packard, Elizabeth Stewart

Staphylococcus epidermidis is an opportunistic pathogen that forms biofilms on medical devices such as catheters, prosthetics, and stents. Current *in vitro* models of biofilm growth do not account for the mechanical environment of the host. Here we evaluate three alginate hydrogel formulations for studying bacterial biofilm growth in soft microenvironments. Additionally, we assess the use of a bacterial growth media reservoir as a nutrient source for bacterial biofilms grown in hydrogels. The mechanical properties of the hydrogels were measured using rheology and bacterial growth within the hydrogels was assessed using confocal microscopy coupled with quantitative image analysis. We determined that alginate hydrogels can be formulated using water or bacterial growth media; however, alginate hydrogel mechanical properties are five times softer when hydrogels are formulated with bacterial growth media. Additionally, a bacterial media reservoir can be utilized as the nutrient source for bacteria grown in hydrogels. Analysis of bacterial growth in hydrogels with a bacterial growth media reservoir showed that bacterial aggregate size was similar in alginate hydrogels across two different stiffnesses, while overall bacterial growth was two times higher in the softer of the two alginate hydrogels. These findings create a basis for using alginate hydrogels in biofilm infection models that mimic the mechanics of the host microenvironment.

Micro-Organoid Generation By Gel-Droplet Platform for Cancer Drug Screening

Angela Taglione, Daniel Montes Pinzon, Fei Fan, Liao Chen, Xin Lu, Sharon Stack, Hsueh-Chia Chang, Donny Hanjaya-Putra

Personalized medicine is an important topic in contemporary healthcare, aiming to tailor medical treatment to individual patients. However, 2D modeling platforms are not as physiologically relevant as 3D models because they cannot reproduce relevant aspects of native tissue such as cell-cell interactions, proliferation, the extracellular matrix (ECM), and tissue development. Spheroids are a type of 3D model that offer a superior representation of solid tumor complexities compared to other models. Despite their advantages, spheroid production is limited due to time consumption and reproducibility. Given that treatment decisions are often required within 14 days of diagnosis, there is a pressing need for a more efficient fabrication method to ensure timely and effective translation to clinical settings. To address these challenges, we evaluated the production of microgels using an elliptical pipette for cell encapsulation by droplet emulsion, for high-throughput drug screening. This pipette droplet technology offers significant improvements over traditional droplet emulsion microfluidics, which typically yield micro-organoids at a low rate, require extended culture times, and restrict matrix composition choices. In addition, this method supports automation via liquid handling instruments and is adaptable to both natural and synthetic matrices.

Norbornene-modified hyaluronic acid (NorHA) polymers were synthesized to mimic the extracellular matrix (ECM) and create the cell-encapsulation platform. Spheroids and microgels were fabricated using the NorHA solution, which included crosslinkers DTT, MMP-sensitive motifs, and RGD for cell adhesion, prepared with Irgacure "I2959" as a photoinitiator. Cell encapsulation was performed via water-in-oil emulsion using a custom-deformed elliptical pipette tip. The droplets were polymerized under UV light to form microgels, which were then cultured in optimized conditions for cell proliferation and drug screening. Cell cultures of ovarian (RFP-OVCAR5) and prostate (LNCaP) cancer lines were maintained under standard conditions and used for experiments at a concentration of 4×10^6 cells/mL. Proliferation and viability assays were performed using mitochondrial activity quantification (WST-1) and confocal microscopy to assess cell growth. For anti-cancer drug screening, encapsulated spheroids were exposed to varying concentrations of doxorubicin and enzalutamide, with IC50 values determined by evaluating cell viability 24 hours post-treatment.

An automated system for pipetting was used to generate microgels and test the effect of pipetting cycles and flow rate on the droplets sizes (**Fig. 2**). In addition, increasing the flow rate did not have an effect on the uniformity of the droplets, while increasing the aspiration/dispensing cycles greatly improved the uniformity of the droplets (**Fig. 3**). The mitochondrial activity assessment exhibited an increase in cell count was ~47% (**Fig. 4A**), with the relative proliferation of encapsulated cells being slightly lower than in monolayer culture (**Fig. 4B**). A combination of MMP-sensitive and DTT crosslinkers were used in the platform. MMP-sensitive crosslinker is nondegradable and therefore leads to less constraint on the cells. This allows for higher rates of cell proliferation because the cells can modify their ECM providing a more natural environment that can promote tissue growth (**Fig. 4C**). Viability assays were performed to assess the health and survival of cells within the gel platform before performing drug screening. The increased viability is attributed to proliferation within the platform. The OVCAR5 and LNCaP had 76.7% and 77.3% viability respectively after 24 hours encapsulation (**Fig. 5**). After 8 days, the LNCaP cell viability increased to 90.98%. The observed increase in viability is attributed to the cell proliferation within the microgels platform. Preliminary cancer drug screening was performed on LNCaP cells using DOX and ENZ at varying concentrations and OVCAR5 cells using DOX. A viability assay and IC50 was constructed. The IC50 values for LNCaP treated with DOX was 0.037 μM, while the LNCaP treated with ENZ had an IC50 value of 1 μM and the OVCAR5 treated with DOX had an IC50 value of 0.3 μM (**Fig. 6**).

Overall, we successfully developed a versatile microgels platform for high-throughput drug screening that can be applied to various cancer cell lines and tumor microenvironments. This technology demonstrated increased cell proliferation and viability and proved its efficacy in screening under different types of tumor models. Significantly, the findings highlighted the impact of tumor size on drug responsiveness. A plateau in drug effectiveness was observed at higher concentrations, particularly in spheroids larger than 200 µm, suggesting limitations in drug penetration that may result in suboptimal therapeutic outcomes for larger tumors. This highlights the need for precision in drug dosing and selection to enhance treatment efficacy. The scalability of this technology, coupled with potential integration with multi-channel systems positions it as a promising tool for generating patient-derived micro-organoids and conducting personalized drug screening.

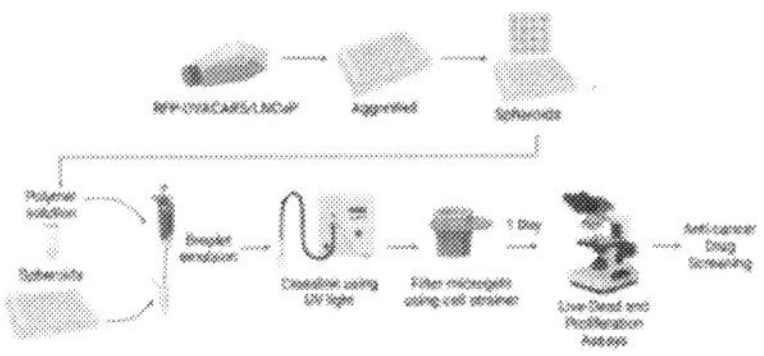

Figure 1. Schematic overview of the experimental setup for granular hydrogel

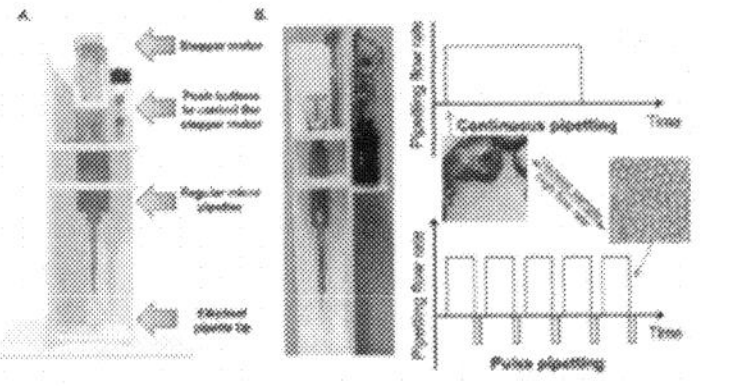

Figure 2. (A) Automated droplet generation system diagram **(B)** Automated droplet generation system photograph

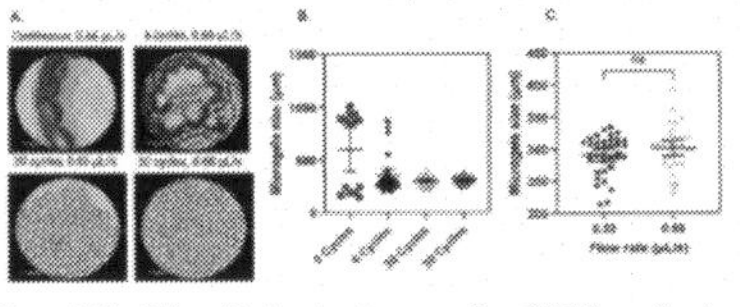

Figure 3. Variables affecting droplet generation. **(A)** Microgels micrographs at a constant flow rate of 0.66 µL/s and increasing cycles (continuous, 6, 20, 30). **(B)** Effect of aspiration/dispensing cycles (0, 6, 20, 30 cycles respectively) on the microgels size. **(C)** Effect of flow rate (0.33 µL/s and 0.66 µL/s) on microgels size.

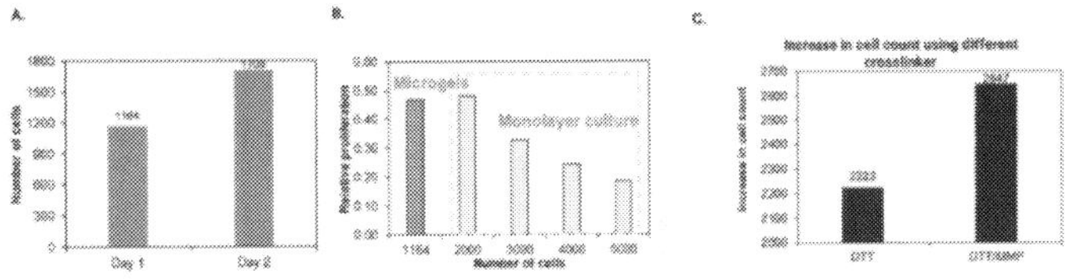

Figure 4. Cell proliferation assays measured by WST-1 cleavage. **(A)** Proliferation quantification based on the calibration curve **(B)** Encapsulated cell proliferation compared in microgels vs a monolayer culture. (C) Optimizing proliferation through ECM degradability. Increase in cell count using different crosslinker DTT alone vs 1:1 ratio DTT:MMP sensitive.

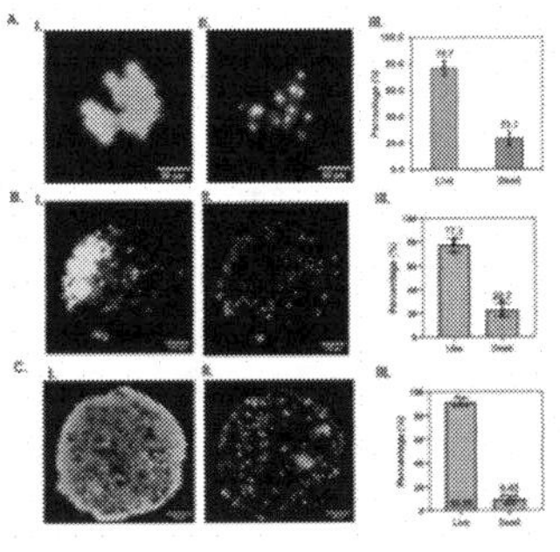

Figure 5. Cell Viability Assays. **(A)** Assessment of OVCAR5 cell viability using fluorescent staining and quantitative analysis. **(B, C)** Assessment of LNCaP cell viability using fluorescent staining and quantitative analysis at day 1 **(B)** and day 8 **(C)**. **(I)** Live cells are stained with Calcein, showing a bright green fluorescence, while **(II)** dead cells are identified using SYTOX with blue fluorescence. Both images indicate cell viability at day 1 with a spheroid width of approximately 120 µm. Scale bar = 50 µm. **(III)** Percentage of live and dead cells within the spheroid population.

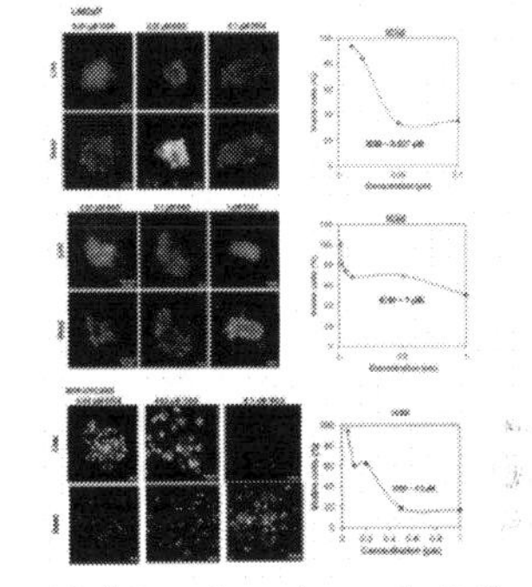

Figure 6. Preliminary anti-cancer drug screening. **(A, C)** Fluorescent images and viability vs doxorubicin (DOX) concentration (IC50) curve showing the effect of DOX at concentrations of 0.01 µM, 0.05 µM, and 0.1 µM on LNCaP cells **(A)** and OVCAR5 Cells **(C)**. Live cells are stained with SYTOX (green), and dead cells are stained with Calcein (blue). Viability was evaluated 24 hours after exposure. **(B)** Fluorescent images and viability vs DOX concentration (IC50) curve illustrating the impact of ENZ at 0.01 µM, 0.1 µM, and 1 µM on LNCaP cell viability. Viability was evaluated 24 hours after exposure.

Repurposing CRISPR-Cas12 for RNA Detection

August Bodin, Piyush Jain, Carlos Orosco, Santosh Rananaware, Michael Baugh, Jordan Lewis, Ian Lange

CRISPR-Cas nucleases have fundamentally been understood as RNA-guided systems, critical for applications in genome editing and diagnostics. The most commonly known CRISPR-Cas system is Cas9 which is commonly used in gene editing due to its high cleavage efficiency. Another commonly used CRISPR-Cas system is Cas12, which are novel enzymes used for diagnostic purposes due to its Trans-cleavage activity. Trans-cleavage is the process when a Cas-crRNA complex binds to a DNA target sequence, resulting in collateral cleavage of ssDNA. If ssDNA with a fluorophore and a quencher are exposed to such cleavage, fluorescence can be measured to determine the presence of a target sequence. This study explores the potential of Cas12 family members to use synthetic small DNA sequences to guide RNA detection for RNA detection. Our findings reveal some Cas12 proteins can activate trans-cleavage in the presence of RNA targets using short DNA "guides", providing a novel mechanism for RNA targeting. We observed that Cas12 could use a spacer-only RNA guide without the traditional scaffold sequence for ssDNA detection. This led to the hypothesis that short synthetic DNA sequences could potentially be employed to detect RNA targets. Experimentation with synthetic DNA guides revealed that adding a loop at the 3'-end of a complementary DNA (cDNA) substantially increased trans-cleavage activity for RNA detection. We demonstrated the efficacy of these guides at detecting a variety of miRNA biomarkers, such as miR-21, miR-122, and miR-155, as well as longer viral RNA sequences from HIV. The broad applicability of DNA guides across different CRISPR-Cas systems highlights their potential in molecular diagnostics. Our results underscore the versatility and cost-effectiveness of DNA guides, offering a transformative approach to nucleic acid detection and diagnostic tool development.

Characterizing the *Aspergillus Nidulans* Kinase Deletion Library for Differential Septation in Response to Cell Wall Stress

Meredith Morse Morse, Mark Marten, Alexander Doan, Joshua Dayie, Feonil Limiac, Raina Miller, Rianna Minter, Greeshma Tarimala, Mael Ndlamba, Shavier Small, Steven D. Harris

Beneficial filamentous fungi are used prevalently in the bioprocessing industry, producing high-value commercial products, while pathogenic species cause significant detriment to the health of crops and humans. Understanding the mechanisms which control the biosynthesis and repair of the fungal cell wall, a critical structure for the survival of fungi, will help us better interact with fungi in diverse contexts. For example, we could modify the cell wall to optimize production goals in industry or identify novel targets to improve the efficacy of antifungal drugs. To maintain cell wall integrity, filamentous fungi employ complex signaling pathways, predominantly mediated by protein kinases. However, many of the 98 nonessential protein kinases in the model fungus *Aspergillus nidulans* have unknown function. Previous research in our lab has shown septation is important for surviving cell wall stress, and we hypothesize that many of the 98 kinases will have connections to both the Septation Initiation Network (SIN) and the Cell Wall Integrity (CWI) signaling pathway. To test this hypothesis, the 98 strains from the *A. nidulans* kinase deletion library, each lacking one specific kinase gene, were grown for 16 hours under two conditions: with and without micafungin, a cell-wall perturbant triggering septation. Fluorescent microscopy was used to capture images of the fungal cells to determine extent of growth and number of septa formed. By comparing the phenotypes of the mutant strains to an isogenic control strain, the involvement of the various kinases in responding to wall stress can be inferred.

Developing Reactors to Improve Oxygen Availability in Cell-Free Protein Synthesis

Elias Gilotte, Chad Sundberg, Vikash Kumar, Govind Rao

In vivo protein production is essential for the pharmaceutical industry and other sectors, but it is limited by the host cell's tolerance to the product and inefficient energy use. Cell-free protein synthesis (CFPS) offers a promising alternative by using cellular components without the constraints of a living cell. CFPS can enhance energy efficiency, as energy can be directed toward specific objectives rather than the multifunctional tasks of living cells. Additionally, CFPS allows the production of biologics too complex or toxic for in vivo methods. However, CFPS remains less efficient than in vivo production.

One of the hypothesized causes of this inefficiency is oxygen availability. Preliminary data suggests oxygen is a limiting factor in E. coli CFPS metabolism, and typical sparging methods hinder protein synthesis. To explore this challenge, we designed a miniature stirred-tank reactor with enhanced gas diffusion and real-time oxygen and green fluorescent protein (GFP) titer sensing. We aim to test how varying oxygen concentrations affect protein titer and metabolism in E. coli CFPS.

While this research is ongoing, we have successfully developed a reactor that supports 4 mL CFPS reactions with real-time measurements of dissolved oxygen and GFP titer. The reactor allows direct sampling to measure adenosine triphosphate (ATP), 1,4-dihydronicotinamide adenine dinucleotide (NADH), and other reaction components.

Characterizing the Black Box of Eukaryotic Cell Free Systems Using Metabolomics

Elizabeth Bennett, Harini Sridharan, Mark Styczynski

Cells naturally adjust their internal processes to survive in changing environments, but their main goal is to stay alive rather than produce specific proteins. This can limit their use in biotechnology, where the focus is on producing large amounts of specific proteins. To overcome this, cell free protein synthesis (CFE), which uses the machinery inside cells but removes the need for living cells, was developed. Instead, DNA templates are added to these cell-free systems to produce proteins for various uses, including medicine and industry.

CFE systems have several benefits over using whole cells, such as being easier to scale up, having longer shelf life, and being more portable. So far, cell -free systems of less complex organisms such as Escherichia coli are being used for several applications such as protein production, biosensors, and biomanufacturing [1]. However, these systems cannot produce the complex proteins needed for many therapeutics due to their lack of the required evolved cellular machinery. Yeast systems have the potential to produce more complicated proteins, but they face challenges, such as lower protein yields and shorter reaction times due to energy loss and toxic buildup [2].

The crude cellular extract is a crucial part of a functional cell free reaction. Its development through lysing whole cells to release their internal machinery, is often the most intensive step due to considerable variation in protocol details from growth medium to lysis methods. Several known and unknown variables, including the presence of endogenous nucleases and the inherent metabolism of the lysate, can affect the performance of the lysate [3]. Thus, this step is often considered a 'black box', due to relatively little knowledge on the processes occurring within the lysate.

Systems biology (proteomics, metabolomics, etc.) is a useful tool often utilized to characterize such an unknown system. It is a 'top-down', systems-scale approach that measures the biological system as a whole. Metabolomics, the analysis of the small molecules present in a system, can be used to identify possible inherent bottlenecks within the yeast cell free system that are restricting protein production [1]. This approach has proven successful on *E. coli* systems, where the adjustment of specific metabolite concentrations has been shown to increase protein yield. These studies have also made it possible to determine the effects of specific steps within the CFE protocol on the final target protein production [4]. By determining possible metabolomic bottlenecks, the system can be fine-tuned to account for these previously unknown metabolic changes leading to an increased protein yield.

We have produced a viable crude cell extract in *Saccharomyces cerevisiae*, or brewer's yeast, and performed cell free reactions to verify a baseline productivity of the system. *S. cerevisiae* was chosen for this experimentation as it is a model organism due to its fast growth, annotated genome, and similarity to more complex mutli-cellular eukaryotic organisms. Originally a luciferase reporter system was chosen, enabling luminescence correlating to protein yield to be read by a microplate reader. Luciferase was chosen as an initial reporter due to its sensitivity towards low outputs, which were expected from an untested system. However, luciferase requires addition of a luciferin substrate to report luminescence, which prohibits the ability to perform a continuous time course reaction. Thus, we switched to a GFP reporter for continuous monitoring, and fluorescence was read by a microplate reader and correlated to protein yield going forward [2].

We have worked to optimize the system to maximize output by varying sonication energy input during lysis and various reactant concentrations including T7 RNA Polymerase, glycerol, and lysate percentage. The effects of different variables affecting the final protein production such as the physiological state of the yeast cells, inherent metabolism of the lysate, and different lysis methods such as sonication and homogenization will be studied. To enable comparison and analysis of the effects of different variables, samples lysed with different methods will be collected at varying stages of growth to determine the effect of the cell's physiological state on the performance of the CFE system. The metabolomics workflow, consisting of various steps such as quenching and extraction, prepares the sample for GC-MS testing [5]. The resultant dataset will be analyzed using various statistical tools such as multivariate analysis using PCA and univariate analysis using ANOVA. As was observed with *E. coli* CFE systems, the resulting data can shed light onto the black box of processes within the lysate and CFE systems. Once one or more of these bottlenecks are determined, it will be possible to further fine-tune the system by harnessing the tools of synthetic biology resulting in a substantially more efficient yeast cell free system.

1. Sridharan, H., F. Piorino, and M.P. Styczynski, *Systems biology-based analysis of cell-free systems.* Curr Opin Biotechnol, 2022. **75**: p. 102703.
2. Hodgman, C.E. and M.C. Jewett, *Optimized extract preparation methods and reaction conditions for improved yeast cell-free protein synthesis.* Biotechnol Bioeng, 2013. **110**(10): p. 2643-54.
3. Khambhati, K., et al., *Exploring the Potential of Cell-Free Protein Synthesis for Extending the Abilities of Biological Systems.* Front Bioeng Biotechnol, 2019. **7**: p. 248.
4. Miguez, A.M., M.P. McNerney, and M.P. Styczynski, *Metabolic Profiling of Escherichia coli-based Cell-Free Expression Systems for Process Optimization.* Ind Eng Chem Res, 2019. **58**(50): p. 22472-22482.
5. Alvarez-Sanchez, B., F. Priego-Capote, and M.D.L. de Castro, *Metabolomics analysis II. Preparation of biological samples prior to detection.* Trac-Trends in Analytical Chemistry, 2010. **29**(2): p. 120-127.

Assay Optimization for a "CRISPR" Understanding of Immune Cell Migration

Andrea Ramirez Hercules

Neutrophils are specialized immune cells that lead the first wave of host defense during an innate immune response, sensing chemotactic signals and rapidly migrating to sites of infection or injury. Migration must be regulated by a series of signals that prevent excessive neutrophil recruitment, which can result in unresolved infections, chronic inflammation, and tissue damage. Abnormalities in neutrophil function can stem from genetic mutations that result in diminished or absent neutrophil response. Understanding genes that play a role in regulating neutrophil migration is crucial for understanding the mechanisms behind this innate immune response.

This project relies on a preliminary screening of genes of interest that have been identified to affect neutrophil migration. Our preliminary screening utilizes a PLB-985 cell line, a model for neutrophil-like cells, in which we have induced silencing mutations of various genes using CRISPR. After identifying genes of interest, knockout lines of individual gene mutations have been generated, and we must now validate their effect on neutrophil migration via a secondary screening. Neutrophil migration is assessed by inducing cell migration through a transwell assay using a chemoattractant, and we expect the silencing mutations of our genes of interest to supress or inhibit this process.

This study investigates the optimization of the transwell assays used in our screenings, which will help ensure accurate results when confirming the effects of various genetic mutations in neutrophil migration. Our optimization of transwell assays explored changes in various components of our experimental procedure, this was done with the goal of obtaining results that align with the expected neutrophil migration trends described by literature. Various factors such as coating and release methods, use of different chemoattractants, and changes in migration period were tested to enhance the reproducibility and accuracy of our screenings, ensure consistency in our migration trends, and establish a reliable system for assessing our individual knockout cell lines.

The primary goal of this project to ensure that the observed changes in neutrophil migration is impacted solely by the effects of various gene silencing mutations, and to prevent our results from being influenced by our experimental conditions. Before we proceed with screening individual knockout lines, we must continue to refine our screenings to align with the migration patterns reported in literature. Exploring candidate migration factors is critical for broadening our molecular understanding of neutrophil function, which will in turn provide a much-needed road map for understanding how and why neutrophil migration defects occur in diseases including infection, cancer, and autoimmunity

CRISPR Facilitated Knockout of RPL39 and RPL39L, Implications for TNBC Metastasis

Gavin DeWitt Graham, Mailin Li, Jenny Chang

Breast cancer stem cells (CSCs) represent a subpopulation of cells within tumors with enriched tumorigenic potential and heightened resistance to both radiation and chemotherapy. Previous work in the Chang lab demonstrated an enriched CSC gene signature post-chemotherapy, suggesting that these cells are selected for by standard therapy. shRNA screening of this 493-count gene signature in MpBC cell lines identified ribosomal protein 39 (RPL39) and its paralog RPL39-Like (RPL39L) as key genes involved in CSC self-renewal, treatment resistance, and lung metastasis. Based on preliminary data showing higher CSC numbers and RPL39 expression in Metaplastic Breast Cancer (MpBC)s compared to non-MpBC Triple Negative Breast Cancers (TNBCs) and considering RPL39 and RPL39L was originally identified from a CSC gene signature, this proposal addresses whether differential expression of RPL39 and RPL39L ribosomal proteins preferentially translates proteins causing the aggressive and chemorefractory nature of MpBC using CRISPR knockout (KO) technology. Crispr-Cas9 was successfully employed, resulting in complete losses of RPL39 and RPL39L function in separate cell lines. Future work involves characterization of stem-cell properties to precisely determine these genes' role in cancer metastasis.

Utilizing Supercoiled Plasmids in CRISPR-Dx for Viral Gene Detection.

Anastasiia Steksova, Noor Mohammad, Qingshan Wei

Having sensitive, specific, and simple diagnostic tools is essential for a well-functioning healthcare system. Among the prevalent diagnostic methodologies, CRISPR-Dx, a technique heavily reliant on costly equipment for fluorescent signaling, has garnered substantial attention. In this paper, we develop less expensive and non-optical CRISPR-based pathogen diagnostic systems using supercoiled ds-plasmid DNA (ΦX174). This new method employs activation of Cas12a protein within a CRISPR-based system that leads to post-nicking relaxation of supercoiled ΦX174 plasmid. We also utilize this new method to detect viral DNA in AVV samples up to 0.2 pM concentration of target DNA. Gel electrophoresis serves to assess the proportion of supercoiled to the relaxed plasmid. The findings indicate that plasmid reporting systems are specific and inexpensive compared to conventional fluorescence reporters.

Detecting PFAS Using CRISPR-Cas12 As a Signal Enhancer

Brynley Platt, Elizabeth Ajibode, Kevin Yehl

PFAS (per- and polyfluoroalkyl substances), referred to as "forever chemicals" due to their inability to break down, are becoming widespread contaminates throughout the environment. Concerningly, their exposure is a major public health concern because it is correlated with cancers, liver disease, and birth defects. The current gold standard for detecting PFAS involves liquid chromatography coupled with mass spectrometry (LC-MS). This approach is highly specific and sensitive, but unfortunately is time-consuming and expensive because it requires trained personnel and a centralized laboratory to perform. Therefore, there is significant interest in developing complementary methods to detect PFAS on-site that do not require trained personnel. Aptamer sensors are a class of functional nucleic acid that are emerging as a potential solution. However, the current sensitivity of aptamers is not low enough to detect PFAS at EPA advisory levels. CRISPR-Cas12 is emerging as a powerful tool for detecting nucleic acids at the point-of-care without requiring sophisticated instrumentation. Therefore, we hypothesize that using CRISPR-Cas12 as a signal enhancer of activated aptamer sensors will increase sensitivity for detecting PFAS. Towards this goal, we investigated several aptamer sensor architectures compatible with Cas12 amplification and examined their corresponding thermodynamic landscape for sensor activation. Several DNA designs were tested having varying 3' and 5' overhang lengths of 1 to 7 nt. While studies are still ongoing, our results show that sensors with 3' overhang lengths between 4 to 7 nt had the largest signal to background enhancement compared to sensors with 5' overhangs. We also investigated buffer compatibility for optimizing aptamer and Cas12 activity. Together, this study is a significant step in designing a more affordable and ultrasensitive biosensor for the detection of PFAS.

Construction and Characterization of *S. Pyogenes* Mutants Influencing CRISPR-Cas9 Memory Formation

Melina Mohammadi

Bacteria employ diverse immunological strategies to detect and eliminate foreign DNA, including that of invading viral genomes such as bacteriophages (phages). Unique among these defenses are Clustered Regularly Interspaced Short Palindromic Repeat (CRISPR) systems, which act as the only known form of adaptive immunity in prokaryotes, generating and retaining genetic memories of previously encountered phages in the form of 'spacers' stored within the CRISPR array. Despite the pivotal role of CRISPR in defense against phage, the process of acquiring new spacers during infection remains exceptionally rare and poorly understood. To identify genetic factors that influence CRISPR-mediated immunity, a transposon mutagenesis screen (Tn-seq) was performed in Streptococcus pyogenes, a native host of a Type II-A CRISPR system. Notably, insertions in recJ, a well-conserved single-stranded DNA exonuclease involved in DNA repair, experienced strong enrichment in our S. pyogenes survivor pool after exposure to lytic phage A1. Generation of a deletion strain, ΔrecJ, validated Tn-seq results with ΔrecJ cells exhibiting a 2-3-fold enhanced CRISPR immunization rate against an untargeted phage. Surprisingly, the ΔrecJ strain also demonstrates increased survival rates against phage already targeted by CRISPR spacers, suggesting that this mutant influences both naive spacer acquisition as well as utilization of spacers once encoded in the CRISPR array. Furthermore, we have isolated a novel phage from wastewater, Faranak, that is not sensitive to the absence of recJ. We aim to leverage this insensitivity to further elucidate the mechanism of RecJ during phage infection and CRISPR immunization.

Cloning, Expression, and Purification of Laccase Enzymes for Biocatalytic Reactions

Kristina Oshiro, Shuyuan Zhang, Ryan Summers

Laccase, a multi-copper oxidoreductase with the ability of oxidizing a wide range of substrates, such as ascorbic acid, 2,2'-azino-bis(3-ethylbenzothiazoline-6-sulfonic acid) (ABTS), 2,6-Dimethoxyphenol (2,6-DMP), guaiacol, and hydroquinone, is found in multiple species including plants, fungi, and bacteria. While it is known that laccase enzymes are capable of naturally oxidizing lignin, a bio polymer found in plant cell walls and provides stability and stiffness to the plant, the reaction between each strain of laccase and lignin will be different. This study aims to study a library of over 40 different laccase, including cloning and optimization of expression and purification, followed by *in vitro* characterization of its reaction with substrates.

To begin, we codon-optimized four different laccases from *Arabidopsis thaliana*, *Bacillus stratosphericus, Bacillus subtilis,* and *Thermothelomyces thermophilus* to fit the bacterial expression system. PCR and Gibson Assembly were employed to produce the laccase coding gene and insert the targeted gene into the bacterial expression vector pET28, which contains an IPTG-inducible T7 promoter and a hexahistidine tag for further *in vitro.* Sequencing confirmed correct construction of laccase-containing plasmids, which were then transformed into *E. coli* expression host BL21(DE3).

Next, BscotA1 from *Bacillus stratosphericus* was picked for the pilot study. In order to maximize the gene expression, we first looked into the study of various cultivation media (LB vs TB) with multiple induction temperature (18C, 30C, & 37C). All of the cultures were pelleted and lysed with sonication, followed by gel electrophoresis to determine the expression level. We observed the most soluble protein using TB medium with 18C induction. We then employed affinity chromatography with a Ni+ column to purify the targeted laccase. With the confirmation of the target laccase protein band in the elution column on the SDS protein page, the project will move to the next phase: 1). further optimize the purification protocol for more pure protein. 2). Apply the pilot kinetic study for such purified laccase to finalize the characterization protocol. 3). Provide both purification and characterization for the other laccase.

Chemical Analysis of Nanoplastics Shed from Tampons

Claire Edwards, Jade White, Joe E. Baio

For many people around the world, tampons are an accessible and easy-to-use period product. Researchers have been increasingly focused on synthetic polymers (plastics) fragmenting into micro- and nanoplastics (<1000 nm) and their environmental impact, but not much work has been done on studying toxicity to humans. A 2021 preliminary study found that tampons release up to 17 billion nanoplastic (<1000 mm) particles per use, with currently unknown adverse health effects. Before determining the potential health implications, the chemistry of components and shed nanoparticles from 5 commercially available tampon brands have been analyzed.

To begin, the chemical analysis of tampon components–such as the outer casing, inner core, and string–were recorded via attenuated total reflectance-Fourier transform infrared spectroscopy (ATR-FTIR). The peaks for notable functional groups were identified in the infrared spectra of the components to determine the composition. Particles were collected through a process meant to mimic physiological conditions and normal wear. Two different solvents were used to limit background interference during FTIR analysis–phosphate buffered saline (PBS) and deionized water. The solvents were filtered to limit potential particles from the solvents and adjusted to between pH 3.8 and 4.2. Samples were taken from different brands of regular absorbency tampons, placed in the pH-adjusted filtered solvent then in a 37°C water bath on a shaker table at 100 rpm for 4 hours. Next, any large particles (>1000 nm) were filtered out and the samples' pHs were adjusted to neutral to prevent excess degradation of the samples. To confirm the presence of nanoparticles in the samples, the concentration and average size of the particles were measured via nanoparticle tracking analysis. Once confirmed, the samples were freeze dried and analyzed via ATR-FTIR.

The FTIR analysis for tampon components was performed by matching peak wavenumbers; this showed that 2 out of 5 brands contain cotton (1030, 1310-1320, 2890-2900, 3330 cm^{-1}), 2 brands contain polyester (721, 872, 1020, 1100, 1240, 1710, 2970 cm^{-1}), 3 brands contain polyethylene (717, 1470, 2850, 2910 cm^{-1}), 1 brand contains polypropylene (1380, 1460, 2840-2950 cm^{-1}), 3 brands contain rayon (1020, 1370, 2890, 3350 cm^{-1}), and 1 brand contains wax paper [believed to be paraffin coated cellulose] (1030, 1320-1460, 2850, 2920, 3340 cm^{-1}). Nanoparticle tracking analysis measured average size between 129.9 and 309.7 nm and average concentration of up to 376 million particles. The FTIR analysis of the particle samples is harder to analyze because of background interference.

The ATR-FTIR analysis performed on the tampon components showed various types of plastics. Furthermore, the nanoparticle tracking analysis confirmed that millions of particles were being shed from tampons. However, future work needs to be done to further investigate the chemical composition of the particles shed.

Evaluating and Validating Protein Biomarkers in Muscular Dystrophy

Sistine Angela Padon, Albert Jiménez Requena, Cristina Al-Khalili Szigyarto

Characterized by the weakness and loss of muscle mass, muscular dystrophy is a rare type of neuromuscular disease where genetic mutations interfere with the production of the proteins necessary to form healthy muscles.

Current methods for the prognosis of muscular dystrophy offer a limited view of the disease's progression. For example, one of the tests conducted to monitor the disease is the 6-minute walk test (6MWT). However, this test relies on the patient's motivation to perform well on the test and ability to understand and follow instructions. As a result, the test may reflect the patient's disposition during the test rather than providing an accurate measure of the disease's progression. In contrast, biomarkers reflect a specific biological assessment of the disease that is more objective and quantifiable.

While testing specific muscle function parameters is an effective method for determining the prognosis of disorders such as muscular dystrophy, validated biomarkers can be more meaningful when improving clinical care and research. These biomarkers not only facilitate early diagnosis and precise monitoring of disease progression but also play a critical role in evaluating new therapies, ultimately improving patient outcome.

This project evaluates the abundance of muscular dystrophy protein biomarkers in serum/plasma through antibody-based protein analysis. In this approach, antibodies specific to the target proteins cover the beads used in Suspension Bead Array (SBA) technology. These beads are then added to the serum/plasma samples, where the target proteins have been biotin-labeled. After the desired antibody-antigen complexes are formed, the detection is carried out using fluorescently labelled streptavidin. This approach also allows for the protein biomarkers to be monitored over time and assess various disease progression states.

Proteolytic Degradation of Amyloidogenic Proteins: Initial Discovery for Transthyretin Amyloidosis

Michael Webb, Raymond Pho, Carl Denard

Transthyretin (TTR) is a highly conserved transport protein in the human thyroid-hormone pathway[1]. Following translation, TTR naturally folds into a homotetrameric protein[2]. However, mutations in the ttr gene and age-related wild-type degradation result in misfolded tetrameric TTR. This instability causes the tetramer to break down and form free monomers. These monomeric units can misfold into aggregates, oligomerize, and form fibril structures which deposit in various organs, including the brain and the heart[1]. Once accumulation in the heart advances, transthyretin amyloidosis (aTTR) can occur and lead to the progressive and deadly disease Transthyretin Amyloid Cardiomyopathy (ATTR-CM). Often under and misdiagnosed, ATTR-CM is characterized by the formation of insoluble TTR fibrils in areas of the heart[3] and is responsible for approximately 20% of heart failure deaths in the United States[4]. The fatal disease occurs most commonly in wild-type TTR amyloidosis (aTTRwt) compared to the genetic variant TTR amyloidosis (aTTRv), and individuals over 60 years of age[5].

Current clinical treatments for ATTR-CM focus on preventing TTR aggregation via protein stabilization, liver transplantation and, more recently, gene editing and silencing[6]. Despite these advances, limited strategies exist to remove misfolded TTR species, including monomers, oligomers and fibrils. Proteases present a promising and novel approach to disaggregation due to their high catalytic activity *in vivo*. The downside to utilizing proteases is their promiscuous nature, which could result in unwanted side effects. As a result, this research aims to identify and characterize candidate proteases to degrade transthyretin aggregates. Currently, we are focusing on the protease HTRA1 as previous research has shown its ability to disaggregate α-synuclein amyloids[7], cleaving after the amino acid valine[8]. Intriguingly, TTR β-strands F and H, which were shown to be responsible for driving TTR aggregation[9], contain two valine amino acids, respectively, introducing the possibility of HTRA1-mediated proteolysis at these sites. Understanding this possible degradation mechanism will allow for broader research into engineering proteases specific to certain amyloids, eliminating the threat of indiscriminate cleaving. Success in this research will open the door to a new novel strategy for treating various deadly diseases caused by aggregating proteins.

References

(1) Liz, M. A.; Coelho, T.; Bellotti, V.; Fernandez-Arias, M. I.; Mallaina, P.; Obici, L. A Narrative Review of the Role of Transthyretin in Health and Disease. *Neurology and Therapy* **2020**, *9* (2), 395–402. https://doi.org/10.1007/s40120-020-00217-0.

(2) Sanguinetti, C.; Minniti, M.; Susini, V.; Caponi, L.; Panichella, G.; Castiglione, V.; Aimo, A.; Emdin, M.; Vergaro, G.; Franzini, M. The Journey of Human Transthyretin: Synthesis, Structure Stability, and Catabolism. *Biomedicines* **2022**, *10* (8), 1906. https://doi.org/10.3390/biomedicines10081906.

(3) Obi, C. A.; Mostertz, W. C.; Griffin, J. M.; Judge, D. P. ATTR Epidemiology, Genetics, and Prognostic Factors. *Methodist DeBakey Cardiovascular Journal* **2022**, *18* (2), 17–26. https://doi.org/10.14797/mdcvj.1066.

(4) Jain, A.; Zahra, F. Transthyretin Amyloid Cardiomyopathy (ATTR-CM). In *StatPearls*; StatPearls Publishing LLC: Treasure Island, FL, 2019.

(5) Nativi-Nicolau, J.; Siu, A.; Dispenzieri, A.; Maurer, M. S.; Rapezzi, C.; Kristen, A. V.; Garcia-Pavia, P.; LoRusso, S.; Waddington-Cruz, M.; Lairez, O.; Witteles, R.; Chapman, D.; Amass, L.; Grogan, M. Temporal Trends of Wild-Type Transthyretin Amyloid Cardiomyopathy in the Transthyretin Amyloidosis Outcomes Survey. *JACC: CardioOncology* **2021**, *3* (4), 537–546. https://doi.org/10.1016/j.jaccao.2021.08.009.

(6) Teng, C.; Li, P.; Bae, J. Y.; Pan, S.; Dixon, R. A. F.; Liu, Q. Diagnosis and Treatment of Transthyretin-Related Amyloidosis Cardiomyopathy. *Clinical Cardiology* **2020**, *43* (11), 1223–1231. https://doi.org/10.1002/clc.23434.

(7) Chen, S.; Puri, A.; Bell, B.; Fritsche, J.; Palacios, H. H.; Balch, M.; Sprunger, M. L.; Howard, M. K.; Ryan, J. J.; Haines, J. N.; Patti, G. J.; Davis, A. A.; Jackrel, M. E. HTRA1 Disaggregates α-Synuclein Amyloid Fibrils and Converts Them into Non-Toxic and Seeding Incompetent Species. *Nature Communications* **2024**, *15* (1). https://doi.org/10.1038/s41467-024-46538-8.

(8) Grau, S.; Baldi, A.; Bussani, R.; Tian, X.; Stefanescu, R.; Przybylski, M.; Richards, P.; Jones, S. A.; Shridhar, V.; Clausen, T.; Ehrmann, M. Implications of the Serine Protease HtrA1 in Amyloid Precursor Protein Processing. *Proceedings of the National Academy of Sciences* **2005**, *102* (17), 6021–6026. https://doi.org/10.1073/pnas.0501823102.

(9) Saelices, L.; Johnson, L. M.; Liang, W. Y.; Sawaya, M. R.; Cascio, D.; Ruchala, P.; Whitelegge, J.; Jiang, L.; Riek, R.; Eisenberg, D. S. Uncovering the Mechanism of Aggregation of Human Transthyretin. *Journal of Biological Chemistry* **2015**, *290* (48), 28932–28943. https://doi.org/10.1074/jbc.m115.659912.

Effects of Protac Mediated BTK Degradation on Platelet Function

Eliana Choi, Alexander Melrose, Helena Ventosa-Capell, Jiaqing Pang, Joseph E. Aslan

Platelets are small, anucleate cell fragments in human blood that function to regulate hemostasis and thrombosis. Bruton's tyrosine kinase (BTK) regulates platelet signaling and activation, and small molecule BTK inhibitors like Ibrutinib act by irreversibly binding to BTK active sites and blocking signal propagation, where scaffolding activities of BTK remain intact. In contrast, proteolysis targeting chimeras (PROTACs) target specific proteins of interest for ubiquitination, resulting in near-complete proteasomal degradation. The PROTAC DD-03-171 specifically targets BTK for degradation to eliminate BTK signaling and scaffolding activities. We compared the effects of DD-03-171 and ibrutinib on platelet adhesion and spreading as well as platelet aggregation. Western blot analysis confirmed a loss of BTK expression in platelets following 2 h incubation with DD-03-171. Loss of BTK expression inhibited platelet aggregation as well as platelet spreading on surfaces of fibrinogen. These observations were comparable to inhibition of platelet BTK ibrutinib, which was used as a control. This study demonstrates that DD-03-171 effectively promotes the targeted degradation of BTK in platelets and highlights the potential of PROTACs experimental as well as therapeutic tools to specifically manipulate platelet signaling and platelet physiological responses.

Artemia As a Testbed for CPA Toxicity Screening Using Machine Learning Analysis

Chloe Johnsen, Jordan Morrow, Crysthal Alvarez, Carla Berrospe-Rodriguez, Guillermo Aguilar

Cryopreservation is a process that requires cryoprotective agents (CPAs) to effectively preserve various cells, tissues, and organs; however, the chemicals used in CPAs can be toxic at certain concentrations, so it is important to establish a biological model to screen the toxicity of CPAs. We propose the common brine shrimp, *Artemia salina*, as a toxicological model to study the toxicity of CPAs at various concentrations, alongside a machine learning tracking program that quantifies the survival rate and swim speeds of the brine shrimp after hatching. For this study, cysts and hatched *Artemia* were placed in three different CPAs at various concentrations with filtered sea water as the base, and ten-second videos were recorded hourly (up to 5 hours) and then every 24 hours to monitor the *Artemia*. This process was repeated for 5 binary solutions at 2.5 mol% and 5.0 mol% of chemicals that are often included in multicomponent CPAs. Each of the videos were then analyzed using the tracking program, which calculated the survival rate and swim speeds of each *Artemia*. A direct correlation was discovered between the concentration of a CPA and the survival rate of the *Artemia*, with a decrease in the survival rate observed with increasing exposure time to the CPA. Additionally, certain chemicals in the binary solutions were more acutely toxic to the *Artemia*, matching previously known toxicity data obtained from other testing methods. Overall, *Artemia salina* has proven to be a promising model for CPA toxicity screening, and when paired with a tracking program that measures the viability and swim speeds, could produce a databank of toxicological profiles similar to those of mammalian cells for a variety of CPAs.

Media Optimization for Biofuels from *Cyclotella Cryptica*.

Alyssa Keptner

Diatoms, distinguished by their silica shells, offer unique potential for biofuel production compared to green algae. This research aims to use diatoms, specifically *Cyclotella cryptica* (*C. cryptica*), for biodiesel synthesis by catalyzing the triacylglycerol (TAG) within them. *C. cryptica* was selected as the subject of study since it is rich in lipids (40% of biomass, 55% TAG). The research focused on determining the optimal silica concentration for maximizing *C. cryptica* growth and its subsequent impact on lipid content throughout the growth phase. Experiments were conducted by varying silica concentrations (0.5x- 5x). Results revealed that a 2x concentration of silica improved *C. cryptica* growth within a 14-day period.

Utilizing Low Volume Samples to Construct Phase Diagrams for Coacervate Systems

Sydney Volheim, Whitney Blocher McTigue

Complex coacervation, a liquid-liquid phase separation phenomenon driven by electrostatic interactions of two oppositely charged polyelectrolytes, is a topic of growing interest in the biomedicine and biological communities. A large part of this investigation focuses on where this complexation occurs, commonly illustrated through binodal curves outlining the two-phase region. With industrial polyelectrolytes, large sample volumes (mililiters or larger) of both the dense liquid phase, the coacervate, and the surrounding dilute phase, the supernatant, can be generated and analyzed. When using biologically relevant polyelectrolytes, such as proteins, enzymes, peptides, etc., small sample volumes (microliters) are obtained. Single large volume samples may be made to generate a handful of points to create a binodal curve, though without the same degree of detail as binodal curves utilizing industrial polymers. Here, we are developing a method to utilize small sample volumes focusing on the supernatant phase with absorbance, emission, and conductivity readings as well as material balances of both phases to create a more detailed map of the binodal curve. Absorbance readings are performed after the polymers are fluorescently tagged. We initially focused on using poly(styrene sulfonate sodium salt) (PSS) and poly(diallyldimethylammonium chloride) (PDADMAC) to form complex coacervates as a point of comparison between what has been done already in the literature and our method. This process will then be applied to more biologically relevant systems where large sample volumes are difficult to make, specifically poly(*L*-lysine) and poly(*D,L*-glutamate). By building these binodal curves, we will have a more accurate representation of where phase separation occurs and can more readily compare how different solution conditions (*i.e.*, salt, pH) change the phase envelope.

Investigating the Effect of Vitamin E on the Viscoelastic Properties of Lung Surfactant

Emmanuelle Ong, Estephanie Nottar Escobar, Madelyn Atkins, Prajnaparamita Dhar

The utilization of e-cigarettes and vapes has become a very prevalent issue in today's society as multitude of adolescents/teenagers have fallen victim to the vaping epidemic. Various studies have been conducted on patients with e-cigarette or vaping product use-associated lung injuries or (EVALI). These studies have observed multiple instances where vaping additives were found in the bronchoalveolar lavages of patients with reported breathing difficulties. It is hypothesized that vaping additives, upon entry to the lungs can interact with lung surfactant, disrupting the stability of the monolayer film, resulting in inflammation of the alveoli and difficulty breathing. The lung's monolayer film exists at the air-liquid interface of the alveoli and is composed of surfactants such as Dipalmitoylphosph-atidylcholine (DPPC), 1-palmitoyl-2-oleoyl-phosphatidylglycerol (POPG), and proteins. These proteins and surfactants work in tandem to reduce the surface tension in the alveoli which enables a person to inhale and exhale properly. Our research focuses on the hypothesis that vape additives such as vitamin E (VitE) can interact with lung surfactant and decrease its ability to reduce surface tension in the alveoli. In this work, a Langmuir-Pockels ribbon trough was utilized to mimic the compression and expansion of the lungs during breathing. Phosphate-buffered saline (PBS) was utilized as subphase in all trials to emulate physiological conditions. Our experiments consisted of spreading a layer of model lung surfactant at the air-liquid interface, followed by compressing and expanding while monitoring the surface pressure. In addition, dilatational rheology was examined via film oscillations at minimal area change to assess the viscoelastic moduli of films. Initial control trials were carried out with pure DPPC to model unaffected lung surfactant. Furthermore, vitamin E a common additive in vape products, was added to DPPC to test the viscoelastic properties of the monolayer film formed on the interface, emulating the exposure to this vape additive. Overall, the plots generated by the expansion and compression did not retain the same shape as the pure DPPC. The phase transition present in pure DPPC compression isotherms, which characterizes the liquid expanded to liquid condensed state changes, was not seen upon the addition of VitE. Rather a higher and smoother surface pressure increase was observed. However, while a pure DPPC film results in maximum surface pressure close to 73 mN/m, the film added by VitE resulted in lower maximum surface pressures of about 64 mN/m. This suggests that VitE addition impacts intermolecular interactions at the interface. More specifically, decreasing packing density at low mean molecular areas. This can also be observed from dilatational rheology results as the viscoelastic properties measured for systems with additions of VitE saw a decrease in the viscoelastic modulus. Particularly the elastic modulus as those values decreased to a value of less than half of the unaffected model modulus. This indicates that addition of VitE significantly impacts the rheology of lung surfactants, causing instability in the alveoli. Further studies will be conducted to investigate the difference between VitE and VitE acetate, as VitE acetate is another common additive found in vaping products.

Egfr Binding Peptide Contrast Agents for Signaling Egfr-Positive Tumors

Mimi Pham, Vivek Kumar, Joseph Dodd-o, Nellone Reid

Triple Negative Breast Cancer, TNBC, is an aggressive and fast-growing subtype of Breast Cancer, which lacks an established therapeutic target. It is characterized by the absence of increased expression of the estrogen receptor, progesterone receptor, and human epidermal growth factor receptor 2 (HER2). Additionally, TNBC affects 13 in every 100,000 women in the United States. Magnetic Resonance Imaging, MRI, is the main noninvasive method to detect these tumors. However, due to TNBC's unique characteristics and nonuniform shape, it is more likely to have inaccuracies in the diagnosis of the tumor and/or tumor morphology. A gadolinium-based contrast agent chelated by an Epidermal Growth Factor Receptor (EGFr) binding peptide is being developed to address these limitations in MRIs. The use of this conjugate will improve the contrast in MRI scans, leading to more accurate detections of EGFr-positive tumors, such as the most common subtypes of TNBC. The conjugate will bind to EGFr, where the paramagnetic gadolinium ions will interact with water molecules in tissues, influencing relaxation rates on MRI scans. EGFr-binding peptides were synthesized by Solid-Phase Peptide Synthesis and coupled with a chelating agent. The main peptide, AEGFr, was found in previous literature and was linked to self-assembling peptides previously developed in KumarLab to develop novel self-assembling multidomain peptides SAP1-AEGFr and SAP2-AEGFr. The peptide conjugates were synthesized and characterized for their chemical, structural, and mechanical properties. The affinity of self-assembling conjugated peptides to the EGFr needs to be further tested with Microscale Thermophoresis, MST to find the best binding affinity. Next, the peptide conjugates are combined with a gadolinium(III) ion solution. The peptide will then self-aggregate to form a translucent soft hydrogel. The formed EGFr Binding Peptide Contrast Agent can then be injected into a patient, increasing the accuracy of EGFr+ tumor diagnosis.

Figure 1: Schematic of Experimental Procedure. Created with BioRender.com

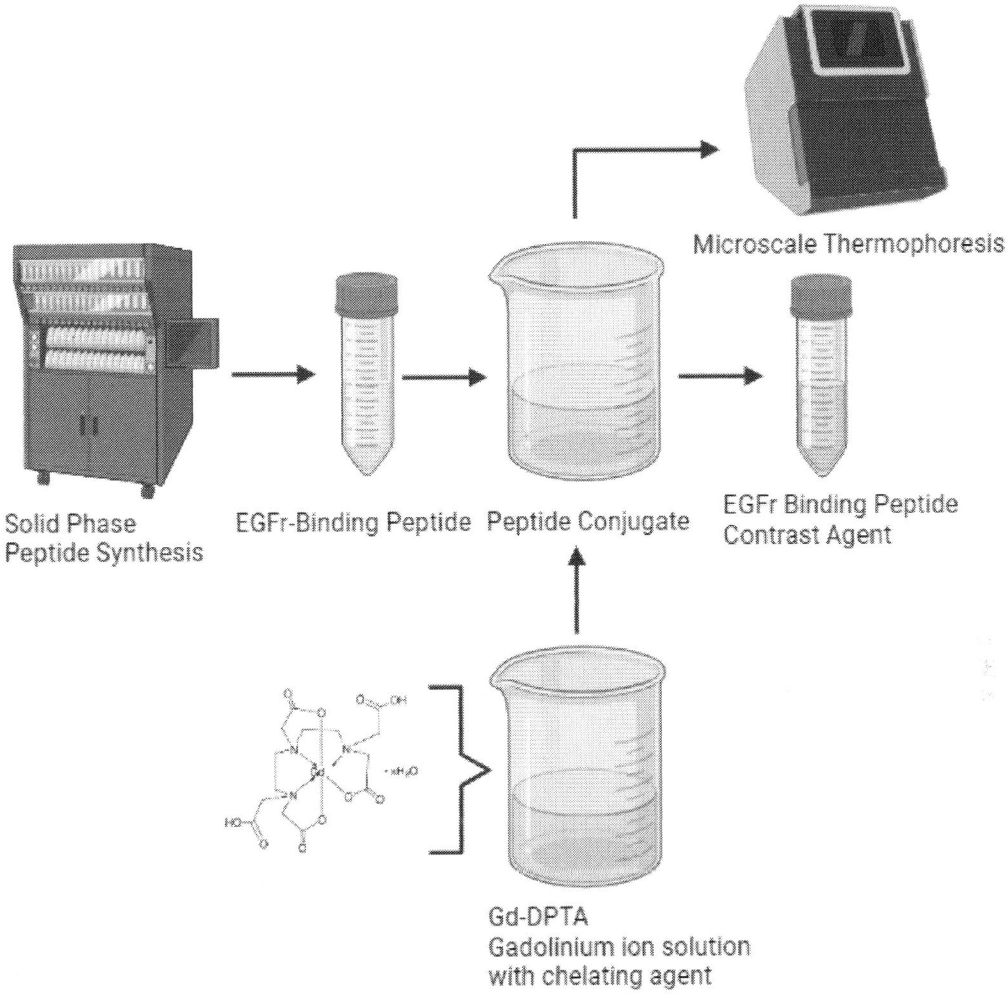
Microscale Thermophoresis
Solid Phase
Peptide Synthesis
EGFr-Binding Peptide
Peptide Conjugate
EGFr Binding Peptide
Contrast Agent
Gd-DPTA
Gadolinium ion solution
with chelating agent

Cooperative β-Sheet Coassembly Controls Intermolecular Orientation of Amphiphilic Peptide-Polydiacetylene Conjugates

Leel Liberty, Tarunya Sudarshan, Sujeung Lim, Jeffrey Li, Alicia Robang, Herdeline Ardoña, Anant Paravastu

In this work, we elucidated the structural organization of stimuli-responsive peptide-polydiacetylene (PDA) conjugates that can self-assemble as 1D nanostructures under neutral aqueous conditions. The amino acid sequences bear positively or negatively charged domains at the periphery of the peptide segments to promote solubility in water while also driving assembly of the individual and combined components into β-sheets. The photopolymerization of PDA, as well as the sensitivity of the resulting optical properties of the polymeric material to external stimuli, highly depends on the structural organization of the assembly of amphiphilic peptide-diacetylene units into 1D-nanostructures. Solid-state NMR measurements on ^{13}C-labeled and ^{15}N-labeled samples show that positively charged and negatively charged peptide amphiphiles are each capable of self-assembly, but self-assembly favors antiparallel β-sheet structure. When positively and negatively charged peptide amphiphiles interact in stoichiometric solutions, cooperative coassembly dominates over self-assembly, resulting in the desired parallel β-sheet structure with a concomitant increase in structural order. These results reveal that rational placement of oppositely charged residues can control β-strand organization in a peptide amphiphile coassembly, which would have implications on the adaptive properties of stimuli-responsive biomaterials such as the peptide-PDAs studied here.

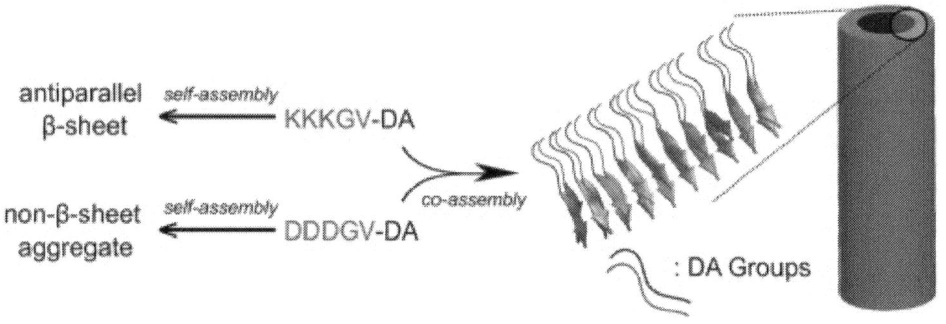

Improving Cardiovascular Tissue Models Via Bio-Ionic Liquid Functionalized Hydrogels

Dominic Pizzarella, Katelyn Neuman, Nolan Burson, Abigail Koppes, Ryan Koppes

Abstract:

Adverse reactions account for 30% of all drug discontinuations with cardiovascular toxicity as the most common complication[1]. Because of this, current drug development projects span around 10-15 years and cost billions of dollars before approval from the Food and Drug Administration is granted[2]. One major contributor to the unpredicted adverse effects of medications in the clinic is the physiological limitations of conventional preclinical models, primarily 2D cell cultures[3]. While these systems have been an industry standard, they display abnormal gene expression and morphology, resulting in poor simulation of native tissues[3]. In response, hydrogels (3D crosslinked networks of hydrophilic polymers) have gained interest as a preclinical screening platform because they can emulate the native extracellular matrix which improves biomimicry[4]. However, many polymers are non-conductive, which has presented complications for modeling excitable tissues like the heart[5,6]. To improve the simulation of the cardiovascular system, the unique electrophysiological characteristics of myocardial tissue should be considered. This work explores the ability of a novel ionically conductive hydrogel to enhance the modeling of the myocardium. In this work, we employ material characterization to tune the biomaterial to cardiovascular mechanical properties. We then evaluate the signaling networks of the culture using a planar microelectrode array (MEA) which is a device composed of electrodes capable of recording extracellular electrophysiology. This research suggests that an 8% (w/v) Gelatin Methacryloyl (GelMA) + 3.5% (v/v) Choline Acrylate hydrogel, nicknamed Gel-Amin, can amplify signaling from a culture of cardiomyocytes (CMs; heart cells responsible for beating) to generate cultures with improved beating synchronicity compared to a GelMA control (Figure 1). The engineered 3D cell culture platform presents a cost-effective, high-throughput accessible device capable of integrating into drug development to better predict cardiotoxicity. By creating an improved model of the heart, this research seeks to decrease drug development costs and timelines, ultimately accelerating the availability of life-saving therapies to patients.

References: 1. Laverty, H. *et al. Br. J. Pharmacol.* (2011). 2. Sun, D. *et al. Acta Pharm.* (2022). 3. Kapalczynska, M. *et al. Arch. Med. Sci.* (2018). 4. Sánchez-Cid, P. *et al. Polymers* (2022). 5. Neuman, K. *et al. Cells Tissues Organs* (2023). 6. Noshadi, I. *et al. Sci. Rep.* (2017). 7. Gadilohar, B. *et al. J. Mol. Liq.* (2016). 8. Kaiser, N. *et al. Biomed. Mater.* (2016). Kreuz, T. *et al. J Neurophysiol.* (2015).

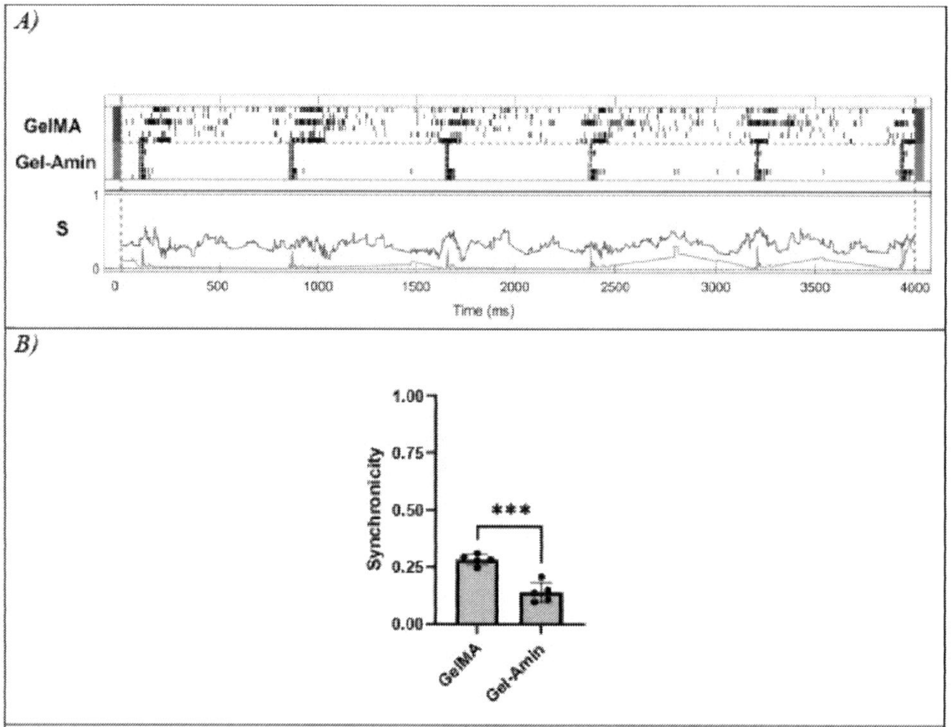

Figure 1. Evaluation of Cardiomyocyte Synchronicity. A) Raster plot showing spike timestamps from individual electrodes. Synchronicity (S) was determined using SPIKY[7], with values near zero indicating strong synchronicity. B) Gel-Amin cultures demonstrate significantly improved synchronicity compared to control (p=0.0011), suggesting enhanced biomimicry of native cardiac tissue.

Development of Agarose-Based Hydrogel Composites for Enterosorptive Removal of PFAS

Ifrah Hammad, Maria Victoria Ximenes Klaus, J. Zach Hilt

Per- and poly-fluoroalkyl substances (PFAS), or "forever chemicals," are persistent anthropogenic chemicals linked to various health disorders and accumulate in the body. Agarose-based hydrogels were investigated as a candidate for removing PFAS from the body to tackle the issue of PFAS accumulation. *In vitro* sorption studies with model dyes were conducted to determine if they could be used for PFAS enterosorption (orally administering sorbents to remove contaminants using undigested sorbents that are excreted). The results of the *in vitro* binding of model compounds suggest that agarose-based hydrogel systems are effective in removing PFAS due to their aqueous sorption behavior. Furthermore, preliminary research shows that hydrogel behavior is maintained within human-simulated conditions. Future research will focus on continuing studies in simulated human conditions and quantifying PFAS removal *in vitro* with the potential to conduct *in vivo* studies.

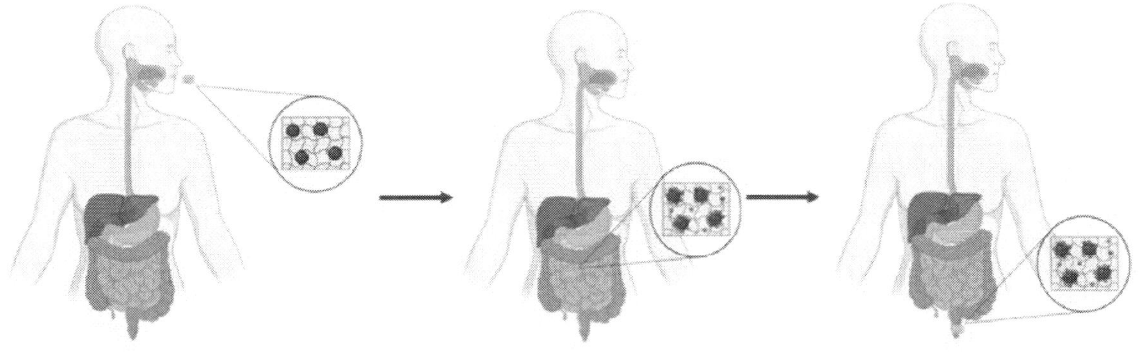

Spheres represent contaminants, which in this case is PFAS.

Alginate and Poly-L-Ornithine Degeling Techniques for Cell RNA Extraction

Brooklyn Lengyel, Philip Glawe, Vivek Prakash, Emmanuel C. Opara, Sivanandane Sittadjody, Justin Saul

Osteoporosis affects approximately 8 million women in the U.S. and is the leading cause of bone fracture with costs expected to reach $95 billion by 2040. Prior to 2002, one widely used treatment for osteoporosis in women who lacked ovarian function consisted of pharmacological hormone treatments (pHTs) that replace the hormones naturally secreted by healthy ovaries. These were found to be effective for reducing osteoporotic fractures but increased women's risks of developing cardiovascular issues and some cancers. Thus the use of pHTs has decreased in use since 2002. New pharmacological treatments are not more effective than hormone based treatments and have the potential for severe, potentially life threatening side effects as well as minor side effects that lead to poor patient compliance.

One reason for the safety risks inherent to pHTs is their inability to regulate the hypothalamus-pituitary ovary (HPO) axis. A cell-based hormone treatment (cHT) that can secrete the hormones produced by healthy ovaries while regulating the HPO axis is a potential approach to safe and effective use of hormone therapies to mitigate hormone-related osteoporosis. This potential treatment involves creating a two-compartment alginate system consisting of an inner core of granulosa cells and an outer core of theca cells taken from a donor. These constructs are coated with perm-selective poly-L-ornithine as a way to mitigate immune attack on the encapsulated cells.

To better understand the molecular mechanisms of action for our cHT approach, we are interested in removing the cells from the construct for RNA extraction. While there are methods of digesting alginate such as alginase and EDTA/ citrate, the presence of PLO makes removal of cells from the constructs challenging. The ability to isolate cells and RNA from the cHT constructs would allow us to determine the mechanisms by which the cells work.

Developing a Fibronectin-Enhanced Elastomer Substrate for the Improved Visualization of 2D Muscle Fiber Morphology

Isabella Weiner, Brandon Rios, Ritu Raman

In vitro tissue engineering allows for faster and more efficient studies of the morphology and function of biological systems compared to *in vivo* models. Skeletal muscle tissue models enable researchers to study electrical signaling and muscle cell growth and differentiation, providing insights into disease mechanisms. Studies have developed 2D methods of engineering muscle fibers and studying their morphology, chemical signaling, and contractility. Muscle alignment, length, and width are key characteristics related to muscle function, necessitating the ability to view and measure whole fibers. However, the common use of extracellular matrix (ECM)-mimicking hydrogels (e.g., fibrin) as substrates results in an uneven surface plane, compromising reproducibility and imaging capabilities. This study develops an alternative substrate for skeletal muscle cell growth using functionalized polydimethylsiloxane (PDMS) with the ECM protein fibronectin, demonstrated in previous studies with cardiac muscle. Various compositions of commercially available PDMS were mechanically tested to match the mechanical properties of fibrin. This blend was then spin-coated onto glass to achieve a thin, clear, uniform layer, and tested first with fibronectin adsorption, then functionalized using aminosilane and glutaraldehyde chemistry, forming stable covalent linkages to fibronectin. C2C12 mouse muscle myoblasts successfully adhered to the substrate and differentiated into measurable muscle fibers, however, they began to detach from the surface by Day 5 of differentiation, possibly due to spontaneous twitching of thickly-formed fibers. Thus, the contractile strength of skeletal muscle fibers may necessitate either a thicker substrate layer or altered attachment geometry.

Optimizing Myoblast Attachment to Shape Memory Biodegradable Elastomers

Ramiya Shelton, Scott Hollister, Sriharsha Ramaraju

Skeletal muscle (SM) has an innate capacity to regenerate its own tissues but struggles to recover from critically sized defects such as Congenital Diaphragmatic Hernias (CDH) in pediatric cases and Volumetric Muscle Loss (VML) and Patch Repair in adults. Tissue Engineering (TE) is a potential therapeutic to address these clinical challenges by integrating scaffolding, tissue architecture, and bioactive factors to stimulate regeneration. This study focused on optimizing tissue architecture for regeneration, using cell-adhesive molecules (CAMs), specifically fibronectin derived RGD to laminin derived YIGSR, on Polyglycerol Dodecanedioate (PGD) substrates. CAMs, which play a vital role in cell and organelle attachment in the Extracellular Matrix (ECM), were used to functionalize the surface of the PGD substrate surface, a biodegradable polymer suitable for minimally invasive implantation due to its thermal manipulability.

We hypothesized that CAMs increase overall murine myoblast attachment, with higher attachment shown in the laminin binding domain YIGSR dominant mass fractions. To test this, PGD was cast into a 96-well plate, sterilized with ethanol, and functionalized with seven mass fractions of RGD:YIGSR (100%-50%). A CBCQA Assay and One-way ANOVA analysis revealed, uniform peptide binding observed across all mass fractions **(Figure 1)**, establishing a consistent baseline for cell attachment.

Murine myoblasts (C2C12s) were seeded onto the peptide-enhanced substrate in triplicates and quantified via fluorescence intensity and a Calcein assay. Results, analyzed with Two-Way ANOVA, show differences in cell attachment between mass fractions at varying time points. At the 24-hour mark, seeding efficiency reached 60% across all mass fractions **(Figure 2)**, with YIGSR dominant mass fractions (100%-50%) having significantly greater attachment at the 6-hour mark. We perceive that Murine Myoblasts attach better to YIGSR due to prior work that indicates their ability to interact with the extracellular matrix laminin binding integrin receptor $\alpha 7$-$\beta 1$.

Despite biological variability, these findings suggest the addition of CAMs, specifically YIGSR alongside RGD, enhances cell attachment and reduces attachment time. Future studies will investigate the use of micropatterned peptide-enhanced substrates to further optimize cell attachment and alignment.

Figure 1

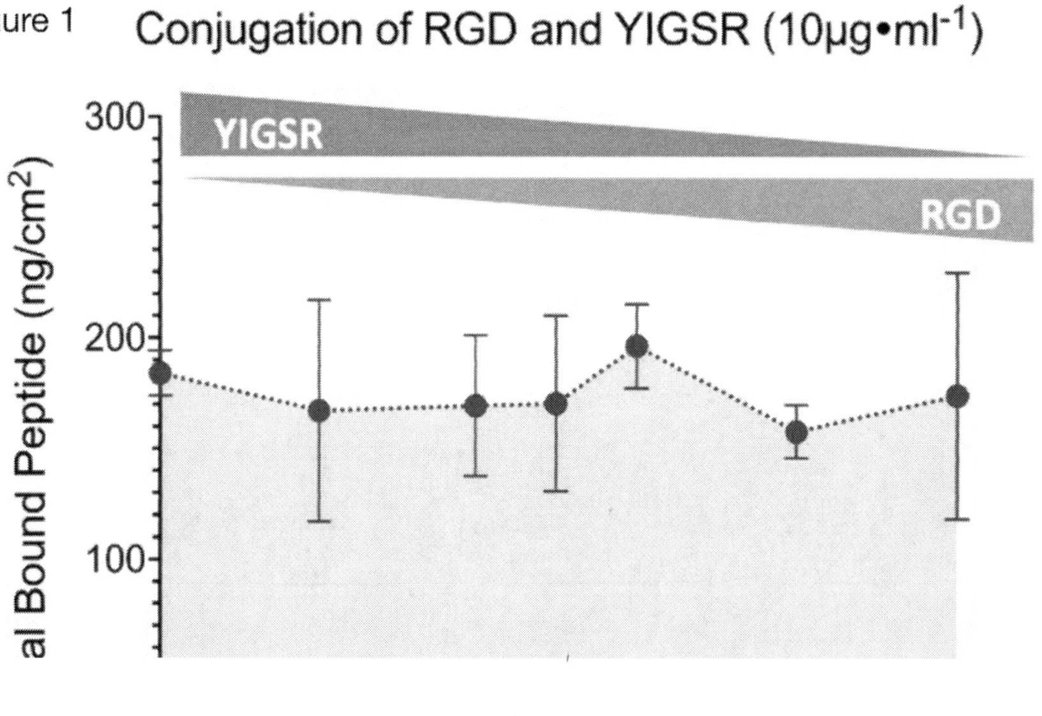

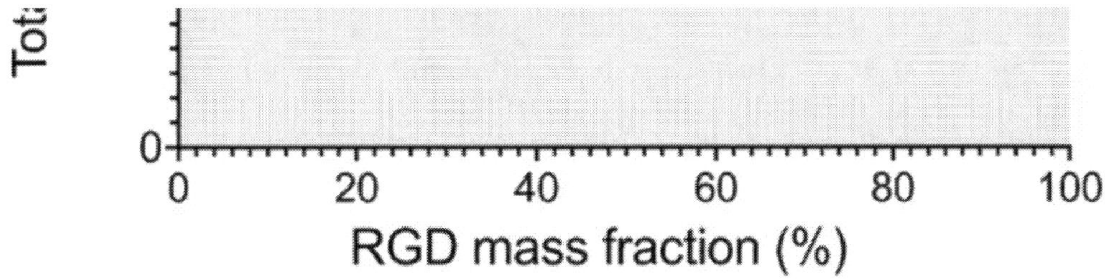
Tot
0
0
20
40
60
80
100
RGD mass fraction (%)

Figure 2

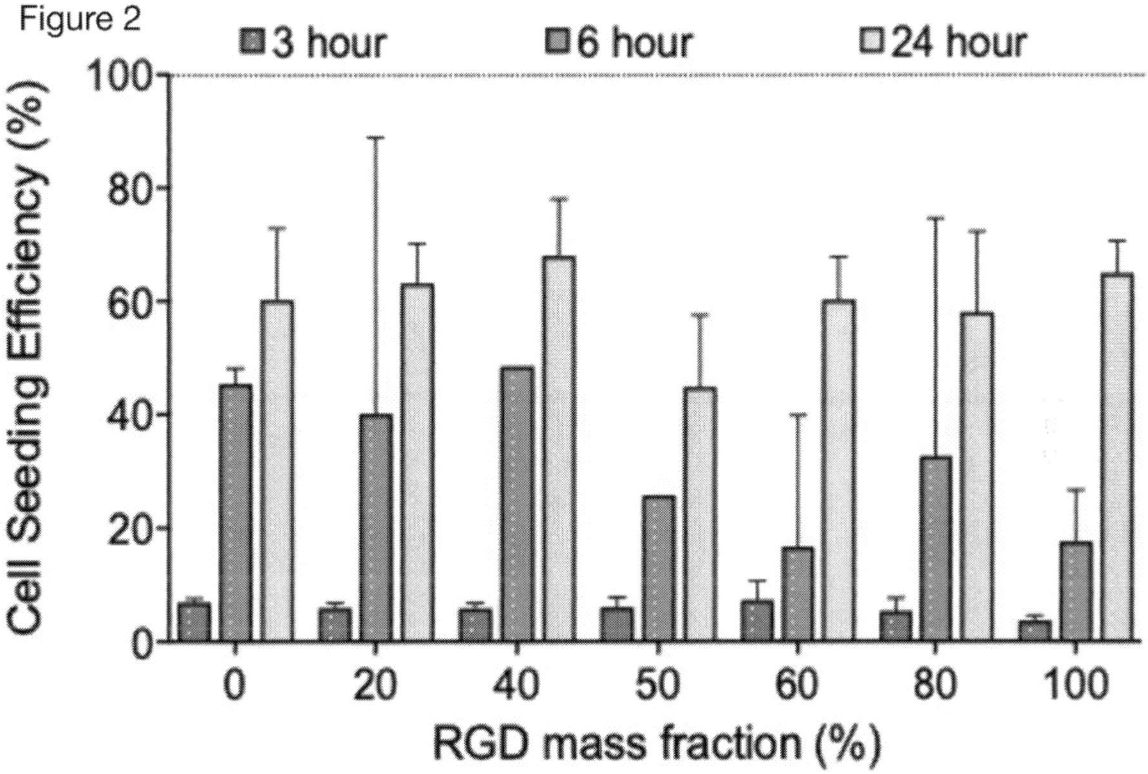
3 hour
6 hour
24 hour
Cell Seeding Efficiency (%)
100
80
60
40
20
0
0
20
40
50
60
80
100
RGD mass fraction (%)

Extrusion of Double-Layer Alginate Constructs for Cellular Hormone Therapy

Philip Glawe, Vivek Prakash, Brooklyn Lengyel, Gabriel Gabrovsek, Sivanandane Sittadjody, Emmanuel C. Opara, Justin Saul

The onset of menopause in women is associated with health risks due to the loss of ovarian hormone production. One of the effects of changing systemic hormone levels is an increased risk of osteoporosis that, in turn, increases vulnerability to bone fracture. One way to combat this is to use pharmacological hormone therapy (pHT) consisting of estrogen alone or in combination with progesterone. However, pHT has been associated with an increased risk of cancer and cardiovascular disease and is, therefore, not currently a first-line approach to preventing osteoporosis. Our approach to potentially enhancing both safety and efficacy of hormone-based therapies is cell-based hormone therapy (cHT). We have used a biomimetic approach in which granulosa and theca cells are encapsulated in a two-compartment alginate system. This construct has the granulosa cells in the interior compartment while the theca cells are in the outer compartment of the construct, providing the juxtaposition of these cell types needed for the paracrine effects seen in native ovaries. We have previously reported the use of this cHT approach in rat models of ovarian failure. The goal of this work is to standardize the procedures used to create the double-layered cHT constructs. We use a syringe-based approach for extrusion of solution-phase alginate into bivalent cations for the formation of spherical particles for the inner compartment followed by a second extrusion of the particles suspended in solution-phase alginate into bivalent cations to form the outer compartment. By standardizing the procedure, the goal is to have greater control over the two-compartment cHT construct diameters so that they can be modified in future tests.

Characterizing Nanoparticle Dynamics in Biorelevant Materials to Predict Epithelial Transport of Orally Administered Medications

Colby Constantine, Daniel Keane, Ryan Poling-Skutvik

This project aims to determine the properties controlling nanoparticle penetration and transport through synthetic porcine gastric mucin and to understand the resulting dynamics within these biomimetic tissue analogues. From this information, nanoparticle-mediated drug delivery systems can be developed and implemented into the pharmaceutical industry, highlighting a rapidly diversifying therapeutic portfolio. To address this challenge, I will investigate the surface charge of both carboxyl and amine-terminated nanoparticles and identify correlations between nanoparticle charge and mucus binding affinity. To quantify the extent of mucus binding, particle tracking and microrheological measurements will be conducted in the tissue analogs, allowing for the determination of sub/super diffusive nanoparticle movement. Conducting microrheological measurements in mucin allows for a more accurate model of the conditions that a drug-encapsulated nanoparticle would encounter once it reaches the duodenum and jejunum, the upper portions of the small intestine, where most drug uptake occurs. It is promising that this biomimetic system will be able to accurately model nanoparticle penetration and paracellular transport as it occurs in biological systems.

Integration of Remote Floating Gate FET-Based Biosensor with Millifluidic Systems for Point-of-Care Diagnostics

Amina Anowara

Methods of developing low-cost biosensors that are rapidly-producible and highly accurate are necessary for effective response to disease outbreaks. Early detection during disease out-breaks can reduce the long-term spread of disease, which reduces the loss of life, and prevents a shutdown of societal functions. Remote-floating gate field effect transistors (RFGFETs) can potentially improve population-based rapid detection capabilities during pandemics. RFGFETs possess many of the necessary qualities of an ideal sensor device to use in emergency scenarios, and have 3 key advantages over other methods of detection: (1) the limit of detection (LoD) is 11 magnitudes lower than the LoD of traditional at-home diagnostic devices, (2) RFGFETs provide results at the minute time-scale rather than hours or days, (3) RFGFETs are low-cost. However, the preparation method that precedes the usage of RFGFETs for pathogen detection is labor-intensive and must be completed in a laboratory setting. Here, we report a simplified procedure as an alternative to the traditional preparation method for RFGFET devices to widen the scope of its use beyond the laboratory setting. This will be accomplished by fabricating a millifluidic device to automate the preparation steps of the sensor platform. Integration of fluidic devices with RFGFET biosensors allows for automation of the preparation process, bringing RFGFETs closer to the target goal of becoming an efficient point-of-care biosensor. The overall preparation process was shortened from approximately eighteen hours to four hours, and concentrations of antigen in the ng/mL range were detected.

Affinity Engineered Living Materials for Protein Immobilization

Milena Zeru, Anjani Chandra, Anna Johnson, Erica Hild, Nathan Soulier, Jonathan K. Pokorski

Engineered living materials (ELMs) are a class of composite materials where a living organism is combined with a synthetic polymer to generate emergent features. One challenge in biosynthetic ELMs, where a protein product is produced from an organism, is containing the protein component within the material. This work aimed to modify the hydrogels' structures to increase the affinity for secreted proteins, therefore containing the proteins of interest.

Sodium alginate was utilized as a scaffold with nickel (II) chloride hexahydrate and nickel sulfate hexahydrate introduced as new cross-linkers that bind to histidine-tagged proteins, preventing them from escaping into the media. This was inspired by immobilized metal affinity chromatography (IMAC), which takes advantage of an amino acid tag's affinity for metal ions to separate the tagged proteins from a solution. Using this idea, the proteins are anticipated to be contained within the hydrogel.

We investigated the effect of polymer and crosslinker concentrations on the structural properties of these hybrid materials. Additionally, this biocomposite was characterized to determine mechanical properties, morphology, and nickel toxicity to cells. As mentioned, only proteins with a His-tag can bind to the gel. Moreover, genetically modified cyanobacteria that secrete a protein of interest with a His-tag were compared to wild-type cyanobacteria. The His-tagged protein is blue, so the gels are expected to have higher absorbance readings at longer wavelengths.

This project aims to better understand how organisms can behave in ELMs, ultimately offering more control over the living material's outputs. In addition, this gel could offer another means for protein purification.

Incorporation of N-Acetyl Cysteine in Polyester Disulfide Materials for Spinal Cord Injury Applications

Marissa Moore, August Hemmerla, Bret D. Ulery

Spinal cord injuries (SCIs) are debilitating injuries commonly caused by traumatic accidents or pre-existing medical conditions. In 2016, it was estimated that the prevalence of SCIs was 721 to 906 per million people in the U.S., highlighting a significant public health concern. These injuries dramatically impact the rest of the body through paralysis or partial loss of function below the injury site. The prognosis for SCIs can vary greatly depending on the severity and location of impact; however, additional damage or complications can follow the primary injury. Consequently, preventing such secondary injuries is a critical component of SCI treatment.

Despite advancements, the gold-standard treatment for SCIs remains invasive spine surgery. While this may provide some relief for patients, these procedures can carry significant risks for patients, including infection and further neurological complications. Additionally, surgery primarily reduces patient pain rather than alters the inflammatory response of cellular and biochemical signaling that leads to secondary damage. Given these limitations, available alternatives only address a singular neural degeneration mechanism and have had limited success in clinical trials. Thus, there is a need for an efficient, localized biomaterial that can limit secondary SCI injury by mitigating the negative signaling cascades responsible for the injury.

In response to this need, this project focuses on treating SCIs by generating a biomaterial that controllably releases simple signaling molecules (SSMs) to encourage neuroprotection. SSMs can be incorporated into monomers that are then polymerized. These degradable polymers provide localized, sustained release, mitigating secondary damage associated with nervous system injuries. In particular, N-acetyl Cysteine (NAC), an antioxidant derived from L-cysteine, was utilized to synthesize polyester disulfide material. NAC has been proposed to influence intracellular reactive oxygen species (ROS) via multiple enzymatic pathways. Studies exploring the delivery of NAC for secondary SCI damage have been limited, reducing the potential for NAC-based treatments.

This work evaluates the neuroprotective capabilities of NAC and its incorporation into a novel polymeric material to evaluate this potential. Initially, the cytotoxicity and neuroprotectant capabilities of NAC were evaluated on the NE-4C cell line utilizing dsDNA and ATP assays. Following this, an NAC-based monomer was synthesized and assessed using Nuclear Magnetic Resonance (NMR) before incorporation into multiple polyester disulfides. Polyester synthesis was achieved through transesterification involving four PEG-based diols. Subsequently, these were evaluated using NMR and gel permeation chromatography (GPC).

Currently, this research is progressing towards integrating these polyester disulfide materials into a hydrogel-based drug delivery platform. This innovative approach aims to provide a multifaceted strategy for neuroprotection in SCIs. This work yields valuable insight into the development of secondary SCI treatments and other SSM-based drug delivery platforms.

A High-Throughput Microfluidic Gut-on-a-Chip for Disease Modeling

Stephen Landry, Bryan Schellberg, Abigail Koppes, Ryan Koppes

Inflammatory bowel disease (IBD), which affects about 2.4 million adults in the United States, is a family of debilitating diseases with pervasive effects on the lives of affected individuals, such as a significant decrease in the quality of life, high hospitalization rates, and expensive treatments [1]. The pathophysiology of IBD is largely unknown, making it difficult to treat the disease at the source. However, the emergence of organs-on-a-chip (OOC), has offered improvement in modeling diseases, such as IBD, by mimicking the structure and function of tissue in vitro.

Although OOC have shown promise, current models are low-throughput, requiring extensive time and resources. To address current limitations, we have developed a high-throughput microfluidic gut-on-a-chip (GOC) that is capable of culturing eight independent samples simultaneously. The GOC features a semi-permeable membrane that supports the cell population and forms distinct apical and basolateral compartments for fluid flow, supplying physiologically relevant shear stress. Additionally, our device distributes fluid flow from a single inlet to all eight culture wells, significantly reducing external tubing and pumping requirements (Figure 1a).

Caco-2 epithelial monolayers were grown and matured on-chip over seven days, then the monolayers were fixed and stained for nuclei, cytoskeletal protein F-actin, and tight junction protein ZO-1. Staining images from a representative well indicate high expression of F-actin, which indicates expected morphology of cells, and ZO-1, indicating epithelial maturity (Figure 1b). In addition, a lucifer yellow assay was performed to quantifiably assess the barrier function of each monolayer. The results of this assay indicated a tight distribution of permeability that was significantly lower than a blank chip and not significantly different from a single-chip control (Figure 1c). Thus, the increased throughput does not compromise the overall physiological relevance of the chip.

In summary, our work demonstrates a novel approach to high-throughput culture in OOC, enabling simultaneous culture of eight independent 'gut samples' in a single GOC device. Specifically, we applied our device to successfully culture Caco-2 cells on-chip, differentiating to mature epithelial monolayers in all eight culture wells. Ongoing work is focused on applying our high-throughput OOC platform to an in vitro model of IBD, as well as additional organ-chip models.

[1] Lewis, J. D., et al. (2023). 'Incidence, Prevalence, and Racial and Ethnic Distribution of Inflammatory Bowel Disease in the United States.' Gastroenterology 165(5): 1197-1205 e1192.

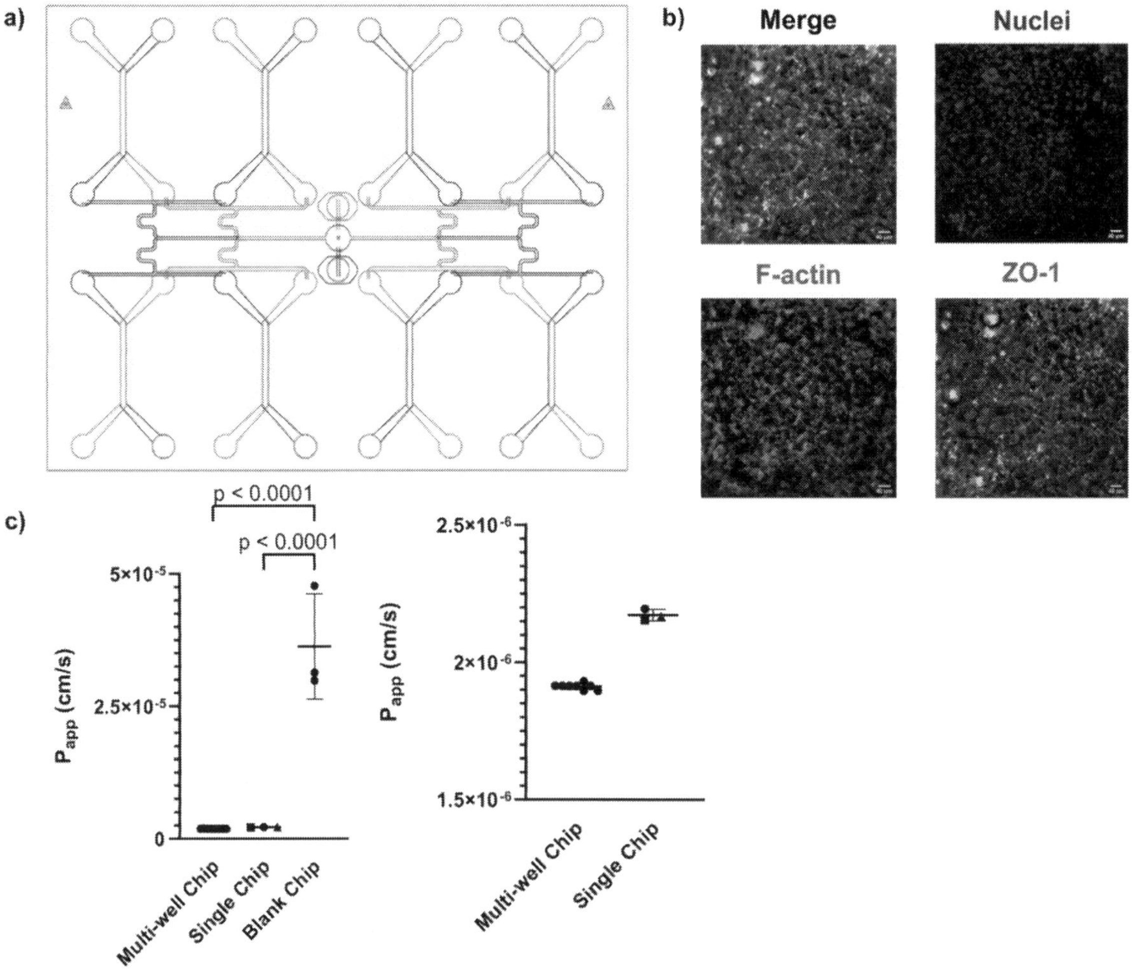

Figure 1. Chip prototype and characterization results. **a)** Top view of final chip prototype. Apical flow layer is depicted in red, and basal flow layer is depicted in black. **b)** Immunostaining results, with DAPI in blue, F-actin in red, and ZO-1 in green. Images were taken at 20X magnification. **c)** Apparent permeability of multi-well chip compared to both single chip and blank chip (left) and only single chip (right). Datasets were compared via one-way ANOVA with Tukey's multiple comparisons. Error bars represent standard deviation.

Production of Bacteriorhodopsin Nanoparticles Isolated from the Haloarchaeon *Halobacterium Noricense* for Biosensitized Solar Cell Development

Vivek Chakraborty, Ruchi Mathad, Mehwish Ali, Sagar Kanekar, Anand Kulkarni

Solar cells present a promising avenue for sustainable energy production by converting light energy into electricity through the photovoltaic effect. Biosensitized Solar Cells are a greener modification of Dye-Sensitized Solar Cells, which utilise eco-friendly compounds like bacteriorhodopsin (bR) protein from a haloarchaeal organism *Halobacterium noricense* MITWPU HA6. This study involves optimising the yield and productivity of bacteriorhodopsin protein by targetting the organism's growth kinetics; extraction and purification of bacteriorhodopsin protein, with a total experimentally recorded yield of 1.28 g/L; studying the electrogenic potential of purified bacteriorhodopsin protein; and developing a protocol to nanocrystallize bacteriorhodopsin for its incorporation into Biosensitized solar cell modules, thus increasing its efficiency. Future studies aim to further enhance the protein's yield and productivity with the help of genetic engineering and metabolic engineering, and fabricate a biosensitized solar cell module using nanocrystallized bR protein, contributing to advancements in bioelectricity and bioenergy.

Development and Optimization of a Modular Microfluidic Device to Study the Effects of Fluid Shear Stress on Metastatic ER+ Breast Cancer

Erica Wood, Emma Grady, Braulio Ortega Quesada, Adam Melvin

Estrogen receptor positive (ER+) breast cancer accounts for approximately 70% of breast cancer cases and relies on the estrogen receptor for proliferation and tumor growth. Of these patients, most will respond to endocrine therapies designed to specifically target the estrogen receptor; however, some develop a resistance to endocrine therapy following metastatic spread. Metastatic cancer is associated with poor patient prognoses and accounts for about 90% of cancer-related deaths. The underlying mechanisms driving endocrine resistance are currently unknown; however, it is suspected that metastasizing cancer cells are exposed to biochemical and biophysical cues that alter their phenotype. One such biophysical force is fluid shear stress (FSS), which prior studies have suggested enhanced proliferation and a pro-survival phenotype at the metastatic site. Most approaches to study the effects of FSS on cancer cells rely on the two-dimensional (2D) culture of cells and interrogation directly after FSS exposure, which does not adequately recapitulate the three-dimensional (3D) growth and cell-to-cell interactions of the tumor microenvironment (TME) or allow for the long-term effects of FSS to be studied. Recent work has shown the potential for the use of 3D spheroids, small aggregates of cancer cells, to mimic what is occurring in the TME. The goal of this study is to develop a modular microfluidic platform capable of first exposing cancer cells to well-defined magnitudes FSS and then growing the cells as 3D spheroids in an array of ~400 circular traps to determine the long-term effect of FSS on metastatic cancer. Initial efforts focused on the optimization of the microfluidic approach including (i) incorporation of a thiol-acrylate (TA) hydrogel scaffold to support 3D growth and (ii) adjusting flowrates of the aqueous and oil phase to ensure proper cell encapsulation. Both sheared and non-sheared MCF-7 cells (a model ER+ cell line) were grown as 3D spheroid for 7 days followed by terminal immunostaining for the proliferative marker Ki67. Both the sheared and non-sheared cohorts were successfully cultured across multiple biological replicates resulting in viable cells in 7 days. Initial results highlight a difference in Ki67 signaling between the sheared and non-sheared cohorts suggesting that exposure to FSS altered the behavior of ER+ breast cancer cells grown as 3D spheroids.

Immunostimulatory Extracellular Vesicle Vaccine for Activation of Immune Responses

Hannah Crane

Extracellular vesicles (EVs) are lipid-based nanoparticles secreted by cells that facilitate the intracellular delivery of biochemical cargoes. EVs are also promising drug delivery carriers as they can be modified and leveraged for an array of treatments. One novel application is the addition of an adjuvant to the EV surface to promote immune activation and antigenicity. To accomplish this, HEK293SF-3F6 cells expressing a fusion of the model antigen ovalbumin (OVA) and the EV-associated protein BASP1 underwent metabolic glycoengineering via N-azidoacetylmannosamine-tetraacylated (Ac4ManNAz) to produce OVA+ EVs with surface azide functionalization. The EVs were then conjugated to a modified version of the stimulator of interferon genes (STING)-agonist diABZI containing a dibenzo cyclooctyne (DBCO) reactive handle using strain-promoted azide alkyne cycloaddition (SPAAC). The efficacy of the diABZI-conjugated EVs was tested in vitro using bone marrow-derived dendritic cells (BMDCs) and compared to that of free diABZI. Results showed a significant increase in antigen presentation and upregulation of BMDC activation markers compared to free diABZI. Interestingly, BMDCs treated with EV-diABZI conjugates secreted significantly more TNF-α than BMDCs treated with diABZI alone. This suggests EV-diABZI conjugates are capable of shifting cytokine secretion from the interferon regulatory factor (IRF) pathway to another immune related pathway, NF-κB, which is conventionally activated by toll-like receptor (TLR) signaling. More investigations are needed to determine the signaling mechanism and downstream implications of this phenomenon. Overall, EV-diABZI conjugates demonstrated efficient BMDC activation and antigen delivery in vitro, suggesting that this platform may serve as an effective cancer vaccine. Future studies aim to optimize this technology for in vivo applications.

Nanobody Conjugated Extracellular Vesicle Vaccine for Adaptive Immune Therapy

Taylor Patino, Hayden Pagendarm, John Wilson

Extracellular vesicles (EVs) which are membrane enclosed nanoparticles secreted by nearly all cell types as a means of intercellular communication exhibit potential as a drug-delivering nanocarrier in a variety of applications. EVs from HEK293SF-3F6 3D suspension cells were isolated through size-exclusion chromatography. During EV biogenesis, HEK293SF-3F6 cells were incubated with $Ac_4ManNAz$, which led to azide incorporation into the EV surface-membrane glycoproteins. A derivative of the STING agonist diABZI, with a DBCO-PEG_{12} handle, was conjugated with to the surface of EVs using strain-promoted azide-alkyne cycloaddition (SPAAC) chemistry to the EV surface azides. The EV conjugates were characterized by nanoparticle tracking analysis (NTA) and western blot. An *in vitro* cell culture experiment was conducted in which murine bone marrow-derived dendritic cells (BMDCs) were treated with the EV-diABZI conjugates and relevant controls. Flow cytometry showed increased median fluorescence intensity (MFI) of the BMDC activation markers CD40, CD80, and CD86, as well as Class II Major Histocompatibility Complex (MHCII) following treatment with EV-diABZI conjugates. Using an enzyme-linked immunosorbent assay (ELISA) assay, we evaluated secretion of interferon-β and the TNF-α, inflammatory cytokines secreted following innate immune activation in BMDCs. We observed an increase in secretion of both cytokines following treatment with EV-diABZI conjugates. Flow cytometry was also used to show that BMDCs presented the Class I MHC-restricted ovalbumin (OVA) epitope SIINFEKL on their surface at a higher-level following treatment with EV-diABZI conjugates that had been engineered to package OVA into their EV lumens (EV-OVA-diABZI), meaning that the BMDCs most efficiently took up and presented OVA when packaged within an EV-diABZI conjugate. Another experiment was done *in vivo* using mice to determine if an antigen-specific T cell response could be generated following vaccination with EV-OVA-diABZI conjugates. Unfortunately, since diABZI is a potent inflammatory agent and the OVA to diABZI ratio was not yet optimized, there was not enough OVA in the EV-OVA-diABZI vaccine to mount a significant T cell response to vaccination as quantified using an H2-kb/SIINFEKL tetramer. Using Shuffle T7 *E. coli*, a camelid VHH nanobody targeting MHCII was produced. The nanobody was designed with a C-terminal sortag motif (LPETGGHHHHHHEPEA) to facilitate sortase-mediated conjugation of amine-containing small molecules to the C-terminus of the nanobody. We used this approach to label the nanobody with NH_2-PEG_3-N_3, followed by sulfo-Cy5-DBCO, to facilitate nanobody tracking. Flow cytometry showed that BMDC treatment with nMHCII-Cy5 demonstrated a profound increase in the MFI of the Cy5 channel compared to controls, suggesting that the nanobody was not only phagocytosed but also bound to the MHCII molecule presented on the BMDC surface. Future work will include ligating NH_2-PEG_3-BCN to nMHCII to facilitate nanobody conjugation to the surface of N_3-EVs to enhance EV targeting to antigen presenting cells (APCs). Additionally, diABZI loading will be optimized to facilitate appreciable OVA delivery without inducing diABZI-mediated toxicities using an EV-OVA-diABZI vaccine.

Determining the Effects of Geometric Confinement on the Fusion of Human Embryoid Bodies

Aman Prasanna, Z. Begum Yagci, Navya Mishra, Albert J. Keung

While human cerebral organoids(hCO) serve as effective 3-dimensional models for experimentation, they often lack reproducibility and precise spatial organization, limiting their utility in research and potential therapeutic applications. Prior literature has suggested that spatial and mechanical factors can influence aspects of hCO development. This study explores the impact of geometric confinement on the fusion and development of embryoid bodies (EBs), to potentially enhance control over the reproducibility and spatial organization of the hCOs they differentiate into. To explore the relationship between geometric confinement and EB fusion, we designed and 3D-printed custom macrowells with different shapes to confine and guide EB fusion. We created an agarose mold of these macrowells and seeded -EBs within them, then following a human cortical organoid protocol. EBs were allowed to develop and fuse within the macrowells for 24 days and notable differences were observed between different geometries and EB developmental stages. We have observed that some geometries worked better for EB fusion than the others. After developing for 30 days within the macrowells, the successfully fused EBs underwent fixation, cryosectioning, and immunostaining for markers such as MAP2, EMX1, and SOX2. We are currently in the process of analyzing the effects of geometric confinement on the organizational differences between these fused EBs using confocal microscopy. If the results suggest that geometric confinement does have an effect on the organizational differences within the fused EBs, researchers may be able to fuse organoids within specific shapes tailored to particular experimental needs or to more closely mimic specific organ structures. These advancements could significantly enhance the utility of human cerebral organoids as a model for studying human development, disease progression, and drug responses.

Effect of Orientation Angle for Needle-Free Jet Injection

Breanna Carruth, Clayton Penttila, Idera Lawal, Jeremy Marston

In this paper, we report on the delivery efficiency of needle-free jet injections using injectors with typical jet speed ≈140m/s, orifice diameter =157 μm, and volume =0.1 mL. The target substrates were either hydrogel tissue phantoms or porcine tissues combined with excised human skin. The novelty of this study is two-fold: First, we investigate the influence of injection angle relative to the skin surface, and second, we also study the influence of the jet path relative to the orientation of muscle fibers. While most commercial jet injectors employ a fitting that would render the device normal to the skin surface, recent studies have proposed oblique injections, which may be beneficial for intradermal or subcutaneous tissue injection. Furthermore, for deeper intramuscular injections, we propose that an angled jet path taking the muscle fiber orientation into account may result in a bolus or dispersion zone that is conducive to increased cellular uptake of the drug.

Plant Virus-Templated Colloidal Nanoparticles: A Sustainable Platform for Pharmaceutical and Biotechnological Applications

Jad Khansa, Michael T. Harris, Che-Yu Chou

Colloidal nanoparticles, known for their tunable properties, are widely used in pharmaceuticals, biotechnology, and other industries. A novel approach involves using plant viruses, such as Barley Stripe Mosaic Virus (BSMV) and Tobacco Mosaic Virus (TMV), as biotemplates for nanoparticle coatings. These plant viruses provide a highly uniform, stable structure, ideal for assembling and coating nanoparticles, enhancing their functionality, stability, and biocompatibility.

This research explores coating colloidal nanoparticles on the surfaces of BSMV and TMV to create virus-nanoparticle hybrids with improved catalytic, optical, and therapeutic properties. In pharmaceuticals, these coated virus particles hold promise for targeted drug delivery and bioimaging, as virus-like particles (VLPs) increase cellular uptake and drug efficacy. In biotechnology, the enhanced surface area and stability of these viral scaffolds make them ideal for biosensors and catalysts.

Using plant viruses as templates for nanoparticle coatings presents a sustainable and scalable alternative to synthetic carriers, making it an eco-friendly solution for various industrial applications. This study underscores the potential of virus-nanoparticle hybrids to drive innovation in medicine, diagnostics, and environmental technologies, paving the way for advanced material design and future biotechnological breakthroughs.

Influence of Particle z-Potential and Experimental Procedure on Protein Corona Formation and Multicomponent Aggregation

Nicole Piccininni, Aida López Ruiz, Mengyuan Xiao, Asmitha Sathya, Guangliang Liu, Noshin Siddiq, Hao Chen, Kathleen McEnnis

Attention has been placed on drug delivery systems (DDSs) with the hope that they can achieve targeted drug delivery all while reducing toxic side effects. Yet it must be noted that a variety of factors reduce DDSs from being optimal. Two such factors are protein corona formation and aggregation, which both result in a decreased circulation half-life of the drug delivery particles and eventually lead to uptake and filtration of the particles by the spleen and liver. For that reason, surface modifications should be made to particles in order to minimize their protein corona formation and aggregation. Reducing these factors will optimize the particles for DDSs. This research modified polystyrene particles utilizing multi-arm and linear polyethylene glycol (PEG) in order to study their influence on aggregation and protein corona formation. It was determined that multi-arm PEGs led to more aggregation than linear PEGs due to the changes in zeta potential from their unreacted end groups. This may lead to multi-arm PEGs having shorter circulation half-lives. Additionally, the formation and composition of protein coronas were studied after different washing techniques to underscore the criticality of studying protein corona formation in undiluted blood plasma.

Effects of Extracellular Matrix Composition on the Neutrophil Response

Tanvi Patil, Christopher Calo, Laurel Hind

Neutrophils are the body's first line of defense in combating infection. The neutrophil response is incredibly effective and powerful when adopted correctly; however, it is prone to dysfunction in diseases. The chemical factors that govern the neutrophil response are well studied, but our understanding of how the physical environment plays a role remains incomplete. Interestingly, diseases worsened by neutrophil dysfunction are often also accompanied by physical changes in the protein composition of the surrounding extracellular matrix (ECM). Specifically, collagen, the predominant protein in the ECM, is responsible for altering its mechanical and structural properties. The impact of the ECM composition is difficult to study in traditional experimental systems. Typical *in-vitro* models are composed of hard plastics and glass with little flexibility that fail to mimic physiological conditions (geometry of arteries and veins, 3-D interactions, etc). On the other hand, most *in-vivo* models do not allow for customization of the ECM composition. As a solution, we are using a biologically inspired microfluidic device that enables us to mimic human physiology while maintaining the tailoring we seek from *in-vitro* models. This project aims to better understand the effects of ECM composition on the neutrophil response through microfluidic devices.

Previous work on the project reveals a biphasic connection between collagen concentration and neutrophil extravasation1. Lysyl oxidase (LOX) was introduced into the system to improve the physiological relevance of this study since both LOX and collagen define the ECM in disease states such as fibrosis, cancer, and general aging. As an added benefit, LOX enables us to decouple the role of ECM mechanical and structural properties on the neutrophil response since the enzyme stiffens matrices without changing its pore size. First, we confirmed LOX treatment does not affect the pore structure within the gels. Using confocal reflectance microscopy (CRM), we found no changes in the major or minor axis pore length of collagen gels treated with LOX. Next, we observed a 1.2-1.4X increase in the stiffness of LOX-treated gels compared to the untreated condition at five days (Figure 1a-b). While this difference is not statistically significant, literature suggests it is biologically relevant on the neutrophil scale. We discovered LOX treatment does not change the extravasation behavior of neutrophils (Figure 1c). Analyzing confocal microscopy images of the neutrophil response to *Pseudomonas aeruginosa* (PAK), we saw no significant differences in extravasation trends between LOX treated and pure collagen gels. Together, the CRM, mechanical testing, and extravasation results suggest mechanical properties such as matrix stiffness have negligible effects on the neutrophil response. Further work on this study involves microscale mechanical testing of LOX gels and quantification of neutrophil antimicrobial activities to better predict the mechanism driving the neutrophil response in varying tissue environments.

1. Calo, C. J., **Patil, T.**, Palizzi, M., Wheeler, N., & Hind, L. E. (2024). Collagen concentration regulates neutrophil extravasation and migration in response to infection in an endothelium dependent manner. *Frontiers in immunology*, *15*, 1405364.

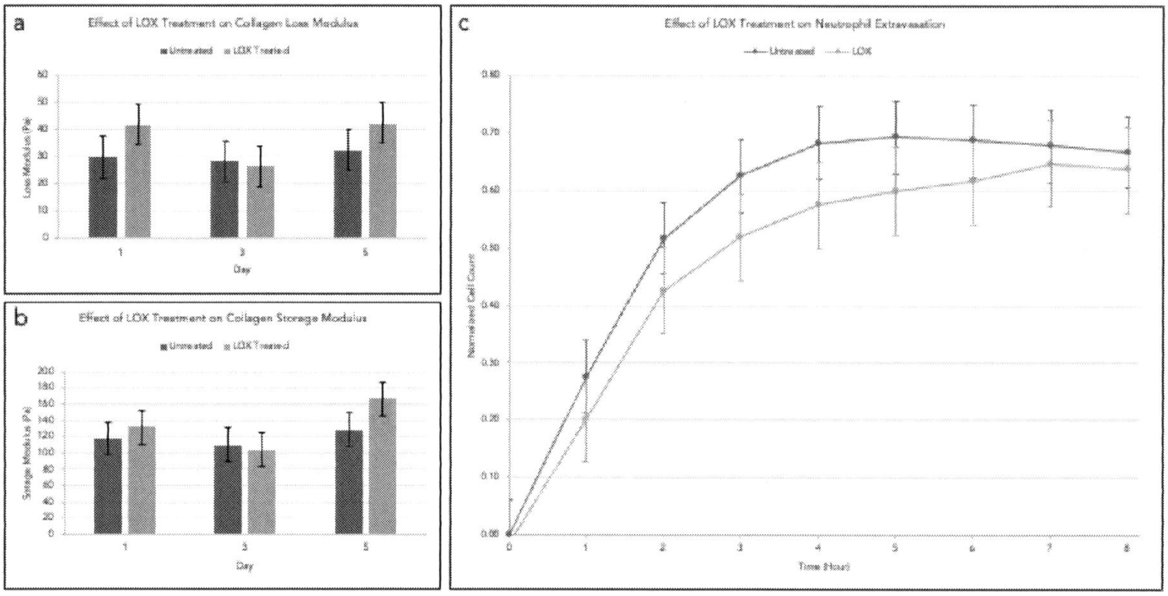

Figure 1: Effects of LOX Treatment a) LOX loss modulus does not become biologically significant from untreated until day 5. **b)** LOX storage modulus does not become biologically significant from untreated until day 5. **c)** LOX treatment has no significant effect on neutrophil extravasation.

Signal and Performance Objective Development in the Bone Morphogenetic Protein Pathway

Jayden Kam, Razeen Shaikh, Gregory Reeves

The Bone Morphogenetic Protein (BMP) pathway, also known as the Smad pathway, is an essential cellular signaling cascade regulating various cellular functions such as differentiation and regeneration across multiple organisms. Its structure is highly conserved across biological species. It consists of a receptor that, upon receiving an input signal, triggers a cascade of biochemical reactions ultimately resulting in the transport of the active complex into the nucleus where it binds to DNA and regulates the transcription of numerous downstream genes [1].

While the structure of this pathway is conserved across species, the system displays flexible behavior, measured as different performance objectives (POs), which are adjusted through various non-conserved parameters (NCPs). As in the engineering world, a perfectly optimal system does not exist and there is a system of trade-offs between competing POs. The system must then be fine-tuned to fit the biological context. The minimum set of POs are the rise time to steady state, noise attenuation ratio (NAR), and the linear sensitivity of the output [2].

Along with the active trimeric complex, (pSmad1)2/(Smad4), there are other complexes of the Smad proteins, including the (pSmad1)/(Smad4) dimer and the pSmad1 monomer, that are transported into the nucleus, forming a multi-layered system where the signal is processed at multiple levels. Unlike a more straightforward scheme, this complex system can give rise to various behaviors desirable for cellular signal processing and interpretation.

To better understand how the multi-tiered system affects signal transduction, we created a mathematical model based on the generic BMP pathway. We used computational tools to analyze the dynamics of the various complexes within. By comparing the POs of several proteins in the pathway, we observed that the performance of the trimer is worse than that of the monomer and dimer. However, the trimer can display a greater range of possible POs in the NCP screen as well as a significantly more varied sensitivity coefficient. This can point to the trimer output as having some unique behavior, not described by the minimum set of POs, desirable for signal processing in the BMP pathway.

[1]Madamanchi A, Mullins MC, Umulis DM. Diversity and robustness of bone morphogenetic protein pattern formation. *Development.* **7**, 148 (2021). Published 2021 Apr 6. https://doi.org/10.1242/dev.192344

[2]Shaikh, R., Larson, N.J., Kam, J. et al. Optimal performance objectives in the highly conserved bone morphogenetic protein signaling pathway. *npj Syst Biol Appl.* **10**, 103 (2024). https://doi.org/10.1038/s41540-024-00430-9

Investigating Interfacial Properties of Major Lipid Components Using a Brain Myelin Model for Multiple Sclerosis

Henry Bair, Estephanie Nottar Escobar, Mara Manolescu, Meredith Hartley, Prajnaparamita Dhar

The myelin sheath is a vital multilayer lipid membrane wrapped around the axons of neurons in the central nervous system (CNS) that protects nerves and facilitates rapid impulse transmission, essential for a proper functioning CNS. Myelin destruction, referred to as demyelination, is the cause of Multiple Sclerosis (MS), a neurodegenerative autoimmune disorder resulting in the interruption of nerve signals and consequent loss of motor and sensory functions. Healthy myelin is composed of an incredibly robust amphipathic lipid membrane, a quality that has been used to relate the properties of the membrane to disease development. Additionally, changes in lipid composition have recently been identified as a signature of demyelination; however, the specifics of these changes and their importance to MS are not yet understood. We hypothesize that changes in lipid composition impact the packing and rigidity of the film thus impacting its stability. To detail the relationship between the structure and function of myelin, this work attempts to develop a realistic biophysical model of the myelin sheath in the brain. Using a Langmuir-Pockels trough, we measured the interfacial properties (surface pressure (SP) and compressibility modulus) presented by monomolecular films of varying lipid compositions, subjected to dilatational interfacial stresses. We also conducted rheological studies by measuring elasticity and viscosity moduli of the films. Lipid mixtures used in these works were composed of major brain myelin lipid components L-a-phosphatidylcholine (Brain PC), L-a-phosphatidylethanolamine (Brain PE), and L-a-phosphatidylserine (Brain PS). While in both studies the differences in lipid concentrations had small effects on the mechanical and viscoelastic properties of the interfacial film, the small increase in SP and elasticity modulus observed corresponded with decreased Brain PC and PE components. This suggests slight changes in surface conformation due to differences in the types of head groups and lengths of alkyl chains present in the lipid mixtures.

Influence of Glia on Tau Propagation Using Patient-Derived Tau Aggregates in a Triculture System

Zander Schwartz, Lauren Drake, Nicole Marguerite, Andrew Kjar, Brian O'Grady, Ethan Lippmann

Tauopathies encompass a spectrum of neurodegenerative disorders marked by the aggregation of the microtubule-stabilizing protein tau. Prominent among these conditions are Alzheimer's Disease (AD) and frontotemporal dementias, characterized by neuronal cell death resulting from the propagation of aggregated tau. There are 26 types of aggregated tau, each of which with a different morphology that determines how it propagates throughout the brain. Understanding the molecular mechanism through which these aggregates spread is essential to designing effective treatment strategies. Previous research has focused on developing animal models to understand the spread of tauopathies. However, this research is often costly while making little progress in treatment strategies: what might be valid in a mouse often fails in humans. *In vitro* human stem cell-based models offer a viable alternative, being a lower-cost and more genetically representative alternative to mouse models that have led to so many failed clinical trials. While no model can adequately represent every function of the human brain, the field hopes that these human cell-based models can fill in gaps that other systems miss.

Many *in vitro* models in the literature utilize neurons, astrocytes, and microglia to provide an accurate representation of the brain cortex's natural physiology. Astrocytes and microglia are types of glia, which provide immunity and support to neurons. Glia are thought to play a role in how tauopathies spread throughout the brain, but evidence of glia-directed propagation is not well known. Understanding the role of glia is essential to designing effective treatment strategies, as glia are responsible for clearance of protein aggregates, and reactive glia are responsible for pruning of neuronal synapses.

We are building a 3D tau propagation model to answer whether glia play a role in the spread of pathological tau through the human brain. This system consists of AD patient brain-derived protein aggregates (with age-matched controls) which validates tau seeding activity of our cohort through a FRET-based tau biosensor. Within a neural triculture of induced pluripotent stem cell (iPSC)-derived neurons, microglia, and astrocytes, we treat glia with patient-derived tau and then re-plate them with untreated neurons to observe whether glia can propagate uptake aggregated tau into neuron bodies. To model a 3D matrix of cortex tissue, we then embed our triculture in Matrigel to observe tau propagation across a 3D network of cells.

Preliminary results have demonstrated that protein derived from patient brains contains seed-positive tau. When dosed in a triculture, this tau impacts the survival and morphology of neurons and glia. Future directions include testing in different hydrogel types, which provide different support matrices for cell growth. Previous work by our group demonstrates enhanced viability and maturation of cortical dopaminergic neurons in GelMA and GelMA-Cad compared to other hydrogels. We intend to explore the growth of our iPSC-derived neurons under similar conditions. We will also test other cell types involved in the neural immune environment, such as T cells and other macrophages.

Investigating Stereoisomers of the Nucleotide Analog Tenofovir As Inhibitors of Sars-Cov-2 RdRp and Exonuclease

Kamonchanok Saejeam, Jingyue Ju

Severe Acute Respiratory Syndrome Coronavirus 2 or SARS-CoV-2, the causative agent of COVID-19, is a member of the Nidovirales order of positive-strand RNA viruses. SARS-CoV-2 possesses two protein complexes responsible for viral genome replication and transcription: RNA-dependent RNA polymerase (RdRp) for replicating viral RNA and 3'-5' exonuclease (ExoN) for proofreading. While several nucleoside and nucleotide analogs have been developed as inhibitors of RdRp with antiviral activity, the proofreader ExoN removes these incorporated analogs, which reduces the drug efficacy. Discovery and optimization of nucleotide analogs which not only inhibit RNA replication by RdRp, but also resist degradation by ExoN, would improve the efficacy of antiviral drugs.

The FDA-approved HIV drug Tenofovir, an acyclic nucleotide lacking a ribose moiety, is a promising candidate to resist ExoN excision. However, another study proved that Tenofovir is inactive against SARS-CoV-2 due to its low incorporation activity with RdRp. Cryo-electron microscopy and X-ray crystallography revealed that the methyl group in (R)-Tenofovir pointing towards the active site of RdRp may cause a spatial clash with the amino group and lead to poor incorporation activity. Therefore, we propose that (S)-Tenofovir with methyl group pointing away from the active site can improve the incorporation activity with RdRp while maintaining the ExoN resistance ability.

To test this hypothesis, we synthesized (S)-Tenofovir diphosphate (DP) and performed RNA extension reactions with SARS-CoV-2 RdRp to assess its incorporation efficiency. Exonuclease reactions were conducted to evaluate the ExoN resistance of (S)-Tenofovir-DP, with MALDI-TOF MS used to quantify incorporation and excision resistance. Results supported our hypothesis, demonstrating that (S)-Tenofovir-DP improves RdRp incorporation and resists ExoN degradation. However, its incorporation efficiency remains lower than that of natural nucleotides.

While (S)-Tenofovir-DP shows promise as an RdRp inhibitor with ExoN resistance, further chemical modifications are necessary to enhance its antiviral efficacy. These findings highlight the potential of Tenofovir derivatives as antiviral agents against SARS-CoV-2 and other coronaviruses, given the conservation of RdRp and ExoN across coronaviruses.

Investigating Neural Cell Behavior in Collagen Scaffolds for Skeletal Muscle Tissue Engineering

Sarthak Pathak, Ryann Boudreau, Steven Caliari

Volumetric muscle loss (VML) is a traumatic injury to skeletal muscle that supersedes the body's natural wound healing capabilities. Current treatments for VML involve surgical operation using muscle from other parts of the body, which often results in poor functional recovery and donor site morbidity. Biomaterials research offers an alternative solution, aiming to guide native regeneration and improve functional recovery in VML injuries. Our research introduces a matrix for cell growth through aligned 3D collagen scaffolds. These scaffolds provide a structure similar to muscle extracellular matrix found in the body and show promising results in mimicking muscle architecture and encouraging new muscle fiber formation. However, the scaffolds alone do not fully restore regular muscle function, nor do they help regenerate other essential connections to the muscle such as the nervous system.

This research builds on our lab's scaffold technology by incorporating neural cells, which have been shown to promote muscle cell maturation and facilitate the formation of critical connections to the muscle. We differentiated rat neural stem cells into motor neurons and seeded them in our collagen scaffolds to test their viability. We looked at nuclei counts and axon growth of the neural cells which revealed the presence of developed neurons in the scaffold. By providing support for neuron viability in the 3D structure, we can begin to work toward a neural-muscle cell co-culture in the scaffold, enabling greater levels of myogenic regeneration as well as stronger neural connections than what is possible with the scaffold alone.

The Membrane Stability of Model Myelin Lipids Modulated By the Presence of Cholesterol

Mara Manolescu, Estephanie Nottar Escobar, Prajnaparamita Dhar, Meredith Hartley

The myelin sheath is a multilamellar lipid membrane wrapped around axons of neurons in the central nervous system (CNS) that facilitates rapid nerve impulse transmission, which is essential for the proper functioning of the CNS. Myelin destruction, called demyelination, causes Multiple Sclerosis (MS), a neurodegenerative disorder that results in the loss of sensory and motor functions due to nerve impulse leakage.

In healthy subjects, the amphipathic lipids that compose myelin form a remarkably stable membrane. The major lipids include the phospholipids L-alpha-phosphatidylcholine (Brain PC), L-alpha-phosphatidylethanolamine (Brain PE), and L-alpha-phosphatidylserine (Brain PS), and cholesterol. Lipid extracts of animal models of remyelination show changes in the lipidome; however, the details of how this lipidome alteration changes the mechanical properties of the myelin sheath is not currently understood. The role of lipids in MS progression has recently been identified; however, the details of these changes and the role of the lipidome change are not currently understood, limiting the treatments for MS. This work addresses the critical need to understand the relationship between structure and function of lipids in the myelin sheath. Using a Langmuir-Pockels trough, surface pressure (SP) versus area isotherms were measured to quantify mechanical properties of lipids extracted from the myelin sheath in the brains of mice with and without MS. Our results showed a significant learning of the compressibility modulus in the lipid films formed from the remyelinated myelin lipid extracts compared to the healthy lipid extracts. In addition, lipidomic studies showed significant differences in the phospholipid and cholesterol composition between the healthy and "sick" animals, establishing the role of the lipidome in MS. To better understand the role of the lipidome alteration on the membrane mechanical properties, further exploration was done using model membrane compositions with varying content of the three phospholipid classes and cholesterol. Careful analysis using several different compositions demonstrate that the largest difference in the monolayer's area compressibility modulus is seen for cholesterol compositions that reflect the native and diseased cholesterol content. Since the area compressibility modulus of monolayers is correlated to the membrane's rigidity, our results conclusively prove the importance of cholesterol in maintaining the structural integrity of the myelin sheath.

The Effect of Cellular Microenvironment on the Mechanical Behavior of Cancer Cells

Dorielis Rodriguez-Suarez, Meenal Datta

The impact of host environment on metastatic cancer cells have been extensively studied. However, the biophysical and mechanical components of the metastatic cascade is poorly understood. In this study, we investigated the effect of different environments on metastatic cancer cell through conditioned media analysis to understand the mechanical differences between brain and breast cancer cells under varying environmental conditions and how these differences influence their properties. We hypothesized that cancer cells would acquire mechanical properties resembling those of the environment they are exposed to. To test this, we cultured brain cancer cells in a normal breast cancer environment and breast cancer cells in a normal brain cell environment. The cells were subjected to conditioned media for twenty-four hours to assess the mechanical properties of their respective environments. Subsequently, a shear assay method was employed to apply stress to the cells and observe their response and resistance. Data analysis was performed using computer programs such as DaVis and MATLAB. The results were utilized to visualize cell deformation and characterize their mechanical properties. Notably, the cells exhibited responses to the different environments, indicating changes in their mechanical properties when placed in unfamiliar conditions. Surprisingly, even cells of the same type cultured in the same environment exhibited varying responses.

This study sheds light on the dynamic nature of cancer cells and their ability to adapt to different environments. By understanding how mechanical properties are influenced by the host environment, we gain valuable insights into cancer disease and its complex mechanisms. Further research in this area is crucial for advancing our understanding of metastasis and developing innovative diagnostic and therapeutic approaches.

Patchy, Locally Multivalent VEGF-Functionalization of Heterogeneous Liposomes May Safely and Effectively Activate Angiogenesis in Vitro

Nicole Korinetz, Semira Kehnemouyi, Xiaofan Ma, Saisumana Peddibhotla, Pooja Hariharan, Stavroula Sofou

Among the leading causes of death, ischemic heart disease results in heart attacks and failure due to insufficient oxygen supply as a result of reduced blood circulation. Current revascularization therapies that stimulate angiogenesis via vascular endothelial growth factors (VEGF) provide only short-term effects due to the short blood clearance half-lives of these agents. While various nanostructures such as hydrogels and peptide amphiphile nanofibers have been applied to prolong the circulation time of proteins, safety concerns limit their potential clinical applications. In addition, to ensure activation of the VEGF receptor, long association times with the VEGF ligand are needed. To address both challenges, we engineered liposomes with stably phase-separated lipid domains, which were functionalized with high local surface concentrations of the VEGF-targeting peptides. This locally multivalent surface architecture, we have previously demonstrated, results in long association times of the functionalized liposomes with a single receptor via continuous exchange of different ligands (from the same liposome) with the same receptor. Additionally, liposomes are an alternative well-studied drug delivery method that have a noticeably lower immunogenic response and longer clearance times. This study uses VEGF-peptides conjugated to lipids which preferentially partition in the stably phase-separated lipid domains of 100nm-in-diameter liposomes. This lipid vesicle topography forms a "sticky patch" on the liposomes that is aimed to effectively activate VEGFR and stimulate angiogenesis that is safe. To evaluate the angiogenic stimulation of these VEGF-functionalized-liposomes, human umbilical vein endothelial cells (HUVECs) were used as an in vitro 2D model. Future evaluation of the liposome safety and angiogenic stimulation may be conducted in 3D in vitro models such as spheroids containing HUVEC.

Modeling Stress Granule Dynamics: A Quantitative Biophysical Framework for Therapeutic Intervention in Neurodegenerative Diseases

Emily Wang, Andrej Košmrlj, Qiwei Yu

Stress granules (SGs) are RNA-protein condensates that assemble in response to cellular stress and play a critical role in regulating mRNA triage during translational arrest. Aberrant SG dynamics have been implicated in neurodegenerative diseases such as Alzheimer's and amyotrophic lateral sclerosis, where protein misfolding leads to chronic endoplasmic reticulum stress. Unlike normal SGs, which disassemble once stress is resolved, pathological SGs persist and disrupt normal cellular functions. Canonically, SGs form through a two-step liquid-liquid phase separation (LLPS) process: proteins like G3BP1, TIA1, and UBAPL2 bind mRNA transcripts to form core nucleating complexes, which interact with low-complexity domains of other RNA-binding proteins (RBPs) such as TDP43 and FUS to create a dynamic outer shell. As pathological SGs age under chronic stress, mutated TDP43 and FUS drive them toward a liquid-to-solid phase transition, forming amyloid fibrils. By gaining insights into and controlling pathological SG assembly dynamics, specifically targeting proteins involved in neurodegenerative diseases and disrupting their interaction landscape, we may prevent or reverse protein aggregation, potentially restoring normal cellular function and slowing disease progression. However, maturation and destabilization pathways of diseased-state SGs and their implications on material-function relations remain largely unexplored.

To address this knowledge gap, we develop a continuum model that interrogates the mechanisms governing SG assembly and disassembly. Our model describes the interaction between SG components with the Flory-Huggins theory and incorporates reaction kinetics that describes their binding and unbinding, including the case of pathological fibrillization. Together, the reaction-diffusion model allows us to simulate phase separation and transitions of SG systems and investigate how the process depends on key parameters, such as the chi parameter, which quantifies molecular binding affinities and interaction energies, along with the concentrations of participating proteins and ribonucleoprotein complexes. We aim to use this model to elucidate how system-level molecular interactions govern the formation and aging of SGs. For example, to probe disassembly mechanisms, we will introduce small molecules or chaperones into the multi-component SG systems to perturb phase behavior. By varying their concentrations and chi parameters, we aim to understand how these new species affect phase boundaries and disrupt protein aggregates. Our simulations will predict how the introduction of these molecules alters the binding interactions of other components, potentially restoring the system to a liquid-like phase that supports normal LLPS function. By developing quantitative models to uncover the fundamental mechanisms driving biological systems and disease, we can explore stress granules as a potential therapeutic target to prevent neurodegenerative disease pathology.

Constructing Astrocyte-Derived Extracellular Matrix to Explore Secondary Tissue Remodeling in Support of Breast Cancer Metastasis.

Dhimitraq Nikolla, Shelly Peyton

Abstract

Metastasis, the spread of cancer cells from the initial site to another region of the body, accounts for 90% of cancer-related mortalities1. Treatment and prevention of metastatic cancer can be more successful if the timing and location of the cancer spread can be predicted accurately. That would allow doctors to make more informed decisions for their patients by minimalizing the risk of unnecessary and difficult treatments. Our team is working on the idea that metastasis is not random and follows the seed and soil notion. Our focus is on astrocyte-secreted remodeling proteins that are initiated by triple-negative breast cancer cues. After a series of trials of injecting mice with conditioned media from breast cancer cells and staining for the presence of remodeling proteins such as tenascin-c, thrombospondin, and connective tissue growth factors, to determine changes in secondary sites, we are shifting to an in vitro approach. By constructing an astrocyte-derived extracellular matrix, our lab aims to accurately measure the response of the astrocytes in the presence of conditioned media or other factors by analyzing the proteins of their ECM. With the lack of biomarkers that distinguish brain metastasis, our research could be the only way to finally understand metastasis and conceivably save the lives of millions of patients who are suffering from the effects of cancer.

1. Barney, L. et al. The predictive link between matrix and metastasis. Current Opinion in Chemical Engineering 11, 85–93 (2016).

Development and Optimization of a Modular Microfluidic Device to Study the Effects of Deformation on Metastatic Breast Cancer

Julia Jacques, Addison Vondersaar, Braulio Ortega Quesada, Adam Melvin

Human cells experience a variety of biophysical forces *in vivo* during metastasis that can result in changes in both biological processes and cellular phenotype, driving pro-survival behavior. Prior studies have shown how matrix stiffness, fluid viscosity, and fluid shear stress enhance the proliferation of metastatic breast cancer. One additional biophysical force cells experience is deformation, which can occur during transendothelial migration during extravasation metastatic events and during transit through narrow capillaries while circulating throughout the body. While the biophysical changes in cells due to deformation have been well studied (e.g., changes in size, shape, recovery time), the phenotypic changes induced by deformation driving pro-survival behavior have not been fully elucidated. The goal of this work was to provide new insight into the biochemical changes, such as altered proliferation or protein phosphorylation during biophysical interrogation in single breast cancer cells. This was accomplished by developing a modular microfluidic device consisting of two unique devices to first mimic the biophysical forces that are endured by the cell during metastasis and to second interrogate single cells using immunostaining for changes in proliferation and protein phosphorylation. Single breast cancer cells were deformed through a venturi-style constriction channel device and collected in a microwell array device for single-cell immunostaining utilizing fluorescence microscopy. Two different breast cancer subtypes, estrogen receptor positive (ER+) and triple negative breast cancer (TNBC), were studied due to prior work that has shown how these different subtypes deform differently, which suggests that the biochemical pathways altered as a result of these biophysical forces could also be variable. Device optimization steps were performed to confirm that 2,051 MDA-MB-231 cells/min were deformed resulting in >98.6% of intact, viable cells after deformation. Single MCF-7 (a model ER+ cell line) deformed cells exhibited decreased levels of the proliferative marker Ki67 and no change in phosphorylated AKT (p-AKT) compared to the non-deformed control cells. Single MDA-MB-231 (a model TNBC cell line) deformed cells exhibited increased levels of p-AKT and decreased levels of Ki67. These preliminary findings suggest that proliferation is diminished immediately after deformation in both subtypes; however, the increase in p-AKT in the TNBC subtype suggests another biochemical pathway is activated due to deformation. These initial findings support that the biochemical changes under deformation appear subtype specific.

Testing Sabotage Attacks Affecting Printability during Bioprinting.

Barry Najarro-Blancas, Muhammad Ahsan, Nastassja Lewinski, Irfan Ahmed

As 3D bioprinting moves from the lab to the clinic, the impacts of cybersecurity vulnerabilities become more consequential. Now is the time to ensure bioprinters are secure by co-developing cybersecurity solutions as 3D bioprinters continue to evolve and become more established in biomanufacturing. In this interdisciplinary project, artificial tissues were printed using compromised and uncompromised G-codes to simulate sabotage attacks. These compromised G-codes affected different printing parameters, such as print speed, layer height, temperature control, nozzle switching, quick-pause-and-print (QPIP), ultraviolet attacks, and contamination attacks. The resulting printed tissues were then characterized for printability and structural integrity by measuring printability ratio (Pr) and parameter optimization index (POI) except for the ultraviolet (UV) attacks and the contamination attacks. Weights of the prints were taken for the QPIP experiments along with Pr and POI values, cell viability was measured for the UV attacks, and colony formation was monitored for the contamination attacks. Results showed visual and quantitative deformation in compromised prints, QPIP demonstrated a significant reduction in weight, no conclusive results were produced from the cell viability tests, and contamination was detected from the contamination attacks. These results illustrate the extent that sabotage attacks can have on bioprinted tissues and the need for bioprinting security frameworks. Moving forward, the consequences of sabotage attacks on cells will be further evaluated.

Using 3D Printing to Correlate Tonsil Size to Aerosol Deposition in the Upper-Airways

Joaquina Somma, Saurav Padhye, Catherine Fromen, Jenna W. Briddell

The anatomy of the upper respiratory tract plays a crucial role in determining local airflow dynamics; this is especially significant in pediatric airways, which continuously evolve and change throughout development (Kolewe et al. AAPS PHarmSciTech 2023). The tonsils, located in the upper airway region, are tissues prone to temporary swelling, which can limit airflow and change how inhaled aerosol medicines reach the lung. This work develops an experimental platform using 3D printing to evaluate the effect of tonsil size on the 1) local deposition of aerosolized therapeutics in the upper-airway and on the 2) amount of inhaled medicine that reaches the rest of the respiratory tract. Tonsil size is categorized by the Brodsky scale, which compares the distance between tonsils against that of the walls of the airway without tonsils. Each category in this scale is referred to as a grade, where a model with less than 25% difference in distance is referred to as grade 1, 26-50% is a grade 2, 51-75% is a grade 3, and 75+% is a grade 4. In this work, CT scans from pediatric patients (ages 3.5-4.7 years old) with various tonsil grades were used to create 3D printable models. Models were digitally split into three different sections according to their placement with respect to the tonsils: top (above tonsils), middle (tonsil area), bottom (below tonsils). After models were split, locking mechanisms were added to ensure easy reassembly, and parts were printed using a Carbon M1 printer. Nebulized doses of a fluorescent tracer (Rhodamine-B) were administered through the model under a fixed flow rate (~15 L/min); deposition in the different upper airway model areas and particle size leaving the model were evaluated. Results from these three patient models show different degrees of deposition as a function of tonsil grade, with generally more deposition in the top and middle segments of the upper-airway compared to the bottom section. On-going studies will continue these experimental evaluations with more models to build a correlation between local deposition and tonsil grade.

The Art of Guided 3D Vascular Channels Via Bioprinting

Linda Liu, Shi Fu, Huiting Luo

The formation of blood vessels are crucial for efficient wound healing as they supply the necessary nutrients and oxygen to the damaged area. At the heart of this process are two critical phenomena: vasculogenesis and angiogenesis. Vasculogenesis is the formation of new blood vessels from progenitor or stem cells, while angiogenesis is the growth of new blood vessels from pre-existing ones. In our previous study, a co-culture of fibroblasts and dental pulp stem cells (DPSCs) with collagen-fibrin ECM was verified as a pro-angiogenesis combination without additive growth factor. Through an integrated co-culture system into our engineered skin tissue, we produced pre-vascularized skin constructs that integrate and form anastomoses with host blood vessels in vivo.

This study aims to optimize the process of creating customized blood vessel patterns using 3D bioprinting. A key challenge in this process is ensuring the formation of a hollow structure with endothelial cells adhering to the walls of the channels. To address this, the thermal properties of gelatin and collagen were utilized to form a small hollow structure mimicking blood vessels. Gelatin was mixed with endothelial cell slurry to serve as a sacrificial material, after 3d printed, gelatin will be dispersed as collagen gelled at higher temperatures, allowing endothelial cells to bind to nearby collagen sites, facilitating vasculogenesis. Enhanced green fluorescent protein-labeled human umbilical vein endothelial cells (EGFP-HUVECs) were used to track the development of these vessels. By carefully controlling the temperature and combining collagen - fibrin as the ECM, the EGFP-HUVECs successfully adhered to the walls of the hollow channels, replicating the endothelial lining found in natural blood vessels. The angiogenesis was observed as early as day 3. By day 11, there is obvious interconnection among channels, suggesting anastomosis in the culture. The current data and results support the feasibility of regulating blood vessel patterns in vitro, which is crucial for accelerating the repair process at injury sites. These findings offer promising potential for future manufacturing efforts to integrate vasculature into bioengineered skin grafts.

Optimization of Bioprinting Protocols for Longitudinal Oxygen Imaging

Lindsey Powers

Bioprinting is a method of constructing biological tissue and organs from bioinks, which are formed from hydrogels containing human cells, nutrients, and other important biomolecules for cell health. Several printing technologies have been developed with the extrusion type being most common. An extrusion bioprinter was built in our laboratory that uses a standard 3 mL syringe and a computer-controlled xyz stage. Three main protocols were followed. The first protocol consisted of using an alginate based bioink into a support bath using the Freeform Reversible Embedding of Suspended Hydrogels (FRESH) method. The second protocol added fibrinogen to the bioink and thrombin to the FRESH. The third protocol used a bioink with alginate, calcium chloride, and collagen without the use of a support bath, but was instead printed into very thin layers that were later stacked after crosslinking. After testing, it was concluded that the best protocol used the alginate and collagen. After this protocol has been optimized, live cells and oxygen probes will be introduced in bioinks. Bioprinted cell-laden constructs will be imaged using a locally built electron paramagnetic resonance (EPR) imaging instrument. 4D images (3D space and time) will be analyzed to extract oxygen consumption rates by the cells. A bioprinting/imaging protocol will be developed and refined using obtained experimental results with the ultimate future goal to ensure better success with tissue and organ implantation.

Developing a High-Throughput Fluorometric Assay for Detecting Enzyme Activity on Ldpe

Mekhi Williams, Ross Klauer, Mark Blenner

Plastic waste processing continues to be a large ecological issue. The current processes for recycling plastic are insufficient to meet the current demand, meaning most of it ends up as waste that accumulates in the environment. Thus a novel and more efficient plastic recycling method must be developed to reduce this impact. Through the analysis of mealworm gut microbiomes, multiple enzymes have been discovered that are capable of oxidizing low-density polyethylene (LDPE) films. An enzyme capable of facilitating this reaction could be used in a complete biodegradation enzymatic pathway for processing LDPE waste, which could be used as a substitute for current plastic recycling processes. These LDPE-active enzymes belong to a large family with large characteristic differences. Knowing which characteristics facilitate LDPE activity would make it possible to engineer an optimized enzyme. Thus, a high-throughput fluorometric assay was developed to conduct a library search for LDPE-active enzymes. 4-Hydrazino-7-nitro-2,1,3-benzoxadiazole Hydrazine (NBD-H) was used as the fluorometric reagent due to its selective ability to detect aldehydes over other ketones. LDPE films were dosed with purified, previously discovered LDPE-active enzymes at optimal conditions and allowed to dry before reacting with NBD-H. A high fluorometric signal was only observed for samples dosed with an active enzyme. Enzymes that were deactivated with large concentrations of cofactor and samples that did not contain the LDPE-active enzymes had a similar baseline fluorometric signal less than that of the active enzyme samples. The success of this assay paves the way for a library search for other LDPE-active enzymes. The screening results will be used to narrow down candidates for future study and optimization.

Optimization and Characterization of 3D Printed Solid Oral Dosage Forms

Yariana N. Bengochea Bonilla

This research focuses on the dosage, characterization, and standardization of three-dimensionally (3D) printed pharmaceutical tablets, highlighting the innovative potential of additive manufacturing in modern drug delivery systems. The study delineates the methodologies, parameters, and formulations required for the bio-ink utilized in the 3D printing process. A robust calibration curve was developed and rigorously validated using UV-Visible Spectrophotometry, facilitating the precise quantification of caffeine uniformity within the printed tablets.

The findings indicate a significant improvement in the content uniformity of 3D-printed tablets compared to those produced via conventional pharmaceutical manufacturing techniques, underscoring the advantages of precision, accuracy, and customization inherent in 3D printing. This advancement suggests a transformative shift in pharmaceutical manufacturing, allowing for tailored dosages and personalized medicine that can better meet individual patient needs.

Future investigations should prioritize a comprehensive evaluation of the mechanical properties of 3D-printed tablets, specifically their hardness, disintegration time, and friability, to further ascertain their quality, efficacy, and reliability for clinical applications. Additionally, exploring the scalability of this innovative technology and its integration into existing pharmaceutical frameworks will be essential for practical implementation and broader adoption. This study lays the groundwork for subsequent research aimed at optimizing 3D printing processes for pharmaceuticals and enhancing their overall effectiveness.

Mechanosensory Constraints in Habituation Learning of Single-Celled Spirostomum Ambiguum

Radeen Dixon, Cedric Kamaleson, Saad Bhamla

Our research targets the fundamental mechanisms underlying non-associative learning in the unicellular organism *Spirostomum ambiguum*. *Spirostomum* is a ciliated protozoa that can contract its body length by 75% when exposed to mechanical stress using a 5 millisecond calcium driven ultrafast contraction. Previous studies have concluded that the protozoa demonstrates a decreasing probability of contraction when exposed to a repetitive vibratory stimulus. These prior examinations of the protozoans' habituation behavior have induced mechanical stimulation through an incident voltage or a mechanical agitation of the organisms. These methods heavily favor binary 'high' or 'low' force measures, which result in a repetitive stimulation that has been correlated to affect behaviors beyond the probability of contraction. Using such a system we employed a low-cost, Arduino-controlled solenoid-based mechanical stimulus, and successfully induced population-level habituation through a contractile-response decrement during the training regime. Building on this, we have designed a microfluidic device that applies precise radially continuous hydrodynamic stresses to probe relationships between stimulus force and habituation—an area previously overlooked by binary stimuli. In order to effectively examine the protozoans' contractions over time we utilize a high speed imaging system capable of capturing high resolution videos over the entire volume of the apparatus. In order to preserve the fidelity of measurements the time and position of each contraction in the apparatus volume will be measured using a machine learning-based neural network for image analysis, leveraging its ability to extract intricate patterns and features from complex datasets. We also utilize a similar computer vision software to analyze video data from a custom-designed apparatus, which induces low Reynolds number flow for real-time cell counting. The system accurately tracks the number of cells passing through the apparatus, ensuring a constant sample size. This cost-effective method, which can be manufactured and implemented for only pennies, provides a scalable solution for cell counting in resource-constrained settings. The hydrodynamic stimuli creates a rigorously controlled and easily adjusted stimulus force. This apparatus is designed to induce a controlled Jeffery-Hammel flow, creating a polar gradient of shear stress experienced by the organism. As the organism progresses radially, it encounters continuously increasing shear forces on its membrane, eventually reaching a threshold that triggers contraction when a critical shear level is attained. To investigate *S. ambiguum* habituation learning under these conditions, we will carefully recycle the cells through equivalent flow fields repeatedly and observe if the critical radial distance decreases with each repetition. Our device's continuous stimulus gradient allows us to explore how *S. ambiguum* integrates environmental information, questioning whether habituation occurs in a continuous or discrete manner. Ultimately, we'll compare learning rates under different mechanical stimuli to reveal any preference in habituation behavior, shedding light on cellular-level learning and its broader biological implications. A thorough understanding of *Spirostomum ambiguum*'s response to hydrodynamic stimuli presents the opportunity to learn more about the mechanism responsible for the organism's contraction. This would yield more insight into basic biophysical concepts and present a new understanding of how single cells induce extreme ultrafast shape changes. Such an understanding presents the opportunity to inspire new engineering applications and could help identify the specific molecules responsible for the protozoans' contractions. Once identified the molecules responsible for habituation can be transplanted into unhabituated organisms in order to better examine the habituation process at the cellular level, laying the groundwork for future studies.

Varying Simulated Body Fluid Formulation to Control Calcium Phosphate Mineralization of Electrospun Fibers

Joshua Kupfer, Ella Starr, Grace Schwarz, Julianne Holloway

The current gold standard in treating large bone defects are autografts which are limited by their availability, donor site morbidity, and other issues. Bone tissue engineering aims to combine cells, biomolecules, and biomaterials to regenerate functional bone tissue. Here, a method of mineralizing electrospun fibers to mimic the hydroxyapatite found in native bone was developed for use as a scaffold for bone repair. Two materials were chosen: polycaprolactone (PCL) and norbornene-functionalized cellulose acetate (nor-CA). PCL was selected because of its common usage in biomaterial scaffolds as a medical grade and FDA approved polymer. Nor-CA was selected because of the combination of handleability and presence of a norbornene group which enables thiol-ene click chemistry. PCL fibers were prepared for mineralization by an activation step using sodium hydroxide and calcium phosphate. Conversely, nor-CA was prepared for mineralization through the conjugation of a calcium-binding peptide to the scaffold. Three types of simulated body fluid (SBF) were chosen for PCL incubation: conventional (C-SBF), revised (R-SBF), and 4x revised (4x R-SBF). For nor-CA, the concentration of peptide on the scaffold was varied and all samples were incubated in C-SBF. Energy dispersive x-ray spectroscopy (EDX/S) was used to quantify the mineral compositions on the mineralized fibers. For PCL, calcium phosphate and sodium chloride-based mineralization was observed for all SBF formulations. The calcium phosphate ratio most like hydroxyapatite was found the 4x R-SBF group. The ratio of calcium to sodium was highest in the C-SBF group. SEM imaging showed distinct regions of calcium phosphate and sodium chloride mineralization for the 4x R-SBF group. In nor-CA, as peptide concentration increased, the amount of calcium deposition also increased. Future studies will explore the mineralization of nor-CA in other SBF formulations and the creation of mineral gradients on the surface of nor-CA through variable peptide concentrations along a singular scaffold.

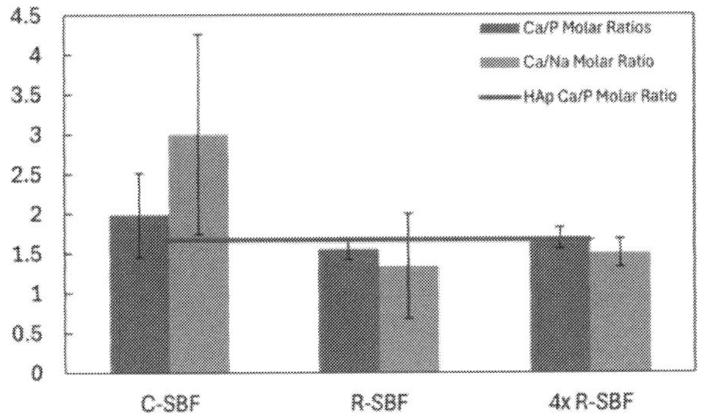

Figure 1: Calcium to phosphate molar ratios on PCL fibrous scaffolds following incubation for seven days in varying SBF formulations. The solid line indicates the molar ratio for native hydroxyapatite.

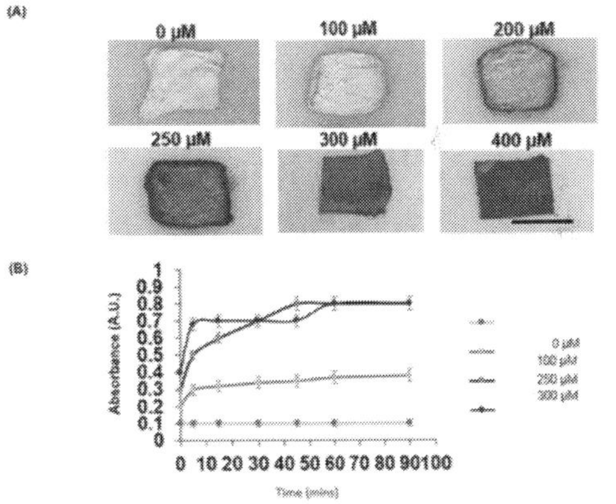

Figure 2: Alizarin Red S staining of nor-CA scaffolds fibrous scaffolds as a function of CBP concentrations and SBF incubation time: (a) macroscopic images of Alizarin Red S stain and (b) absorbance of Alizarin Red S stain.

Hydrophobins for Improved Plastics Binding and Bio-Deconstruction

Hanna White, Ross Klauer, Rachel Silvestri, Mark Blenner, Kevin Solomon

Microplastic pollution is rapidly accumulating in the world, leading to significant environmental and human health issues. For example, plastic waste makes up 80% of all marine pollution. A surgery study with 200 people also noted that 60% of patients had microplastics in their main artery and were 4.5 times more likely to experience a heart attack, a stroke or death in the 34 months after the surgery. Current waste management methods such as mechanical or chemical recycling are inefficient and lack product specificity. Microbial degradation has been leveraged as a way to degrade plastics but falls short due to the poor access microbes have on the plastic. In this work, we utilize a strain of *Aspergillus fumigatus* to bind to microplastics and pull them out of the environment with nearly 100% recovery. We identified hydrophobin proteins as the predominant biomolecules driving efficient plastic binding flocculation. 7 different hydrophobin genes, RodA-RodF, were present in the yeast genome. RodA is reported as the most expressed hydrophobin responsible for cell wall surface hydrophobicity. By knocking out each gene individually then testing the plastic recovery, the RodA knockout resulted in significantly less plastic recovery. Therefore RodA is the main hydrophobin gene responsible for plastics binding and it can be used to subsequently improve plastic deconstruction. We are developing a proof-of-concept hydrophobin and plastic-oxidizing enzyme surface display system for improved plastic oxidation in *Yarrowia lipolytica*. This system can then be tied to a full bio-deconstruction cascade, allowing for efficient and sustainable plastics deconstruction.

A Cross-Species Study of X-Y Chromosome Pairing in *Drosophila* Male Meiosis

Kiran Kuriakose, Alina Salagean, Yukiko Yamashita

Meiosis is a specialized cell division responsible for the production of sperm and eggs. The goal of meiosis is to reduce the chromosome number by half, a crucial step in ensuring genetic balance across generations. When this reductional division fails to segregate the chromosomes properly, aneuploidies may arise which can lead to a variety of genetic diseases. In order to ensure accurate chromosome segregation, homologous chromosomes must pair at the onset of meiosis. Canonically, chromosome pairing is stabilized by the synaptonemal complex (SC) and the repair of double strand DNA breaks through recombination. However, *Drosophila* males undergo a non-canonical meiosis that proceeds without an SC or recombination, presenting a unique model to study chromosome pairing mechanisms. Our research explores how meiotic cells in *Drosophila* males achieve stable pairing by focusing on the X and Y chromosomes, which pair by using a 240bp repeat in the intergenic spacer (IGS) of the rDNA locus–the only known pairing site for any chromosomes in *D. melanogaster*. Previous research has shown that IGS, in combination with the pairing proteins SNM (stromalin in meiosis) and MNM (modifier of mdg4), is both necessary and sufficient for X-Y chromosome pairing. Interestingly, some of the *D. melanogaster* sister species (*D. simulans*, *D. sechellia*, and *D. mauritiana*) have lost the rDNA locus from their Y chromosome, yet retain IGS, implying that IGS may also be utilized for X-Y pairing in sister species. Using DNA fluorescence in situ hybridization (FISH) and confocal microscopy, we visualized chromosome pairing dynamics in *D. melanogaster*, its sister species, and hybrids between these species. Whereas wild type flies (parental species) were highly competent in X-Y pairing as expected, hybrid species exhibited a considerable increase in X-Y nondisjunction. These data suggest that X-Y pairing mechanisms between species are not compatible, despite the fact that their IGS sequences are highly conserved. Interestingly, a comparative analysis of SNM protein structure across sister species revealed a truncated form of SNM in *D. mauritiana*, raising a possibility that SNM, which is known to mediate X-Y pairing, may be incompatible between *D. mauritiana* and other sister species. Our finding may hint at how IGS and its binding proteins are utilized to facilitate X-Y pairing in non-canonical pairing mechanisms. Understanding pairing mechanisms may help future therapeutic use of artificial chromosomes that can be efficiently transmitted through the germline.

Integrated Stress Response Alters Energy Metabolism in Brain Cells

Harin Sim, Samantha O'Keeffe, Junyoung Park

Energy metabolism is critical for the survival and function of human cells and organs. When cells encounter stress, such as infection, nutrient deficiency, or toxins, they alter their metabolism to adapt to the perturbation. The integrated stress response (ISR) is an intracellular pathway that rewires metabolism to help cells adapt to these disruptions. Dysregulation of the ISR in the brain is associated with neurodegenerative diseases, including Alzheimer's, making it a potential therapeutic target. Despite its significance, the specific metabolic changes induced by ISR activation and its mechanisms remain poorly understood. Here, we aimed to elucidate these metabolic changes by using optogenetic H4 neuroglioma cells. Using light, optogenetics allows precise triggering of the ISR. By employing stable isotope tracers and liquid chromatography-mass spectrometry, we observed rapid metabolic reprogramming within the first hour following ISR activation.

Our findings revealed altered glucose utilization, shunting it away from glycolysis towards the pentose phosphate pathway (PPP). Glycolysis produces the cellular energy currency ATP, while the PPP generates reducing power. Downstream of glycolysis, cells utilized anaplerosis, which produces carbon backbones for biomass instead of ATP generation. A common phenomenon observed was the decrease in ATP production, matching the reduced energy demand for protein synthesis from ISR activation. These rapid changes suggest the presence of regulatory mechanisms beyond the known transcriptional and translational effects of ISR, which typically cause changes over longer timescales. This knowledge provides new insights into fast-acting metabolic adaptations driven by ISR and underscores its potential as a therapeutic target for treating neurodegenerative diseases.

Engineering a Mesophilic Prokaryotic Argonaute for Transcriptional Regulation

Ashley Kalan, Brett Graver, Kevin V. Solomon

Prokaryotic Argonautes (pAgos) are programmable endonucleases that use a small oligonucleotide guide to target a nucleotide substrate. Currently, pAgos have been deployed in diagnostics and have the potential for *in vivo* gene editing. Active pAgos can nick a nucleotide substrate with a catalytic tetrad (residues DEDX, where X is D, H, or N) which is responsible for its endonucleolytic cleavage. However, one potential biotechnology application of catalytically inactivated pAgos is transcriptional regulation. Transcriptional regulation is central to biomolecular engineering, enabling diverse applications such as sustainable energy production, synthesis of medicinal compounds, treatment of disease, and development of climate resilient crops. A common strategy uses the catalytically dead version of the CRISPR Cas9 protein usually fused with a transcriptional activator or repressor. This Cas fusion can be programmed to regulate a selected gene of interest; however, the effectiveness of the Cas9 system is restricted by a protospacer adjacent motif (PAM) site requirement for target recognition. Because PAM sites are not evenly distributed across genomes and are less abundant in GC-biased organisms, it is difficult to optimally target genes in certain cases. pAgos, on the other hand, are a promising alternative to the CRISPR Cas system because they have no known recognition restrictions *in vivo,* allowing for optimal gene targeting. Here, my work investigates the application of the pAgo from *Clostridium butyricum* (dCbAgo) for transcriptional regulation. To better understand the properties of each residue in the catalytic tetrad, I made point mutations in CbAgo to determine which residues alter protein activity. To further investigate the degenerate 'X' position in the tetrad, I made mutations with other metal coordinating residues to determine changes in activity. Assessing these CbAgo mutants, I found that non-native residues at the 'X' positions results in different activity levels, yet not abolished, in comparison to the wild type aspartic acid. Additionally, I determined single and double alanine point mutations within the catalytic tetrad which inactivate CbAgo. These catalytically dead mutant pAgos can be further characterized *in vitro* to determine binding strength and targeting restrictions. To determine the transcriptional interference capabilities of the catalytically dead version of CbAgo (D541A, D611A), I built a fluorescent reporter assay which utilizes a green fluorescence protein fused with a C-Terminal LAA degradation tag. This system allows the study of the current transcription rates of the protein by this active protein degradation turnover. Through this, I demonstrated dCbAgo mediated interference when targeted to the promoter region of the fluorescent protein gene, achieving a 30% knockdown. Future work will determine optimal targeting regions as well as targeting restrictions with the potential for implementing transcription factor fusions to dCbAgo. Continuing to engineer pAgos as a tool for transcriptional regulation will provide the synthetic biology field with a more flexible, alternative biotechnology with applications across biomolecular engineering.

Metabolic Kinetics of T Cell Activation with Soluble Antibodies and Antibody-Coated Magnetic Beads

Janet Johnson, Meghana Kalluri, Melissa Skala

Human T cell activation is a crucial area of study as it provides insights into the adaptive immune response and T cell cytotoxicity. T cell activation requires rapid changes in T cell metabolism to support proliferation and effector functions. This study investigates differences in T cell metabolism kinetics between two activation methods: antibodies (CD3/CD28) that are soluble vs. coated onto magnetic beads (Dynabeads). The Dynabeads are hypothesized to more closely mimic the mechanical features of antigen-presenting cells. Studies were performed in human plasma like medium (HPLM), which is designed to mimic the metabolic profile of human plasma. First, we collected a sample of healthy human blood and isolated the CD3+ T cell population. T cell metabolism was imaged using two-photon fluorescence lifetime imaging microscopy (FLIM) of an autofluorescent metabolic co-enzyme before activation and over three days post-activation. Specifically, FLIM measured the fluorescence lifetime of the reduced form of nicotinamide adenine dinucleotide (phosphate), or NAD(P)H, a crucial metabolic coenzyme in cells. NAD(P)H exists in two conformations: a free form and a protein-bound form, the bound conformation of NAD(P)H has a longer lifetime than the free conformation. Therefore, NAD(P)H FLIM measures T cell metabolism related to NAD(P)H protein-binding activity. Since NAD(P)H is autofluorescent, this technique is a label-free method to probe cellular behavior. Following baseline imaging, the T cells were divided into two groups: one activated with the soluble antibodies and the other activated with the Dynabeads. Single cell segmentation was performed using Cellpose to generate masks, followed by automatic manual corrections using Napari. SPCImage was used to extract the lifetime decay parameters including the short and long lifetime and their relative weights. The mean lifetime was also calculated, which is the weighted average of the short and long lifetime. The mean lifetime of NAD(P)H decreased progressively each day following activation, indicating decreases in the proportion of bound NAD(P)H with respect to free NAD(P)H, for both soluble antibody and Dynabead groups. This suggests that activation in HPLM with Dynabeads and soluble antibodies produce similar changes in T cell metabolism over three days. Future research should explore whether these trends persist with different cell culture mediums to determine if Dynabeads and soluble antibody activation results in similar metabolic changes across other conditions.

Harnessing the Biosynthetic Machinery of Natural Products for Biomanufacturing

Margaret Guilarte-Silva, Yash Chainani, Keith Tyo, Linda J. Broadbelt, Aindrila Mukhopadhyay, Jay Keasling, Hector Garcia Martin, Tyler Backman

Biomanufacturing offers a sustainable alternative to petrochemical processes for producing commodity chemicals, such as biofuels, dyes, and monomers as it relies on renewable biomass-derived feedstocks rather than rapidly depleting and pollutive fossil fuels. To date, most engineered biosynthetic pathways for manufacturing bulk chemicals have relied exclusively on monofunctional enzymes. However, nature has also provided us with another type of underutilized enzymes known as polyketide synthases (PKSs). Type I polyketide synthases, in particular, are modular and versatile enzymes capable of producing an extensive array of natural products such as erythromycin, an antibiotic used to treat bacterial infections. PKSs are unique in their ability to generate compounds with elongated carbon chains and uncommon functional groups, such as terminal alkenes or alkynes. By leveraging their biosynthetic capabilities alongside monofunctional enzymes and synthetic chemistry, new pathways for synthesizing valuable chemicals can be unlocked.

Despite existing tools for retrobiosynthesis and *de novo* metabolic pathway design, no current platform effectively integrates PKSs with both biological and chemical systems. To address this limitation, we developed TridentSynth (tridentsynth.ese.lbl.gov), an open-source web tool that combines the chimeric PKS design software RetroTide with the pathway design tool DORAnet, providing a unified platform for constructing complex biosynthetic pathways. We deployed TridentSynth on the web by utilizing a stack of modern web development technologies for both ease of use and broad compatibility. Streamlit's simplicity and user-friendly interface allows users to easily interact with the application and input their own data into the program. On the backend, Django manages jobs based on users requests, chosen for its robustness and scalability. To facilitate communication between the front and back ends, Celery serves as an asynchronous task queue and Redis as a message broker. The entire application is containerized using Docker, providing a consistent development environment and ensuring compatibility across different systems. This microservice-based architecture enabled the creation of a scalable and easily deployable application. By streamlining the process of pathway exploration and analysis, TridentSynth facilitates the discovery of novel chemical routes and supports more efficient and innovative biomanufacturing practices.

The Unique Metabolite Landscape of the Aged Microenvironment Dictates Melanoma Migratory and Invasive Capacity

Payal Patel, Gretchen Alicea, Meihan Wei, Vito Rebecca, Denis Wirtz

Metastatic burden and organ failure are the root causes of mortality in patients with advanced melanoma; however, there is little known about the molecular drivers that enable early organ colonization and the formation of overt metastases. Patients over the age of 55 have a higher incidence of metastasis compared to younger patients independent of other prognostic factors. Metabolic dysregulation has been identified as a hallmark of aging; however, not much is known about how host cell metabolic reprogramming during aging contributes to melanoma motility, invasiveness, and metastatic capacity. Following metabolomics of host cells and melanoma cells in the context of aging and secondary validation, here we report that changes in bioavailable nutrients to melanoma cells occur in the aged tumor microenvironment (TME), which confer elevated melanoma cell motility and invasiveness. We identified novel transcriptomic and differential proteome signatures that expound the migratory and invasive differences observed in cells under distinct age-dependent metabolic environments. This work has important implications for our understanding of the metabolic determinants of melanoma migration and metastatic colonization during aging.

Exploring Metabolic Adaptations in Treponema Pallidum through Genome-Scale Metabolic Modeling

Edward Stone, Nabia Shahreen, Niaz Chowdhury, Elle Knobbe, Rajib Saha

Treponema pallidum, the causative agent of syphilis, exhibits unique metabolic characteristics, including the absence of a fully functional TCA cycle, which raises questions about how it sustains its high motility with limited ATP production. Moreover, *T. pallidum* is difficult to culture in vitro This project aims to reconstruct a genome-scale metabolic model (GEM) of T. pallidum to gain a better understanding of the bacteria, specifically addressing how it achieves high motility with limited ATP generation and whether its uptakes lactate as a carbon source to compensate for its lack of TCA cycle. To build this model, we first conducted a literature review to elucidate the bacterium's metabolism. Using Kbase, we drafted an initial model refined by adding and removing reactions to accurately represent *T. pallidum's* metabolic traits. After developing the model, we validated it using Memote, achieving a score of 92%. To further validate the model, we predicted the essential genes for *T. pallidum*. Since no information on *T. pallidum's* essential genes is available, we used BLAST to compare the predicted genes with those of *E. coli* and *Neisseria gonorrhoeae*. BLAST revealed that 41 out of the 42 predicted genes had orthologs in E. coli, and 37 had orthologs in *N. gonorrhea*. Of the *E. coli* orthologs, 26 were essential for growth, while 28 orthologs were essential in *N. gonorrhea*, further demonstrating the robustness of our model. Additionally, we used transport DB to verify the transporter proteins and tested the model's ability to metabolize various carbon sources such as glucose, pyruvate, and mannose. The resulting GEM provides insights into *T. pallidum's* metabolism and offers a foundation to explore key questions about the bacteria's high motility and its potential to uptake lactate as a carbon source. In the future, it can serve as a powerful tool for studying the bacterium's metabolic adaptations and responses to various environmental conditions.

Effects of Embryonic Hyperglycemia on Early Stage Embryo Development

Ariana Adams, Ananya Lanka

Pre-gestational diabetes mellitus (PGDM) refers to diabetes diagnoses received before pregnancy [1]. As of 2010, there are around 300,000 reproductive age women living in the United States who could have diabetes but have not been diagnosed [1]. In cases of pregnancies with PGDM, higher levels of glucose (hyperglycemia) in a pregnant person causes embryonic hyperglycemia [1]. Maternal hyperglycemia increases the risk of developmental defects, including cardiac malformations, in the baby [2].

To further explore the effects of embryonic hyperglycemia on the development of the embryo, avian models were exposed to high levels of glucose before incubation. A 500mM D-glucose solution was injected into the yolk at a volume of 300uL (D300) or 600uL (D600) before the egg was incubated. Control groups were similarly dosed: 600µL of a 500mM L-Glucose solution (L600) served as the osmotic control, and 600µL of saline (saline) was used as a vehicle. Untreated (Control) eggs were pierced with a needle but were not injected with any solution to account for the effects of the injection itself.

Eggs were opened on day 3, then the embryonic heart was imaged and velocity of blood through the primitive heart outflow tract of the heart was measured with optical coherence tomography (OCT). The viability was tracked daily through day 9.

At Day 3, D600 embryos died at a higher rate than control embryos (Table 1), suggesting that glucose additionally impacts their development early on. By day 7 of incubation, physical defects were observed in D600 embryos compared to control embryos. Defects include not developing organs or being under-developed.

Viable embryos were imaged and staged at day 3 and their development observed (Figure 1, Part A). Using Hamburger-Hamilton (HH) Stages, a common method for staging chicken development, as a guideline, a trend was found with D600s embryos generally showing underdevelopment (Figure 1, Part A) [4]. Both control embryos are at expected development stages. D600 embryo #1 was slightly underdeveloped and D600 embryo #2 was too deformed to determine how developed it was (Figure 1, Part A). Both L600 embryos showed to be maldeveloped (Figure 1, Part A) but only those two day 7 L600 embryos were photographed so it is difficult to determine a trend in L600 from this.

From the OCT scans, the L600 scans both show additional noise and malformed curves (Figure 1, Part A). D600 embryo #1 did show a clear outflow tract (Figure 1, Part A) and D600 embryo #2 showed extreme malformations and no clear outflow tract. Both controls show the expected smooth curve of primitive heart outflow tract, which is the expected development (Figure 1, Part A).

Both control embryos showed normal outflow velocity graphs when normalized (Figure 1, Part B) and outflow velocities (Table 2) within the same range as those found by previous studies [3]. The L600 embryos both showed abnormal graphs, with the graph trends being inconsistent (Figure 1, Part B), and abnormal velocities (Table 2). For the D600 embryos, both graphs showed to be fairly normal (Figure 1, Part B) and the outflow velocity of embryo #1 was within the normal range, but embryo #2 had a higher, abnormal velocity (Table 2).

The effects of embryonic hyperglycemia in avian models become apparent by day 3 of incubation. Future exploration of this topic could include running trials using D300 instead of D600 to determine if the same defects would still occur. If the results are similar, the D300 would offer a higher viability method to study the defects caused by embryonic hyperglycemia. Regardless of how the topic is explored in the future, prioritizing the understanding of PGDM pregnancy risks could ultimately aid both expectant parents and their offspring.

Figures:

Day 3 Embryo Scan Chart

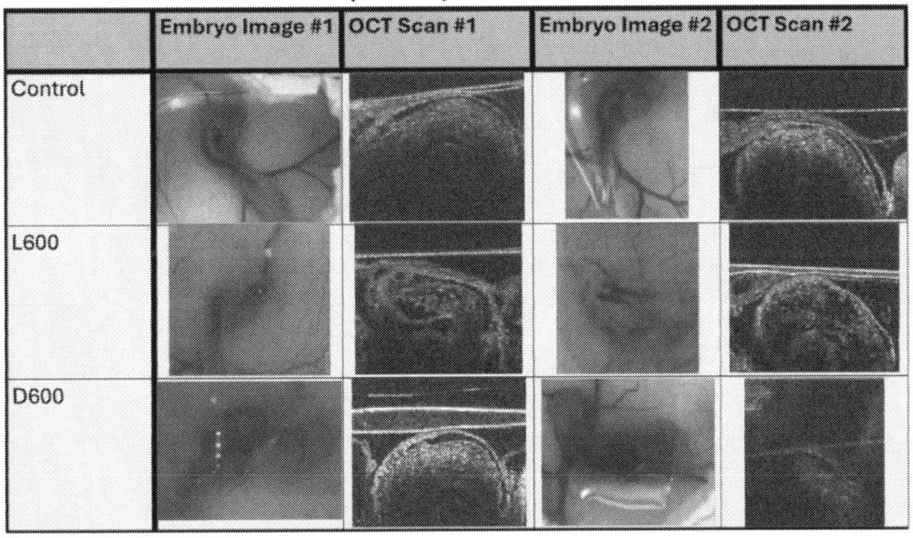

Figure 1, Part A: Aphotograph and OCT scan are each shown for six different day 3 embryos: two of each treatment type. These embryos best represent their treatment's trends. For Control embryo #1 and #2, both photographs show the expected level of curve in the developmenting embryo. This is contrasted with the D600 embryos, #1 of which shows only a slightly mis-curved embryo while #2 has a very abnormally curved form. Similarly, the OCT scans of Control #1 and #2 show the healthy, half-circle outflow tract while D600 #2 OCT scan shows no developed curve, indicating severe developmental defects. The L600 photographs both show abnormal curves in the embryos and no smoothly curved outflow tract in the OCT scans.

Day 3 Embryo Velocity Chart

	Velocity Graph #1	Velocity Graph #2

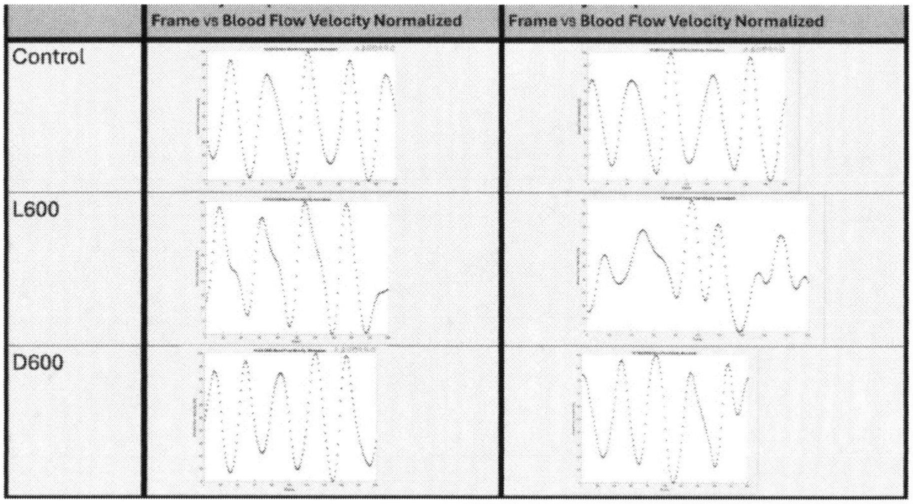

Figure 1, Part B: The graphed normalized curves of the velocity of blood through the primitive heart outflow tract are shown for each of the embryos presented in Figure 1, Part A. The Control embryos show smooth, consistent heartbeats while D600 embryo #2 shows a slight disfunction towards the far side of its graph with a lower velocity. Both L600 embryo graphs show inconsistent and asymmetrical curves.

Tables:

Embryo Viability

Treatment #of	Control	Saline	L600	D600	D300
Embryos #of	28	17	22	42	11
Embryos Alive at Day 3	16	6	7	15	5
Survival Percentage at					

Day 3	57.1	35.3	31.8	35.7	45.5
#of Embryos Alive at Day 4	9	6	7	11	0
Survival Percentage at Day 4	32.1	35.3	31.8	26.2	0.0
#of Embryos Alive at Day 7	7	1	4	2	0
Survival Percentage at Day 7	25.0	5.9	18.2	4.8	0.0

Table 1: The percentages of embryos alive at day 3, 4, and 7 of incubations based on the treatment type. The controls show the highest viability at day 3, with D300 being the next highest while D600 was the second lowest. A drop in viability can be seen between day 3 and day 4 for the Control and D600 embryos, with another viability drop being observable between day 4 and day 7 for all treatment types.

Outflow Velocities

	Outflow Velocity (mm/s) Embryo #1	Outflow Velocity (mm/s) Embryo #2
Control	34.7 71.6 34.3	33.2 29.7 45.3
L600		
D600		

Table 2: The velocity of blood through the primitive heart outflow tract of the six different day 3 embryos shown in Figure 1, two of each treatment type. The Control embryos show the expected outflow velocity which can be used as a comparison for the other embryos. For the D600 embryos, embryo #1 is within the normal range, but embryo #2 had a much higher, abnormal velocity. The outflow velocity for L600 embryo #1 was extremely high while embryo #2 was only slightly low.

References:

1. Razzaghi H, Marcinkevage J, Peterson C. Prevalence of undiagnosed diabetes among non-pregnant women of reproductive age in the United States, 1999-2010. Prim Care Diabetes 2015;9(1):71-3. DOI: 10.1016/j.pcd.2013.10.004.
2. Priest JR, Yang W, Reaven G, Knowles JW, Shaw GM. Maternal Midpregnancy Glucose Levels and Risk of Congenital Heart Disease in Offspring. JAMA Pediatr 2015;169(12):1112-6. DOI: 10.1001/jamapediatrics.2015.2831.
3. Lansford R, Rugonyi S. Follow Me! A Tale of Avian Heart Development with Comparisons to Mammal Heart Development. J Cardiovasc Dev Dis. 2020 Mar 7;7(1):8. doi: 10.3390/jcdd7010008. PMID: 32156044; PMCID: PMC7151090.
4. Hamburger, V. and Hamilton, H.L. (1951), A series of normal stages in the development of the chick embryo. J. Morphol., 88: 49-92. https://doi.org/10.1002/jmor.1050880104

Quantitative Understanding of Pmad at Different Stages of Cell Division in Germline Stem Cells

Matthew Ball

Stem cell decision making is not fully understood, which makes the use of stem cell therapy inconsistent. More research in understanding the pathways by which stem cells differentiate can provide better control on determining the outcome. A key part of this understanding is the Bone Morphogenetic Protein (BMP) Pathway, which is a major biological signaling pathway that plays a part in stem cell decision making. In particular, the BMP pathway occurs in the ovaries of fruit flies, where germline stem cells divide into differentiated cells called the cystoblast. Dpp, the main ligand in the fruit fly BMP pathway, is secreted by the cap cells, which are adjacent to the germline stem cell. Secreted Dpp then forms an extracellular diffusion gradient that extends to the end of the germline stem cell. Dpp activates receptors on the surface of the stem cells resulting in the phosphorylation of the protein Mad. Phospho-Mad (pMad) binds with protein Medea (Med) and acts as a transcription factor in the nucleus of the cell, downregulating bam, the factor that causes differentiation in germline stem cells. Despite its central role in regulating differentiation, there is no quantitative measure of pMad in relation to stem cell differentiation/self-renewal, especially at different stages of cell division. Having a quantitative understanding of this system would improve the efficiency and reliability in stem cell therapy. The goal of this project is to document how pMad changes at different stages of cell division. This goal will be achieved by performing ovary dissections and immunostaining to preserve the spatial and concentration of proteins of interest. The results of this project would be a collection of pMad concentrations in the germline stem cell and cystoblast at different stages of division. The results can be analyzed to begin to understand the role pMad has in making the stem cell decision making distinct to either differentiation or self-renewal.

Solution Blow Spinning of Complex Coacervates for Wound Healing Applications

Allison Courtenay

Surgical site infections (or SSIs) are typically caused by bacteria such as Staphylococcus, Streptococcus, and Pseudomonas. Most SSIs—depending on severity—can be treated with antibiotics, however delay in treatment of SSIs can lead to sepsis and septic shock which can be deadly. According to the CDC, the number of SSI events associated with in-patient surgery has continued to increase from 2015 to now, despite advances in infection prevention practices, suggesting a need for solution in diagnosing the infection. This project demonstrates proof of concept for a rapid, visual indicator of bacterial infection to detect SSIs faster and in-situ. We use a complex coacervation—liquid-liquid phase separation phenomenon—system of a poly(diallyldimetheylammonium chloride) (PDADMAC) and poly(stryrene sulfonate) (PSS) in potassium bromide solution. Via a technique called solution blow spinning with a commercial airbrush, we can spray the complex coacervate onto a substrate mimicking a bandage sprayed on a wound in-situ. Variables including distance to the substrate, air pressure, salt concentration, and static versus sweeping airbrush motion impact the morphology of resulting droplets. We also test the encapsulation of small molecule cargo like fluorescent dyes and pH indicators in the system, for the proposed application of visual indication of pH flux due to bacteria presence. Future research using carboxymethyl cellulose (CMC) as a polyanion to complex with PDADMAC can improve the biocompatibility of the system. Our research on solution blow spraying complex coacervates with pH indicator encapsulation proves the possibility of a sprayable bandage to improve effective diagnoses of surgical site infection.

In Vitro Pancreatic Tumor Microenvironments using DNA-Directed Patterning

Lily Towery, Soheyl Mirzababaei, Molly Kozminsky

As one of the most lethal malignancies worldwide, pancreatic cancer underscores the urgent need for advancements in effective treatment. The pancreatic tumor microenvironment (TME) plays a crucial role in tumor progression, metastasis and immunosuppression, making it an important target for pancreatic cancer research. Among the various techniques applied to explore TME dynamics, DNA-directed patterning stands out as a promising approach through leveraging photolithography-based templating and the unique properties of DNA and Watson-Crick base pairing to engineer intricate cellular arrangements at the microscale. This technique involves two distinct steps: initial patterning of a substrate with short single-stranded oligonucleotides; followed by cell labeling with complementary oligonucleotides and their placement onto the patterned substrate, thus generating precise cellular configurations. These patterns may contain various components of the TME, such as cancer-associated fibroblasts (CAFs), different macrophage phenotypes, and tumor cell nests. Subsequently, by introducing different mixtures of extracellular matrix (ECM) components, researchers can further mimic the native tumor microenvironment with excellent precision. This modulated spatial organization and iterative nature enables investigations into cellular behaviors, including migration, proliferation, differentiation, and responses to therapeutic agents. DNA-directed patterning thus provides a versatile and high-throughput platform to study cellular behaviors in response to varying cues, allowing for a more accurate representation of *in vivo* conditions.

Waste Plastics to Fuel: Analytical Techniques to Classify Plastics and Evaluate Fuel Quality

Laura Osborne

Plastic is not infinitely recyclable, but little end of life processing has been developed outside of landfilling. The **Plastics to Fuel** project aims to change this standard and create an alternative use for plastic once it reaches the end of its life cycle. To do this, the Plastic to Fuel project starts with an unknown mixed plastic sample, characterizes it through Differential Scanning Calorimetry (DSC), Thermogravimetric Analysis (TGA), and Fourier-Transform Infrared Spectroscopy (FTIR). Then, using a single step slow pyrolysis kiln reactor, a liquid carbon fuel (diesel) is produced. By performing Gas Chromatography (GC), and Nuclear Magnetic Resonance (NMR) on the product, a comparison can be made to a model HDPE pyrolysis diesel, as well as with farm (red) diesel.

Through these analysis methods, it has been shown that the pyrolysis diesel is comparable to commercial farm diesel fuel. However, if fossil fuel-based electricity to power the reactor it diminishes the energy savings and makes it necessary to have a source of electricity. In summer 2024 the project team tested the pyrolysis kiln reactor powered by solar panels supplied by the OSU Solar trailer provided by the OSU Sustainability Coalition. The results were comparable to those obtained with conventional electricity sources.

Moving forward, the focus will be to capture and analyze the off-gas (< C6) to close the mass balance and determine the process efficiency. Recirculating the off-gas to supplement heating the kiln air will also be studied. The ultimate goal of the project is to provide a low-cost, simple to operate and maintain pyrolysis reactor to underserved, island, and rural communities that struggle with disposing of their plastics and have a need for diesel fuel for their tractors, boats , generators, etc. With further development, this project has the potential to reduce waste in landfills and improve the quality of life while also reducing local cost for fuel in rural communities.

Futile to Utile: Waste Plastics to Fuel

Abigail Marshall

Oregon onion farmers are encountering a growing waste crisis from accumulating agricultural plastics. i.e, irrigation supplies such as drip tubing, plastic film coverings, buckets, etc. Unfortunately, mechanical recycling is cost-prohibitive and unavailable in these rural Oregon communities, resulting in the accumulation of tons of agricultural plastics with no plan to manage them. Chemical recycling technologies, like pyrolysis, could provide a reliable means to convert waste plastics into fuels for tractors, generators, etc. off-setting diesel needs and reducing farm operating costs.

The **Plastic to Fuel (PTF)** project is bringing together experts from transdisciplinary fields and backgrounds, in academics, industry, and the community to assess the techno economic, environmental, and social impact for use of small-scale chemical recycling units to efficiently convert waste agricultural plastics into a farm-fuel, i.e., a blend-stock for red diesel. A common agricultural waste product in the greater Malheur community is drip-tape which is used for irrigation of many crops, the most prominent of which is onions. The drip-tape is polyethylene, which has been experimentally confirmed with the Differential Scaling Calorimetry (DSC). In Summer 2024, four OSU chemical engineering students provided a technology demonstration on two field days at the OSU Malheur Ag Experiment Field Station. A bench scale kiln pyrolysis reactor was used to convert the drip-tape to a liquid hydrocarbon product (farm/red diesel). The pyrolysis diesel was run in a tractor at a 1:3 blend of pyrolysis diesel to farm diesel. In addition to the demonstration, oral and poster presentations were made and surveys were handed out to gauge the local interest and concerns. There were about 150 people in attendance between both events and students fielded questions from regional economic development, growers, vendors, and community members. At one event, one of the students donned an onion costume (provided by the station head) to discuss the process with curious participants! Upon returning to the lab, Gas Chromatography Flame Ionization Detection (GC-FID) and Nuclear Magnetic Resonance (NMR) were used to analyze the diesel and confirm it compared favorable to standard farm diesel. The current plan is to construct a small-scale reactor to place permanently at the Malheur Field Station and train local Community College students give demonstrations more frequently to spark interest in supporting a full-scale reactor (500kg capacity) for the region. This will be accomplished through collaboration with a local pyrolysis company.

Durability and Performance Improvement of Anion Exchange Membrane Water Electrolysis with High Ionic Strength Anolyte

Chenyu Li, Habin PARK, Paul Kohl

Green hydrogen is emerging as a cornerstone of the global energy transition, providing a pathway to decarbonize industries that are otherwise difficult to electrify, such as chemical production, heavy transportation, and steel manufacturing. Water electrolysis, especially when powered by renewable energy sources, has been considered a cost-effective method for green hydrogen generation. Anion exchange membrane water electrolyzers (AEMEL) are a promising technology that combines the advantages of conventional liquid water alkaline electrolyzers (AEL) and proton exchange membrane water electrolyzers (PEMEL). The alkaline operation environment allows the use of earth-abundant catalysts for the oxygen evolution reaction (OER) and non-perfluorinated polymer membranes, while the compact design of AEMEL reduces the device cost compared to AEL. However, long-term durability of AEMEL still remains challenging under high-pH electrolytes (pH>13) and high current density (>1A/cm^2) due to the degradation concerns. While low hydroxide concentration is desirable for improved chemical stability, increased cell overpotential in low-pH environments has been observed, compromising the overall efficiency of the system.

This study presents high ionic strength anolyte as a promising strategy to improve the performance of AEMEL operating at a low hydroxide concentration. The research delves into the mechanisms behind these improvements, particularly addressing the cathode hydration issues associated with low pH alkaline anolytes in a dry-cathode AEMEL configuration. The findings demonstrate that cations in the high ionic strength anolyte significantly reduce the cathode overpotential by enhancing water transport from the anode to the cathode via cation migration and diffusion through the AEM. This approach allows for the use of neutral salts as the main cation source, which effectively reduces the anolyte alkalinity and mitigates the potential degradation of cell parts. Different alkali cation species and concentrations were tested consistently in a membrane electrode assembly. The result shows that the improvement on the overpotential depends on the mobility and dynamic ionic radius of the cation. With the implementation of the optimized low-pH high ionic strength anolyte, the AEMEL achieved durable and energy-efficient performance at 60°C and 1A/cm^2 for over 1,000 h.

Circular Technology Framework for Biodiesel Production from Non-Edible Cassia Fistula Seed Oil

Aditya Kashyap

This poster presents a circular technology framework for biodiesel production using Cassia Fistula seed oil, an abundant and non-edible feedstock in India. The framework integrates high-selectivity biocatalysts derived from Cassia Fistula to enhance the transesterification process, boosting yield and minimizing reliance on conventional catalysts. This innovation promotes sustainable fuel production and aligns with circular economy principles.

Captured CO_2 emissions are converted into methanol through a catalytic process, which is then utilized in the transesterification reaction. This closed-loop system effectively repurposes CO_2, reducing greenhouse gases and contributing to the sustainability of biodiesel production. Methanol acts as a key reactant in this cycle, promoting a more sustainable fuel supply chain.

The framework also explores biodiesel blends with fossil fuels to improve engine performance and reduce emissions, with the potential to incorporate emissions from biodiesel combustion into the petrochemical industry. The use of Metal-Organic Frameworks (MOFs) for CO_2 adsorption and separation improves emissions capture and enables more efficient methanol production.

Additionally, nano-additives are investigated for their role in enhancing fuel stability and combustion efficiency, further reducing emissions. This circular approach offers considerable benefits for the automotive industry, helping meet stringent emissions standards while advancing energy efficiency and sustainability.

Keywords: Biodiesel, Cassia Fistula, CO_2 conversion, circular economy.

Reaction Study of Guaiacylglycerol-Beta-Guaiacyl Ether

Jordan Sexton, Yiqi Xu, J. Will Medlin

Lignin, a biopolymer found in woody and other plant biomass, is a potential energy source that can offset the demand for fossil fuels by providing a new pathway for making reagents—such as bio-aromatics and olefins—and aviation fuel. However, Lignin is historically underutilized primarily due to a plethora of recalcitrant chemical bonds that inhibit lignin from depolymerizing. One such prolific bond is the beta-O-4, connecting various compounds within Lignin.

For the project, Guaiacylglycerol-beta-guaiacyl ether (GGE) was used as the model compound, containing the beta-O-4 bond. The research regards GGE and the reactionary steps needed to depolymerize it into its monomer components, coniferyl alcohol and guaiacol. By identifying the best reaction conditions for depolymerizing GGE, this research can support the eventual creation of a techno-economically sound, lignin-focused biorefinery.

For experimentation, solvolysis reactions were conducted, consisting of GGE in various concentrations of water and methanol and at varying lengths of time. These reactions occurred in batches utilizing a fluidized sand bath reactor held at different temperatures. After spending time in the reactor, the samples were cooled in an ice bath, and a vial of the solution was taken for gas chromatography analysis afterward.

The gas chromatography data from the differing solvent concentration experiments indicate positive relationships between temperature and GGE conversion and another between solvent polarity and product selectivity. The data from the differing times experiment displays a positive correlation between reaction time and GGE conversion but a negative one between reaction time and product yield.

In addition, a byproduct of the reaction processes was identified; a solid precipitate was present in the final samples of some of the reactions. The concluding hypothesis is that coniferyl alcohol, and potentially guaiacol, are repolymerizing into unknown oligomers at some stage in the reaction process. This is leading to various experiments to isolate the solid product and identify its molecular structure through techniques, such as mass spectrometry (MS) and nuclear magnetic resonance (NMR) imaging.

Development of Safe, Efficient, and Cost-Effective Solid State Hydrogen Storage

Savannah Murt

The demand for alternatives to fossil fuels is rapidly growing due to the production of greenhouse gas emissions and the progression of climate change. Using hydrogen as an energy source is cleaner for the environment than hydrocarbon-based systems. Sodium borohydride (SBH, NaBH4) has been attracting great attention as a hydrogen storage material because of its relatively high hydrogen storage capacity and low cost compared to other materials. Also, this hydride produces hydrogen, but it does not release harmful byproducts into the environment. Here, our research group proposed the solid-state hydrolysis of SBH and water formed by thermal dehydration of sodium metaborate tetrahydrate. This approach offers a higher H2 yield compared to conventional hydrolysis reactions based on the amount of the sample. In addition, this approach operates at much lower temperatures and pressures to offer improved safety. In the present work, we aim to develop a storage that can safely and efficiently produce hydrogen on site. This hydrogen storage equipped with a fuel cell can be used for a variety of applications, such as unmanned and small-sized motorized vehicles, military devices and equipment, and emergency energy storage.

Renewable Hydrogen Production: In-Plane Electrical Conductivity of Polymer Electrolyte Membrane Water Electrolysis (PEMWE) Catalyst Layers for Performance Study

Chloe Cheng, Iryna Zenyuk, Florian Chabot, Hung-Ming Chang, Nadia Tolouei

Rising carbon dioxide emissions due to fossil fuels raise the importance of developing renewable energy alternatives for energy conversion and storage. Proton exchange membrane water electrolysis (PEMWE) has been proposed as a promising energy-dense method for green hydrogen production. However, PEMWE is associated with high costs, prominently due to the iridium anode catalyst. Ultra-low iridium loading (~.1 mg/cm^2) with added supports can reduce electrolyzer costs and optimize performance. However, ultra-low-loading catalyst layers may display less uniformity and lower in-plane conductivity. This research aims to understand and correlate the relationship between in-plane electrical conductivity, anode iridium catalyst loading, adding iridium supports, catalyst uniformity, and cell performance. Methods were first developed for in-plane electrical conductivity measurements of anode catalyst layers using a four-point probe. Experimentally, catalyst-coated membrane (CCM) and decal in-plane conductivity were measured and compared for TiO_x-supported IrO_2 (IrO_2/TiO_x), NbO_x-supported IrO_2 (IrO_2/NbO_x), mixed TiO_x-NbO_x-supported IrO_2 and Tanaka 110 (TKK110, pure iridium). Results confirmed an inverse relationship between iridium catalyst loading and sheet resistance as predicted, with higher sheet resistance at ultra-low loading that may be attributed to catalyst non-uniformity. IrO_2/TiO_x featured greater conductivity than IrO_2/NbO_x and much higher conductivity than TKK110. Ultra-low loading IrO_2/TiO_x and IrO_2/NbO_x samples displayed durability and minimal degradation under Accelerated Stress Test (AST) protocol compared to TKK110, correlating increased in-plane conductivity with strong cell performance. This research on in-plane electrical conductivity with supported ultra-low loading anode catalysts progresses PEMWE performance characterization for future large-scale implementation of PEMWE and lower-cost renewable energy.

Anhydrous Polybenzimidazole-Based CO_2 Electrolysis Using Computational Fluid Dynamics

Mia Alvarez, Dhairya Shah, Kris Likit-anurak, Benjamin H. Meekins, Sirivatch Shimpalee

CO_2 electrolysis has become an increasingly well studied method of using captured CO_2 from the atmosphere as a source of popular chemical feedstocks.[1] Additionally, it can be more time- and cost-efficient to model CO_2 electrolysis using computational fluid dynamics (CFD) before applying the calculated predictions to a lab setting. Current CO_2 electrolysis simulations use a NafionTM brand proton exchange membrane (PEM) and a liquid flow model[2], but this research investigates using a polybenzimidazole (PBI) membrane in a gas phase model. This membrane's ability to function in high-temperature, anhydrous environments[3] makes it a promising next step in the area of CO_2 electrolysis due to its possible increased efficiency. Furthermore, a gas phase simulation could allow for faster diffusion rates and less overpotential than a multiphase model[4]. Using Star-CCM+ as the CFD software, a geometry and meshing based on an actual, densified and 24-hour reimbibed, PBI electrolyzer cell were developed. Then, parameters from previous models[3,5], the literature[6], given constants, and prediction calculations were input into the simulation. The electrolyzer's performance in terms of species concentration distribution, potential distribution, current density distribution, temperature distribution, and pressure distribution were plotted and analyzed, as well as its polarization curves. Using this data, experiment design and predictions in the lab setting can become more precise; in return, lab results can be used to create a more accurate model for CO_2 electrolysis.

References:

1. CO2 Electrolyzers. Colin P. O'Brien, Rui Kai Miao, Ali Shayesteh Zeraati, Geonhui Lee, Edward H. Sargent, and
   David Sinton Chemical Reviews 2024 124 (7), 3648-3693 DOI: 10.1021/acs.chemrev.3c00206.
2. Lin, S., Zhao, S., Dou, H., Wei, G., Li, Y., Gao, S., Gao, J., & He, Y. (2024). Mass-transfer mechanism of Nafion
   proton-exchange membranes in fuel cells—a review. Energy Technology, 12(6). https://doi.org/10.1002/ente.202400164.
3. Likit-anurak, K., Karki, I., Howard, B. I., Murdock, L., Mukhin, N., Brannon, J., Hepstall, A., Young, B.,
   Shimpalee, S., Benicewicz, B., & Meekins, B. (2023). Polybenzimidazole membranes as Nafion
   replacement in aqueous HCl electrolyzers. ACS Applied Energy Materials, 6(10), 5429–5434. https://doi.org/10.1021/acsaem.3c00522.
4. Hou, P., Wang, X., Wang, Z., & Kang, P. (2018). Gas phase electrolysis of carbon dioxide to carbon monoxide
   using nickel nitride as the carbon enrichment catalyst. ACS Applied Materials & Interfaces,
   10(44), 38024–38031. https://doi.org/10.1021/acsami.8b11942
5. Star-CCM+ PEM Fuel Cell tutorial.
6. Pérez-Rodríguez, S., Barreras, F., Pastor, E., & Lázaro, M. J. (2016). Electrochemical reactors for CO 2
   reduction: From acid media to gas phase. International Journal of Hydrogen Energy, 41(43), 19756–

19765. https://doi.org/10.1016/j.ijhydene.2016.06.130.

Optimizing Oil Use during Pipeline Flushing and Changeover Operations

Emma Padros, Thien An Pham, Amarelys Rios, Andrew Malec, Joel Patterson Jr., Barnabas Gao, Kirti Yenkie, Robert Hesketh, C. Stewart Slater

Pipeline flushing operations are vital for ensuring product integrity in multi-product petroleum facilities where different grades of petroleum products are transported using the same pipeline network. During product changeovers, pipelines must be thoroughly flushed to remove residual oil from previous batches. Traditional flushing methods, which involve using the next product as the flushing oil, generate commingled oils—a mixture of residual and flushing oils that do not meet the required product specifications. These oils are sold at a lower value, leading to high economic losses as petroleum facilities conduct thousands of flushes annually. Furthermore, emissions associated with these flushing operations contribute to environmental concerns due to limited repurposing and reuse. With the quantity of commingled oil produced across the industry, there is a critical need for more efficient flushing processes that lead to resource minimization while improving sustainability.

This research aims to address these challenges by introducing an optimized flushing process. To better understand the flushing dynamics, we explore two flushing operations of direct oil to oil displacement and an enhanced air blowing technique. The direct oil to oil displacement process involves the introduction of a moving plug of the flushing oil to displace the residual oil, thus an interplay of a two-oil system is produced. The enhanced air blowing technique involves partially displacing the residual oil using air, followed by introducing flushing oil to completely remove both the air pockets and the remaining residual oil. Optimal conditions of the air blowing technique in terms of air flowrate and duration of air blowing were attained through the design of experiments. The experiment analyzed the flushing dynamics between a transition from a higher to a lower viscosity product known as a step down and a transition from a lower to a higher viscosity product known as a step up. The product purity is determined through the monitoring of the instantaneous real-time viscosity during the flushing process as detected by an inline viscometer. With a data acquisition system seamlessly integrated, data from the experiment is collected and analyzed. The effectiveness of these two flushing operations is analyzed for optimal conditions of time and volume requirement to achieve product purity. These experiments were conducted using a pilot plant scaled down to 1/5th of a generic lubricant oil plant. The effectiveness of the pilot plant in simulating the operations at the industrial plant was validated through dimensionless residence time distribution (RTD) studies[1]. This study demonstrates how an industrial-scale pilot plant was used to explore and optimize flushing techniques during product changeovers involving oils of varying viscosities. The findings provide valuable insights for improving flushing operations at the industrial level, offering a scalable solution that promotes resource efficiency and sustainability.

References

[1] B. Gao *et al.*, "Improved Design of Flushing Process for Multi-Product Pipelines," presented at the Foundations of Computer-Aided Process Design, Breckenridge, Colorado, USA, Jul. 2024, pp. 137–144. doi: 10.69997/sct.171679.

Additional References:

1. Jerpoth, S. S.; Hesketh, R.; Slater, C. S.; Savelski, M. J.; **Yenkie, K. M.*** 2023 Strategic Optimization of the Flushing Operations in Lubricants Manufacturing and Packaging Facilities. ACS Omega, 8(41), 38288-38300.
2. Gao, B.; Slater, C. S.; Curtis, S.; Fracchiolla, M.; Theuma, D.; Roth, S.; Padros, E.; **Yenkie, K.**

M.; Hesketh, R.P.*Enhancing efficiency and quality in oil pipeline flushing: a novel laboratory scale pilot plant study. Accepted to the 2024 ASEE Annual Conference, Portland, OR.

3. Jerpoth, S. S.; Hesketh, R.; Slater, C. S.; McClernan, R.; Savelski, M. J.; **Yenkie, K. M.*** Application of discretized non-linear programming to minimize mixed oil formation in flushing operations of lubricants pipelines. FOCAPO 2023 conference, January 8-12, 2023

Equivalent Circuit Modeling from Electrochemical Impedance Spectroscopy for Aqueous Redox Flow Battery Characterization and Performance

Edward Chen, Jorden Corpuz, Blake Smith, Lia Stokes, Noah Keith, Andrew Gatzke, Ethan Fergel, Jeanine Mantooth, Samuel V. Cowart, Corey James, Simuck Yuk, Enoch A. Nagelli

Electrochemical equivalent circuit models for redox flow batteries are key in assessing current state of charge and predicting overall charge-discharge performance. The electrochemical characterization and performance analysis of separators and porous carbon felts for their application in redox flow batteries will help inform the selection of components based on properties such electronic and ionic resistance and capacitance associated with the charge transfer and the mass transfer. From analysis of the Nyquist plots obtained with electrochemical impedance spectroscopy (EIS), an RLC model is developed to identify the ohmic and ionic resistance along with capacitance in both flow and no-flow conditions in symmetric full flow cell hardware with hydrogel coatings onto Celgard^{TM}. A fundamental understanding of the contributions of each component will also allow for further experimentation in electrolyte chemistries with specific permeable electrode and composite separator combination for redox flow battery applications.

Rechargeable Aluminum-Organic Batteries with Sulfonamide-Based Electrodes

Blerina Sehitaj, Harrison Asare, George John, Robert Messinger

Abstract

As the demand for electric-powered technologies increase, energy-dense batteries composed of earth abundant, sustainable, and safe materials must be developed to ensure resource availability on a global scale. Aluminum-organic batteries combine earth abundant and energy dense aluminum (Al) metal with renewable, tunable, and sustainable organic materials to provide an environmentally friendly energy storage device. However, to date, Al-organic batteries exhibit low operating voltages (typically, <1.5 V vs. Al/ Al(III)) and significant capacity fade during galvanostatic cycling, decreasing energy density and cycle life.1

In this study, we design and synthesize a derivative of a previously reported conjugated sulfonamide2 that was used as a high-voltage (>3.0 V vs. Li/Li+) cathode material in lithium-organic batteries. The new organic material, oxidized ortho-Phenylene do methane-sulfonamide (oxy-OPDSA) was then used as a cathode material in a rechargeable Al-organic battery. The electrolyte was a Lewis acidic mixture of aluminum chloride (AlCl3) and 1-ethyl-3-methylimidazolium chloride (EMImCl). Al-oxy-OPDSA cells achieved a median voltage of ca. 1.9 V vs. Al/Al(III), the highest yet reported in an Al-organic battery, with specific capacities of 180 mAh/g. The absence of labile protons that would otherwise incur deleterious side reactions and impede efficient binding of the electroactive caiton contributes to cycling stabiltity. The effects of mass loading of oxy-OPDSA on specific capacity were also investigated. This research demonstrates that organic structures can continue to be tuned at a molecular level to enable higher potentials and cycling stability in multivalent battery chemistries, thereby paving the way for more efficient and environmentally friendly energy storage technologies.

References.

(1) Kim, D. J.; Yoo, D.-J.; Otley, M. T.; Prokofjevs, A.; Pezzato, C.; Owczarek, M.; Lee, S. J.; Choi, J. W.; Stoddart, J. F. Rechargeable Aluminium Organic Batteries. Nat. Energy 2019, 4 (1), 51–59. https://doi.org/10.1038/s41560-018-0291-0.

(2) Wang, J.; Lakraychi, A. E.; Liu, X.; Sieuw, L.; Morari, C.; Poizot, P.; Vlad, A. Conjugated Sulfonamides as a Class of Organic Lithium-Ion Positive Electrodes. Nat. Mater. 2021, 20 (5), 665–673. https://doi.org/10.1038/s41563-020-00869-1.

Batch Vs. Slug-Flow Continuous Production of $Ni_{1/3}Co_{1/3}Mn_{1/3}C_2O_4$ Precursors for Li-Ion Battery

Aardra Sakpal, Jethrine Mugumya, Sourav Mallick, Arjun Patel, Michael L. Rasche, Ethan Huchler, Sunuk Kim, Mo Jiang, Ram B. Gupta

Growing demand for clean energy for various applications necessitates the advancement of energy storage devices such as the lithium ion battery (LIB). The cathode material is one of the determining factors that influences the battery manufacturing cost and its electrochemical performance. The suggested strategies to rationally address these factors are the judicious selection of the production method and optimization of the synthesis parameters of the cathode precursor material. Lithium Nickel Cobalt Manganese Oxide (NCM 111) is a cathode material known for its good structural stability, high theoretical capacity and high energy density. However, the traditional stirred tank-based batch manufacturing process leads to inhomogeneity in reaction environments, leading to non-uniform particle morphology of the cathode precursor particles. Therefore, a slug-flow based reactor is used for the continuous production of NCM111 oxalate precursors and comparative study with the batch derived product is performed. It is found that the product of the continuous slug-flow production method, when compared with the batch product, yielded bigger particle size, narrower particle size distribution (PSD) and higher tap density, suggesting a successful improvement of homogeneity due to the slug-flow micro environment. An NH_4OH concentration of 0.08 M and a residence time of 2 minutes followed by lithiation of the precursors shows cycling stability of more than 80% both at 0.5 C and 1 C, and better electrochemical performance of the NCM 111 based LIB.

Automated Quantification and Failure Analysis for Li-S Batteries

Matthew Miyagishima, Saurabh Parab, Shen Wang, Y. Shirley Meng

Lithium-sulfur (Li-S) batteries are highly promising for next-generation energy storage due to their high theoretical energy density and cost-effectiveness. These attributes make them attractive candidates for applications such as electric vehicles and grid energy storage. However, their widespread practical implementation is hindered by several critical issues, including capacity loss, limited cycle life, and reduced efficiency—all of which are driven by complex degradation mechanisms. Key challenges like polysulfide shuttling, sulfur dissolution, lithium dendrite formation, and inactive lithium accumulation significantly affect battery performance and longevity. As these degradation processes overlap, they complicate the diagnosis of failure modes, making it difficult to identify the underlying causes and develop effective mitigation strategies. Thus, a rapid, accurate diagnostic approach is essential to propel the development of high-performance, durable Li-S batteries.

To address these diagnostic challenges, we have developed an advanced analytical framework, supported by an automated software tool named Dr. HUGS. Dr. HUGS represents a major advancement in data analysis for Li-S battery research by automating the interpretation of raw data from various analytical techniques. By eliminating the traditionally time-consuming manual data processing, Dr. HUGS enables researchers to obtain comprehensive diagnostic insights within minutes. The software is designed with a user-friendly interface, allowing users to quickly input experimental data and retrieve crucial diagnostic parameters. Dr. HUGS rapidly processes raw data through peak identification, fitting, and quantification, linking peak areas to calibration curves to determine species concentration and capacity. The results are then visualized in clear formats, such as vector plots and capacity storage charts, providing actionable insights into the battery's health and performance. This automated process drastically reduces data analysis time while maintaining high accuracy and consistency, thereby allowing researchers to quickly diagnose failure mechanisms and iterate on cell designs more effectively. By cross-validating Dr. HUGS' outputs with conventional manual processing, we have demonstrated its reliability, with any discrepancies falling within negligible ranges.

Moreover, Dr. HUGS is versatile and adaptable to different battery testing conditions. It offers detailed insights into how different configurations, such as variations in pressure setups, influence the compositional uniformity and behavior of Li-S batteries. This capability enables researchers to better understand the impact of testing environments on battery performance, guiding the optimization of cell designs and operating conditions. By automating and standardizing the data analysis process, Dr. HUGS provides a powerful platform for accelerating Li-S battery research, ultimately advancing the understanding and development of next-generation energy storage technologies.

Alloy Assisted Rechargeable Zinc Oxide Anodes for Aqueous Nickel-Zinc Batteries

Tanishq Moondra, Zhitao Chen, Nian Liu

In the past few years, the lithium-ion battery (LIB) industry has experienced remarkable growth, along with a significant increase in demand. However, the high cost of raw materials combined with the flammability in LIBs poses challenges to manufacturing and safety. An alternative solution lies in rechargeable aqueous zinc-ion batteries, which are made of more abundant materials (zinc is ~300 times more abundant than lithium), are more environmentally friendly, and a have high specific and volumetric capacity (820 mAh/g and 5845 mAh/cm^3). There are drawbacks, however, particularly concerning the parasitic side reactions and irregulated metal deposition of zinc anodes in aqueous solution. Irregulated metal deposition can result in dendrite formation on the electrode, which can pierce through the separator and cause a short circuit, thereby limiting the lifespan of the battery.

In our research, we focus on engineering a nickel cathode and a zinc oxide anode coated with a layer of silver nanoparticles, which is then coated with a layer of ion-sieving carbon nano-shells. Silver is a zincophilic alloy with a low nucleation barrier and can regulate the deposition of zinc as well as minimizing the dissolution of zinc. The ion-sieving carbon nano-shell layer acts as a physical barrier to further prevent active material dissolution and achieve a higher coulombic efficiency.

Zinc- and Copper-Ion Intercalation and Plating on Chevrel-Phase Mo_6Se_8 Cathodes in Aqueous Electrolytes

Sara Avraham, Brian Chen, Robert Messinger

One possible failure mechanism for ion intercalation-based energy storage systems (e.g., lithium (Li)-ion) is uncontrolled metal electrodeposition (plating), where metal ions from the electrolyte reduce on the electrode surface to deposit metal instead of intercalating. This work investigates zinc (Zn) and copper (Cu)-ion intercalation and metal plating on chevrel-phase Mo_6Se_8 cathodes in aqueous electrolytes. The objective is to develop a better understanding of the competition between electrochemical intercalation and plating of these metal cations into a model battery electrode; the results will be analyzed with respect to lithium cation intercalation versus lithium metal plating on graphite anodes to yield a general understanding of how ion charge density and size affect these competing electrochemical processes.

We studied Zn and Cu-ion intercalation and metal plating in two-electrode coin cells comprised of either Zn and Cu metal anodes and Mo_6Se_8 cathodes in 1 M $ZnSO_4$ and 1 M $CuSO_4$, respectively. We demonstrate for the first time that Zn and Cu metal electrochemically plate on Mo_6Se_8 cathodes. The presence of both metallic Zn and Cu metal deposits was determined using scanning electron microscopy (SEM), energy dispersive X-ray spectroscopy, and X-ray diffraction on galvanostatically discharged Mo_6Se_8 electrodes. Zn- and Cu-ion intercalation and metal plating on Mo_6Se_8 was investigated further through variable-temperature (+50 °C down -10 °C), constant current (10 mA/g) discharge of Zn/Cu-Mo_6Se_8 coin cells. Specifically, the results show that the discharge of Zn-Mo_6Se_8 cells reveal a temperature-dependent interplay between Zn^{2+} intercalation, hydrogen evolution reaction (HER), and Zn plating on the Mo_6Se_8 electrode surface. However, the discharge of Cu-Mo_6Se_8 suggests that an interplay between Cu-ion intercalation and Cu metal plating occurs in the absence of HER even at +20 °C. Arrhenius analysis on an empirically defined rate constant () for Zn metal plating shows that the Zn plating process possesses a negative apparent activation energy, indicative of a multi-step plating process that includes Zn^{2+} adsorption on the Mo6Se8 electrode surface.

This work enables the study of Zn and Cu metal plating as an analog to Li plating and establishes a connection among the three metal plating processes. Building a mechanistic understanding of Zn, Cu, and Li plating as failure modes could improve each metal's respective battery chemistries in more demanding cycling conditions including fast- and low-temperature charge/discharge.

3D-Printed Carbon Scaffold for Structural Lithium Metal Batteries

Joanne Hui

Focus on advancement of energy storage has now turned to curbing carbon emissions in the transportation sector by adopting electric vehicles (EVs). Technological advancements in lithium-ion batteries (LIBs), valued for their lightweight and high capacity, are critical to making this switch a reality. Integrating structurally enhanced LIBs directly into vehicular design tackles two EV limitations: vehicle range and weight. In this study, 3D-carbon (3D-C) lattices, prepared with an inexpensive stereolithography-type 3D printer followed by carbonization, are proposed as scaffolds for Li metal anodes for structural LIBs. Mechanical stability tests revealed that the 3D-C lattice can withstand a maximum stress of 5.15 ± 0.15 MPa, which makes 3D-C lattices an ideal candidate for structural battery electrodes. Symmetric cell tests show the superior cycling stability of 3D-C scaffolds compared to conventional bare Cu foil current collectors. When 3D-C scaffolds are used, a small overpotential ($\approx$0.075 V) is retained over 100 cycles at 1 mA cm−2 for 3 mAh cm−2, while the overpotential of a bare Cu symmetric cell is unstable and increased to 0.74 V at the 96th cycle. The precisely oriented internal pores of the 3D-C lattice confine lithium metal deposits within the 3D scaffold, effectively preventing short circuits.

Hybrid Organic-Inorganic Fast-Charging Structural Batteries

Coby Scrudder, Suyash Oka, Alexander Albrecht, Chen Wang, Jodie Lutkenhaus

As lithium-ion battery specific energy and power nears its ceiling, other approaches must be taken to increase the system-level specific energy. One such approach is structural batteries, batteries which can bear mechanical load and can thus be used as body panels of UAVs, cars, or satellites. In this goal, carbon-fibers are coated with a cathode slurry to make a carbon-fiber composite cathode. The cathode slurry contains a mixture of an organic active material and an inorganic active material. Organic radical polymers – such as PTMA (poly(2,2,6,6-tetramethyl-1-piperidinyloxy-4-yl methacrylate)) – are a promising cathode material due to their straightforward synthesis, availability of raw materials, and fast kinetics. However, they fall short of inorganic materials in their specific capacity. To solve this issue, a hybrid organic-inorganic cathode using PTMA and LFP (lithium iron phosphate) was investigated due to their similar reaction potential. Using this hybrid system, the capacity of the cathode was greatly increased and maintained much of the fast-charging capability due to a beneficial charge-transfer reaction between the PTMA and LFP. This system demonstrates capacity of up to 105 mAh/g at 1 C-rate and 25 °C – a 57% improvement over the pure PTMA system. In addition to this, the mechanical properties of the carbon fiber are not deteriorated by the cathode coating, enabling a high specific modulus. The combination of high specific power and high specific modulus make this hybrid system a promising evolution in structural batteries.

Carbon Electrode Surface Treatments and Novel Electrode Materials for Energy-Efficient, High-Power Vanadium Redox Flow Batteries.

Anna Burson, John Harris, Matthew McDowell

Efficient storage and rapid delivery of clean energy are critical issues facing our energy grid. Vanadium redox flow batteries (VRFBs) offer a solution to these challenges through long-term, large-scale energy storage and a decoupling of power and energy. VRFBs store energy in vanadium-ion redox couples that are soluble in aqueous electrolytes. Spatial separation of the electrolyte and electrode after charging minimizes self-discharge and capacity loss over time. Charging or discharging of the battery occurs when the liquid electrolytes flow past or through porous electrodes. Consequently, the electrode dictates VFRB efficiency and power by the reversibility and rate of redox reactions at its surface as well as mass transport of electrolyte within its volume. VRFB electrodes have been limited to carbon-based materials like carbon paper and graphite felt because of their stability in the acidic electrolyte. However, vanadium redox reactions are slow on carbon surfaces, and carbon's hydrophobicity hinders mass transfer of the aqueous electrolyte. To overcome these shortcomings, surface treatment is a widespread practice to introduce functional groups and defects that increase redox kinetics and aid mass transfer. In this work, we explored atomic layer deposition (ALD) as a novel method of carbon electrode surface treatment. With an example of aluminum oxide, we illustrated the ability of ALD to coat thin films of electrocatalyst onto the carbon fibers within carbon paper. We then quantified vanadium redox reaction reversibility on ALD-modified carbon paper with cyclic voltammetry and described mass transport within these electrodes with contact angle measurements. These findings were compared to a baseline of thermal oxidation as a surface treatment. Finally, we contrasted cyclic voltammograms of surface-treated carbon paper to novel electrode materials like porous copper to understand if carbon materials are the most optimal VRFB electrodes. The knowledge gained from this work will guide the design of electrodes for more efficient and more powerful VRFBs. Advancing this energy storage system will importantly enable the widespread use of clean energy and reduce greenhouse gas emissions through effective grid-scale energy storage and fast delivery of clean energy for off-production use.

Integrated Electrochemical-Acoustic Modeling for in-Silico Characterization of Lithium-Ion Batteries

Zoe Herman, Elizabeth Katzman, Gunnar Thorsteinsson, Dan Steingart

Lithium-ion batteries are a key focus in the drive for energy transition due to their modularity and high energy density. Recent research efforts have been directed at studying the acoustic characterization of cells as a non-invasive probe for chemomechanical dynamics to complement electrochemical signals. However, a challenge in the progress of this novel characterization technique has been a disconnect between the two domains. Here, we address this *in silico* by coupling the Python Battery Mathematical Model (PyBaMM) with the acoustic k-Wave model. By using the electrochemical model as a feedstock for the mechanical one, we achieve continuity in the interpretation of acoustic features. This hybrid model is tuned and validated against experimental data from commercial-grade pouch cells. The framework developed in this study highlights the importance of integrated modeling solutions for battery acoustics for enhanced diagnostics of lithium-ion batteries.

Techno-Economic Study of an e-Methanol Production Plant Using Solar Energy and Carbon Dioxide Capture from a Cement Plant in Peru.

Claudia N. Sierra Huallpayunca, Denis Cristian Alfonte Lupaca, Piero Enrique Varela García

Techno-economic study of an e-methanol production plant using solar energy and carbon dioxide capture from a cement plant in Peru.

Methanol is a widely used raw material in the industrial and chemical sectors of many countries (Cieza & Ugaz, 2018). Despite its local demand, there is no industrial production in Peru that can meet it, so it is imported from countries such as the USA, Trinidad and Tobago, and China. A promising route for the production of e-methanol is the utilization of CO_2 from greenhouse gas emissions in cement plants in Peru, which account for 4-8% of total global emissions (INEI, 2021). These emissions are concerning due to their large volume and potential impact on air quality at local and regional scales (Yarlequé, 2022).

In this productive route, the use of hydrogen is required, and considering Peru's natural resources, green hydrogen produced with photovoltaic solar energy was chosen. This research aimed to economically and technologically evaluate the design of a green methanol production plant. With this objective set, the selection of the plant's location was initiated. The plant's location considers areas suitable for renewable energy sources (RES), specifically photovoltaic solar energy. The La Joya region has one of the highest solar radiation indices in the country, presents adequate atmospheric characteristics, and the energy resource averages 4.89 kWh/m^2/day (SENAMHI, 2024). Regarding logistical aspects for the construction of the plant, its proximity to ports such as the Port of Matarani, located 90 km from the cement plant, its accessibility to major roads, and flat terrain were considered to ensure technical feasibility.

The capacity of our methanol plant is proposed based on the production of green hydrogen generated from a solar plant with 300 MW photovoltaic panels, which has a capacity similar to the San Martín project currently under construction in the La Joya region of Arequipa. The solar plant, the electrolyzer, the carbon dioxide capture plant, and the methanol synthesis plant were simulated using AVEVA Process Simulation Software.

The selection of the most suitable CCU technology began, with post-combustion technology chosen for its operational advantages and commercial maturity (Rodríguez, 2022). This technique can be classified based on the possible mechanisms employed: adsorption, absorption, membrane separation, and chemical reaction. Among these options, the most commonly used in the industry is the use of absorption/desorption towers. In these, a solvent, usually monoethanolamine (MEA), is employed (Mukherjee et al., 2019).

For the production of green hydrogen, photovoltaic panels with an installed capacity of 300 MW and an electrolyzer with a capacity of 196.4 MW were used, where water will be chemically separated into oxygen and hydrogen. Regarding methanol, according to Calo (2021), the chemical route for methanol production from CO_2 consists of a direct hydrogenation of the CO_2 molecule. At the same time, the reverse water gas shift reaction (RWGS) occurs. This reaction is undesirable and should be avoided as it consumes hydrogen. Therefore, a catalyst with high selectivity is needed, and copper (Cu) has been identified as the most suitable, promoted by another metal, mainly some zinc or zirconium oxide, or commercial preparations based on $Cu/ZnO/Al_2O_3$ (Duyar et al., 2020).

Considering all the aforementioned factors, the simulation resulted in an input of approximately 2,000,000.00 tons per year of flue gas, from which approximately 105,120.00 tons of MEA per year are required to capture 350,400 tons per year of CO_2 with a purity of 97.7%. Regarding green hydrogen, 36,226.5 tons per year were produced at a temperature of 80°C and a pressure of 8.9 bar. With the captured CO_2 and produced hydrogen, 179,600.00 tons of methanol with a purity of 75% were produced annually.

An economic analysis was conducted using CAPEX and OPEX variables, yielding values of 854.59 MUSD and 41.338 MUSD, respectively.

Although the process aims for the sustainability of green methanol production, aspects of the technologies used for transporting the generated hydrogen and its proper use should be considered. Additionally, based on the economic analysis and using variables such as WACC and TCI, the project is viable in the long term but not in the short term. Improvements in these economic aspects can be sought based on methanol production. Finally, possibilities for using methanol to produce various chemical compounds such as formaldehyde and methylamine should be explored.

Evaluation of Asphaltene Deposition Using Seawater/Oil Emulsions in a Porous Media Microfluidic Device

Sam Nance, Omran Taqi, Thao Vy Nguyen, Maura Puerto, Sibani Lisa Biswal

Asphaltenes are the heaviest, most complex, and most polar component of crude oil. Asphaltenes often aggregate and deposit in oil pipelines, plugging flow and reducing production. This presents a severe flow assurance challenge in the oil & gas industry, with up to millions of dollars in lost production and treatment costs. Asphaltenes are amphiphilic and thus stabilize water in oil emulsions. While this is known, there is limited knowledge on how the presence of water impacts asphaltenes behavior. This study aims to use microfluidic methods to investigate the effect of emulsified crude oil-sea water solutions on asphaltenes' behavior. The methods of the experiment are as follows. Synthetic sea salt was mixed in deionized (DI) water to make synthetic seawater. Seawater was emulsified with crude oil using a probe sonicator for 5 minutes. The seawater to crude oil ratio explored in this project ranged from 0 (control) to 30%. The experiments were conducted in low porosity porous media microfluidic devices to mimic behavior in a near wellbore region. The microfluidic devices were fabricated using NOA-81 (Norland Optical Adhesive), a thiolene photopolymer. NOA-81 has been proven to be resistant to organic chemicals and low/high temperatures, making it suitable for asphaltene experiments. Water-emulsified crude oil and n-heptane are co-injected at a 50 μL/min flow rate through the microfluidic device at the onset ratio to promote the formation of asphaltenes deposits. Asphaltene deposition is observed and recorded through an inverted microscope connected to a camera. The pressure differential across the microfluidic device was also recorded throughout the experiment. Following the deposition phase, heptane is disconnected from the syringe pump. Water-emulsified crude oil is pumped through the microfluidic device at an increasing flow rate from 50 to 200 μL/min to shear off the asphaltene deposits. Image analysis is leveraged to calculate the percentage of porous media area blocked by asphaltene deposits. Different concentrations of water emulsified in crude oil had varying impacts on asphaltene deposition. It was found in all water emulsion cases except the 5% case that there was less deposition than in the control case. The results show that at an optimal salinity concentration (OSC) of 1% water emulsified in crude oil, the asphaltene coverage on the porous media decreased by 63% from the control after the deposition period. It was also found that the asphaltene deposits were less resistant to shear in the water-emulsified crude oil cases than in the control case.

Evaluating the Efficacy of Ideal Solution Theory for Biodiesel Mixtures of Varying Complexities

Michael Senra

Biodiesel is a promising replacement for petroleum-based diesel in a number of different applications. However, one of its major disadvantages is its inferior cold flow properties. Previous work has shown that the challenge with the cold flow properties is strongly correlated to the composition of the biodiesel, which primarily contains fatty acid methyl esters or ethyl esters (FAME/FAEE). Researchers have used a wide variety of thermodynamic models of varying complexities to conduct solid/liquid equilibrium analysis to estimate the cloud point of the biodiesel. This work comprehensively looks at cloud point data from a wide variety of sources and analyzes them using ideal solution theory. Sensitivity analyses were conducted to determine what variables caused the most variation in Cloud Point Prediction.

Development of Electrochemically Active WO3/PVDF Membranes for Intensifed Hydrogenation Reactions

Daniel Kesler, Evan Miu, James R. McKone

The chemical industry currently relies on fossil fuels to produce hydrogen, contributing to global climate change through the emission of greenhouse gases. In this work, we are developing a fossil-free method of hydrogen utilization to limit carbon emissions currently tied to hydrogenation reactions. In our approach, hydrogen is activated from water using electrochemically active tungsten-oxide (WO_3) membranes in a reactor system. During this presentation, the design, fabrication, and characterization of WO_3/PVDF composite membranes will be discussed. Cyclic voltammetry measurements demonstrate that these WO_3/PVDF composites could intercalate hydrogen similarly to benchmark WO_3/Nafion inks. Analysis of the reduction peaks exhibited a similar magnitude of the current response, correlating to simmilar redox reaction rates for the hydrogenation of composites. Another telling feature of the intercalation behavior was the change in color [green to blue] of the composite membrane. Further, chronoamperometry measurements reveal the ability of the membranes to conduct hydrogen over micron-scale distances. Finally, WO_3/PVDF composites were coated onto carbon felt gas diffusion electrodes and placed into a prototype hydrogenation reactor for gas and liquid permeability testing, finding limited permeability. Present work includes testing the full hydrogen reactor for hydrogen production and incorporating further reactions for carbon upcycling. In summary, this work covers the full cycle of design and fabrication of a WO_3/PVDF composite that is capable of electrochemically activating hydrogen from water.

Optimization of Desalting of Crude Oil

Joshua Scott

The escalating demand for oil as an energy source has led to the depletion of light crude oil reservoirs, prompting exploration of alternatives such as heavy crude oil. However, heavy crude oil's complex composition, including salts, poses challenges during processing, particularly in refining. To get the salt out water emulsifiers are used to mix the water with the crude oil. Effective separation methods to separate the water retaining the salt out of the oil include, demulsification using silicon-based compounds, electroporation, and heat.

Enhanced Performance of Lithium-Sulfur Batteries with Trace Water Electrolyte

Joshua Brown, Karam Eeso, Nian Liu

With the growing demand for energy storage systems capable of delivering high energy densities, lithium-sulfur (Li-S) batteries have emerged as a promising solution. These batteries offer a significantly higher theoretical specific capacity of 1672 mAh/g, which exceeds the capacities of many conventional battery technologies. The reason behind this high capacity lies in the two-electron transfer per sulfur atom, which enables efficient energy storage. However, the practical realization of this potential is hindered by the polysulfide shuttling effect. This phenomenon occurs when intermediate lithium polysulfides (Li_2S_x, $4<x<8$) formed during the discharge process dissolve into the electrolyte and migrate to the lithium anode. These polysulfides can react with the lithium anode, leading to capacity loss and reduced battery efficiency. One of the strategies to mitigate the polysulfide shuttling effect is to modify the electrolyte. Some works add lithium nitrate to protect the lithium anode to stop the direct reaction of the polysulfides with the lithium anode while others change the solvents to decrease the solubility and mobility of the polysulfides. In our research, we explore an approach by adding trace amounts of water to the electrolyte to restrict the mobility of polysulfides. Through this modification, the shuttling of polysulfides is significantly reduced, preventing their unwanted interactions with the lithium anode. Initial charge-discharge tests conducted on coin cells demonstrate a promising outcome: the introduction of 1000 ppm (parts per million) of water into the electrolyte results in a 25% improvement in specific capacity. However, the amount of water added must be carefully controlled. While trace amounts enhance performance, excessive water can degrade the battery's lifecycle by promoting side reactions that impact stability and longevity. Achieving the right balance between improving specific capacity and maintaining battery longevity is critical to advancing Li-S battery technology. Our research contributes to the ongoing optimization of electrolyte formulations, potentially bringing Li-S batteries closer to commercial viability and supporting future energy storage needs.

Electrochemical Characterization of Composite Separators in Aqueous Redox Flow Batteries for Grid Scale Energy Storage

Woodson Squier, Samuel Bedor, Blake Smith, Lia Stokes, Jeanine Mantooth, Noah Keith, Andrew Gatzke, Ethan Fergel, Samuel V. Cowart, Corey James, Simuck Yuk, Enoch A. Nagelli

Redox flow batteries (RFBs) meet the increasing demand for more effective and scalable energy storage technologies to implement renewable energy sources into the grid. The objective of this study is to determine the most effective combination of composite polymer separators and carbon felts within a aqueous flow battery. The electrochemical characterization of separators and porous felts is critical to overall flow battery performance. Through electrochemical impedance spectroscopy (EIS), cyclic voltammetry (CV), and chronoamperometry (CA) we can determine the charge transfer resistance, mass transfer resistance, and electronic and ionic conductivity in a custom flow cell hardware with various supporting and redox electrolyte chemistries at varying flow rates and at stagnant no-flow conditions. Composite separators for electrochemical characterization including Celgard™ and Daramic™ coated with poly(acrylic acid) and poly(vinyl alcohol) hydrogels are studied. Porous carbon felts with varying electronic conductivities are studied to assess the impact on voltaic and Coulombic efficiency performance. The overall electrochemical characterization will help inform the design of aqueous redox flow batteries with cost-effective separators.

Nano-Invaders: Unraveling the Secrets of Nanoparticle-Lung Interactions

Jonathan Arredondo, Monica Iepure, Younjin Min

Airborne nanoparticles, both natural and engineered, are becoming an increasing concern due to their potential impacts on human health and the environment. The increasing frequency and intensity of wildfires driven by climate change are generating higher levels of airborne carbonaceous particles. The inhalation of fine particulate matter from wildfires has been linked to adverse pulmonary health outcomes and there is a critical need to better understand the interactions between airborne nanoparticles and the human respiratory system, particularly the effects of inhaled particles on the lung surfactant system - the first line of defense against foreign materials in the alveolar space. The alveolar boundary in the lungs was modeled with 50 nm silica nanoparticles (SNPs), which were modified with trimethoxy silanes with varying carbon tail chain length groups dispersed onto dipalmitoylphosphatidylcholine (DPPC), the main component of the pulmonary surfactant. Utilizing surface and interfacial tension measurements through the optical shape analysis of a pendant drop apparatus, the interfacial behavior of the DPPC and modified SNP monolayers were examined through dilational rheology experiments and surface pressure-area isotherms. Our studies demonstrate that with increasing carbon tail lengths of the modified nanoparticles the DPPC monolayers were subject to stronger intermolecular interactions. This is due to strong van der Waals interactions between the hydrophobized nanoparticles and the acyl chains of the DPPC molecules. These interactions lead to more condensed and densely packed DPPC/SNP monolayers as carbon tail length increases. Despite the increasing radial length of the coupling agents, our studies show that the larger increase in hydrophobicity is the stronger force condensing the DPPC monolayer. This is in contrast to the unmodified SNP, which with no carbon tail functionalization, was attracted to the hydrophilic region of the monolayer, decreasing the density of the monolayer. The intermolecular interactions caused by the presence of nanoparticles in the monolayers were attributed to the surface pressure-area isotherms. Additionally, dilational rheology experiments determined the packing density and viscoelastic properties of the DPPC/SNP monolayers. As the carbon tail lengths of the modified silica nanoparticles increased, resulting in increasing surface hydrophobicity, reduced the mean molecular area of the DPPC monolayers. These findings provide important insights into the interactions between airborne nanoparticles and the LS system, which is essential for maintaining proper lung function, and improve our understanding of how the physicochemical properties of nanoparticles influence their impact on LS. These findings shed light on the potential health risks associated with exposure to airborne nanoparticles from natural sources.

Enhanced Chemical Resistance in Elastomeric Gloves through Graphene-Based Nanomaterial Coatings

Grey Small, Aidan Stone, Zidan Yang, Jiaman Wang, Maria Louiza Dimtsoudi, Aicha Sama, Rebecca Martin-Welp, Indrek Kulaots, Francesco Fornasiero, Robert Hurt

Graphene-based nanomaterials possess unique properties that can effectively address contemporary challenges in personal protective equipment (PPE) development. A promising application is the engineering of ultrathin, chemically resistant barrier coatings for elastomeric gloves. We hypothesize that these coatings offer a range of benefits, including effective chemical resistance, compatibility with elastomeric glove scaffolds, durability for extended use, and retention of dexterity—all while adhering to green chemistry and manufacturing principles.

These graphene-based films are produced by casting and drying graphene-based inks, which vary in concentration, type of graphene suspension, and crosslinkers. The primary material tested was graphene oxide, noted for its mechanical flexibility, crosslinker compatibility, and barrier properties. Nitrile rubber was chosen as the glove scaffold due to its established commercial use. Casting methods explored included drop casting and dip casting, which align with nitrile glove manufacturing practices. The solutions were cast on pre-stretched scaffolds, which induced a microscale crumpling of the films upon relaxation, granting greater mechanical flexibility and droplet repellency.

The coated films underwent rigorous testing for molecular permeation, mechanical wear resistance, and solvent immersion stability. Coatings applied to pre-stretched scaffolds maintained integrity during stretching, folding, and rubbing. Coatings treated with crosslinkers demonstrated enhanced solvent stability. Most significantly, the permeation data indicates that these graphene-based coatings substantially reduce the molecular permeability of the coated scaffolds.

Ligand Exchange on Oleylamine Passivated $BaZrS_3$ Nanocrystals for Thin Film Creation

Paul Gramelspacher

As the global transition towards renewable energy accelerates, perovskite materials are emerging as promising candidates for photovoltaic applications. Among these, the chalcogenide perovskite $BaZrS_3$ stands out due to its low toxicity and high stability when compared to common lead halide perovskites. Its direct band gap of 1.75-1.85 eV also makes it well-suited for some solar cell applications. $BaZrS_3$ nanocrystals have been colloidally synthesized at temperatures between 330°C and 365°C using molecular precursors. However, to be used in photovoltaic devices, these nanocrystal dispersions must be turned into high-quality films. Unfortunately, the surface of colloidally synthesized $BaZrS_3$ nanocrystals are typically passivated by oleylamine and oleic acid. These long bulky ligands hinder charge carrier mobility and reduce nanocrystal density on films, ultimately lowering efficiency. To address this, we propose procedures for exchanging these ligands for smaller inorganic ligands. In this work, we systematically screen various combinations of inorganic ligands, small organic ligands, and solvents to achieve well-dispersed, inorganic-capped $BaZrS_3$ nanocrystals. These nanocrystals are then spin-coated into thin films and characterized using scanning electron microscopy (SEM), UV-Vis spectroscopy, and energy-dispersive x-ray spectroscopy (EDX). Our goal is to produce films with improved quality, allowing for a more detailed investigation of their electronic properties and optimization for photovoltaic applications.

Quantifying Dissipative Losses in Nanoparticle Films for Acoustic Gas Sensors

Yacob Melman, Ethan Moran, Jose Bahamonde, Stylianos Siontas, Sean McSherry

The integration of chemical sensing materials (i.e., receptors) with thin-film bulk acoustic resonators (FBARs) enables sensitive and selective gravimetric gas sensors. This approach offers devices with significantly smaller size, power, and cost compared to conventional resistive and optical-based approaches, making it a promising technology for Internet-of-Things sensing applications. One barrier to the application is the difficulty of achieving high quality receptor films that propagate short-wavelength acoustic waves ($\lambda = 2.5$ μm) without substantial energy loss. In this study, we quantify the dissipative energy losses in proxy receptor films made from SiO_2 nanoparticles that are deposited onto 3 GHz FBARs. Our findings demonstrate that decreasing the nanoparticle diameter from 100 nm to 20 nm decreases dissipative energy losses by a factor of $\approx$ 25 for drop-casting, and $\approx$ 3 for dip-coating deposition methods, respectively. Scanning electron microscopy reveals this reduction in energy losses is due to improved surface coverage for drop-casted films. For dip-coating, where surface coverage is more consistent across different nanoparticle diameters, we hypothesize that films made from smaller nanoparticles (possessing smaller ratio of nanoparticle diameter to acoustic wavelength) are closer to an 'effective-medium' and are therefore less likely to scatter and/or attenuate acoustic waves. In practice, these findings can be applied as design guidelines to implement high-quality chemical sensing materials on high-frequency FBARs for gas sensing applications.

Fundamental Study of Germanium Nanowire Growth and Deposition

Rachel Lee, Stephen Maldonado

The growth of individual Ge nanowires on ultramicroelectrodes (UMEs) through the electrochemical liquid-liquid-solid (ec-LLS) crystal growth mechanism was studied. Ge nanowires have become an emerging area of study with the growing interest in semiconductor nanowires for battery anode applications. The formation and development of individual micro- and nanowires was studied to improve understanding of Ge crystal behavior and growth from liquid metal electrodes. Previous work in the Maldonado lab has observed the nanowires exhibiting a coil-like structure and eutectic gallium-indium cap. We thoroughly characterized the shape, structure, and length of Ge nanowires grown from Pt and Cu ultramicroelectrodes dipped in eutectic gallium-indium using optical and electron microscopes. Understanding of the fundamental behavior of Ge nanowire growth can be applied to improve efficacy and promote the appearance of desired characteristics (e.g. length and thickness) when growing Ge nanowires at larger scales.

Optimized Co-Encapsulation of Colloidal Nanocrystals and Small Molecules into Polymeric Nanoparticles *Via* Sequential Nanoprecipitation (SNaP)

Jacques Zacharie Thaddeus Ponce, Rachel Pollard, Nouha El Amri, Parker Lewis, Ashley Han, Xiuxian Li, Nathalie M. Pinkerton

Polymeric core-shell nanoparticles (NPs) are widely used as effective carriers for active pharmaceutical ingredients (API) in various biomedical applications. The formation of NPs carrying colloidal materials, including iron oxide nanocrystals (IONC) and quantum dots (QD), has drawn substantial interest due to their extensive physical and optoelectronic properties as medical contrast agents. However, while the parameters that control small, hydrophobic molecule formation in NP precipitation have been thoroughly investigated, there is limited knowledge on the assembly process of colloidal materials. In this work, we encapsulate IONCs (5-20 nm) and QDs (5 nm) alongside a small molecule fluorophore into stable NPs *via* two different precipitation processes, Flash Nanoprecipitation (FNP) and Sequential Nanoprecipitation (SNaP), and compare their control over NP size, API loading, and cargo uniformity. FNP is a well-established, one-step process that involves the simultaneous core formation of hydrophobic molecules and adsorption of the stabilizing amphiphilic block-copolymer through rapid mixing. SNaP is an emergent synthesis process which utilizes similar mixing principles as FNP, but incorporates a two-step sequential mixing modality, creating the ability to delay the time points at which components are incorporated into the NP. Characterization of the FNP particles showed a lack of control of NP size with generic formulation variables such as percent cores and total mass concentrations. There was also notable nonuniformity in API and nanocrystal distribution among formed NPs, which was visualized using Transmission Electron Microscopy (TEM). Through SNaP, the nanocrystals aggregated prior to block-copolymer stabilization, yielding increased control over NP size, ranging from 100-250 μm in diameter, and a 2-fold improvement in API uniformity compared to FNP. These results highlight the application of SNaP as a robust, scalable method for the synthesis of multimodal nanoparticles with both drug delivery and bioimaging capabilities. Future work will continue to optimize the manufacturing processes of theranostic nanoparticles.

Dynamics of Polymer-Grafted Nanoparticle Suspensions

David Amirsadri, Nehal Nupnar, Masoud Abdi, Patrick Barrett, Michael Hore, Kiril Streletzky, Ryan Poling-Skutvik

Polymer-grafted gold nanoparticles (AuNPs) are a widely studied class of material that combine polymer and colloidal physics. Gold nanoparticles provide structural integrity and impart desirable optical properties; polymers grafted to the nanoparticle surface provide stability in a colloidal suspension. Their unique physical properties give rise to a broad range of applications ranging from drug delivery, to biomedical sensing, to catalysis. This poster explores the dynamics of polymer-grafted nanoparticle suspensions from two perspectives: (1) evaluating the effects of anisotropy on chain dynamics, and (2) determining how polymer molecular weight controls the colloidal stability of AuNP solutions. We explore the effects of nanorod geometry on chain conformation, relaxation dynamics, and rotational and translational diffusion using depolarized dynamic light scattering (DDLS). We observe an increase in the hydrodynamic dimensions of nanorod AuNPs grafted with 10 kDa thiolated PEG, with a larger increase in the radial direction than in the longitudinal direction. This increase is consistent with our understanding of grafted polymer chain distribution on the surface of geometrically anisotropic AuNPs. Similarly, we modulate the rate of aggregation through grafting and characterize aggregation kinetics in solution. We can examine the effect of different molecular weights of polymer on any structural hysteresis in the relationship between particle size and temperature. Additionally, a DLS temperature ramp was performed to explore aggregation and dispersion behavior. DLS revealed a kinetic effect from slow aggregation for rods functionalized with a 57 kDa thiolated polystyrene and in spheres functionalized with a 233 kDa polystyrene. Future experiments will characterize the effect of varying both polymer molecular weight and nanorod radius on chain conformation and the diffusive properties of the grafted nanoparticles. Additionally, measurements will explore the effect of varying temperature on the absorbance of UV light by the particles, to tailor the aggregation kinetics to optimize aggregation in the presence of environmental contaminants and generate a novel class of nanoparticle-based sensors.

Small Angle X-Ray Scattering of Metal Halide Perovskite Nanocrystals

Jalen Pryor, Dylan Ladd, Jessica Kline, Shaun Gallagher, Ben Hammel, Gordana Dukovic, David S. Ginger Jr., Michael F. Toney

From photocatalysis to optoelectronics, metal halide perovskite nanocrystals are an incredibly promising material. Notably, they have a high photoluminescence quantum yield, narrow emission linewidth, and precise tunability of their band gap by adjusting both size and halide composition. A major issue with the colloidal synthesis of these materials is the occurrence of a size distribution of nanocrystals, driving inhomogeneous broadening. In a quantum confined regime, different sized particles emit different frequencies, causing the overall emission spectra to broaden - ultimately impacting the purity of the energy emitted by the material. Due to this major size dependency on the uniformity of the material produced, precise size control of colloidal nanoparticle synthesis is integral. One technique for characterizing the size and shape of nanocrystals is small angle x-ray scattering (SAXS) which can provide robust insights into their morphology. In this study, SAXS was used to investigate two main questions related to colloidal $CsPbBr_3$ nanocrystals: the origin of an asymmetric red-tail shift observed in its photoluminescent spectra and the impact of different surface ligands on the size, size distribution, and shape of nanocrystal solutions. By analyzing SAXS data with different shape models, these factors were compared across a series of samples. While the symmetry of the size distribution did not have a significant impact on presence of a red tail, different surface ligands, such as lecithin, affected both the size and shape of the nanocrystal solution in a distinguishable manner - such as trends in the emission linewidth and photoluminescent quantum yield. Future work involves more robust characterization, modeling, and analysis of additional samples to further connect structural properties of CsPbBr3 nanocrystals to optoelectronic performance.

Rapid Phase Separation Driven Formation of Biomimetic 2D Nanosheets

Katherine Kimball, Ronald J. Vogler, Dominic Bujanos, Benny D. Freeman, Manish Kumar

Biomimetic materials, such as membrane proteins, exhibit intriguing potential as membrane materials due to their fast transport rates and high selectivity. When membrane proteins are self-assembled into 2D nanosheets with block copolymers and deposited onto membranes, the resulting membranes have exhibited exceptional performance for nanofiltration. However, the methods required to create these self-assembled 2D nanosheets can take up to multiple days to complete. In this work, we leverage detergent solution phase separation to achieve a rapid self-assembly of membrane proteins and block copolymers. To create the 2D nanosheets, Outer Membrane Protein F (OmpF), poly(butadiene-b-ethylene oxide) (PB-PEO), and n-octyl-oligo-oxyethylene (Octyl POE) detergent were mixed and then heated to 60°C for varying durations. After heating, the solution was diluted to lower the Octyl POE concentration below its critical micelle concentration. To analyze the structures that were formed, Transmitting Electron Microscopy (TEM) was utilized. Fast-Fourier-Transform analysis of TEM images revealed the formation of 2D nanosheets with a trigonal packing of OmpF proteins in PB-PEO, the same structure that was formed using previous methods. We hypothesize that the 2D nanosheet formation was caused by the aggregation of Octyl POE micelles containing OmpF and PB-PEO at 60°C, which happens within seconds. These sheets were formed in as little as five minutes, which is about 2,100 times faster than the conventional method for 2D nanosheet synthesis. Additionally, many of these nanosheets were found to be nearly an order of magnitude larger in size than the largest OmpF nanosheet synthesized by the conventional method. This method was scaled up in volume 30x and continued to produce similar quality 2D nanosheets, indicating scalability potential. Ultimately, the method reported here eliminates the time and material constraints previously associated with 2D nanosheet formation from membrane proteins while maintaining nanosheet quality, critical factors for achieving scalability of 2D nanosheet biomimetic membranes.

Effects of X-Ray Irradiation on Self-Regulating Hyperthermia Magnetic Nanoparticles of Gadolinium Silicide

Samantha Smith, Santiago Bermudez Naranjo, Jessika Rojas Marin, Ravi Hadimani

In the development of nanomedicine, materials for magnetic hyperthermia treatment, such as gadolinium silicide (Gd_5Si_4), have been proposed as potential candidates due to their self-regulating heating nature, which utilizes a tunable transition temperature to prevent overheating [1,2]. Given the biocompatibility of Gd_5Si_4 nanoparticles, their potential exposure to X-ray irradiation for medical imaging or compensatory irradiation therapy raises questions about possible changes in magnetic properties that could affect self-regulating heating safety. This study investigates the effects of high-dose (120 Gy/min) X-ray irradiation on Gd_5Si_4 nanoparticles.

Gd_5Si_4 nanoparticles were prepared using the procedure reported in [1]. An X-RAD 225XL Precision X-ray system was used to irradiate the sample for 5 hours at a dose rate of approximately 120 Gy/min (ASTM E 1026-04). This study produced results that showcased observable effects on the magnetic properties of the material when exposed to X-ray irradiation. XRD analysis was carried out on the samples before and after irradiation, and there were no observable changes in the crystalline structure.

A vibrating sample magnetometer (VSM, Quantum Design, 3 T VersaLab) was used to measure the magnetization (M) of Gd_5Si_4 with applied magnetic fields ranging from -3 T to 3 T, and magnetization was measured from 200 K to 400 K at an applied magnetic field of 100 Oe. Fig.1(a) shows M-T data from 200K to 400K before irradiation and Fig.1(b) shows M-T data after irradiation. A small change is seen in the transition temperature of 1K-3K, and a change in magnetization (M) of the samples is present. Fig.2(a) and Fig.2(b) present hysteresis graphs at 300K. Change in saturation is present in post-irradiation graphs with a decrease in magnetization in irradiated samples. These results indicate that irradiation affects magnetic hyperthermia treatment by changing the magnetization of the samples and the transition temperature without having an effect on structural properties.

Acknowledgments: Authors thank the funding from the Commonwealth Cyber Initiative Grant, NSF #2304513 and #2349694.

References:

[1] Z. Boekelheide, Z. A. Hussein, S. M. Harstad, A. A. El-Gendy and R. L. Hadimani, 'Gd5Si4 Micro- and Nano-Particles for Self-Regulated Magnetic Hyperthermia,' in IEEE Transactions on Magnetics, vol. 53, no. 11, pp. 1-4, Nov. 2017, Art no. 5400204, doi: 10.1109/TMAG.2017.2708688.

[2] A. A. El-Gendy et al., 'Ferromagnetic Gd5Si4 Nanoparticles as T2 Contrast Agents for Magnetic Resonance Imaging,' in IEEE Magnetics Letters, vol. 8, pp. 1-4, 2017, Art no. 1507504, doi: 10.1109/LMAG.2017.2728503.

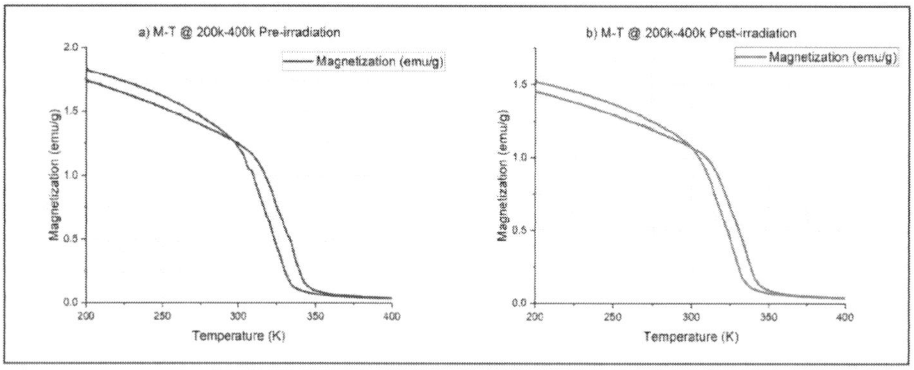

Fig.1 (a) MT Data from temperature range of 200 K to 400 K at 100 Oe applied magnetic field before irradiation and (b) MT Data from post irradiation demonstrating change in magnetization (M) and an small change in transition temperature.

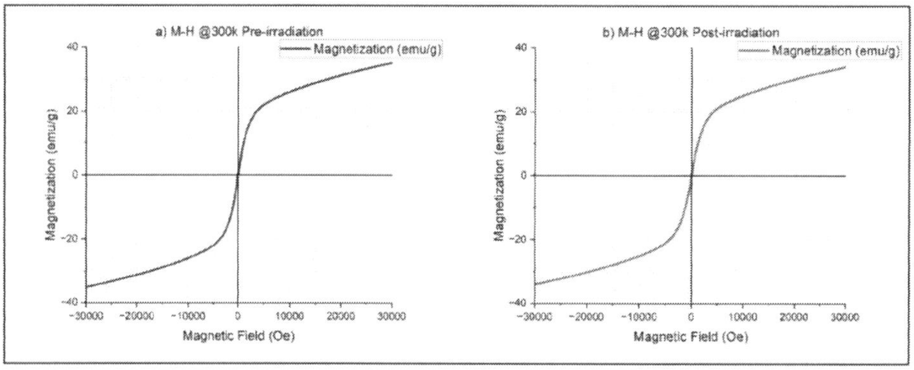

Fig.2 (a) MH hysteresis data from -3T to 3T at 300 K before irradiation and (b) MH hysteresis data after irradiation showcasing decreased magnetization and a changed saturation.

Iron Aluminate Reticulated Porous Ceramic Fabrication for Use in Solar Thermochemistry

Jessica Connell, Linnea Helenius, Liam Taylor, Kent Warren, Alan W. Weimer

Solar thermochemistry involves cycling redox materials to split CO_2 and H_2O into CO and H_2, which can then be converted into hydrocarbon fuels and syngas. This process occurs at high temperatures (~1500°C), so to directly irradiate the active material, a fixed form must be used as opposed to powder. While many use ceria as an active redox material, we propose that iron aluminate offers substantial benefits when compared with ceria, and thus has the potential to increase the efficiency of the process.

One advantage of using iron aluminate is its high productivity under isothermal or near-isothermal operating conditions. Ceria requires a temperature swing from ~1500°C for reduction to ~900°C for oxidation to effectively split water, which causes heat loss and decreases process efficiency. Iron aluminate has demonstrated productivity substantially higher than that of ceria while operating isothermally at 1400°C with a slight pressure-swing (Tran et al., 2023) and has demonstrated thermal stability over 2000 redox cycles.

To compare iron aluminate's water and CO_2 splitting capacity with that of ceria, we have developed a novel process for fabricating iron aluminate reticulated porous ceramics (RPCs). RPCs are solid, porous structures. Their mechanical stability and porosity make them ideal candidates for solar thermochemistry because they are robust through cycling and optimize volumetric absorption of radiation. Fabricated iron aluminate RPCs will be demonstrated in the ETH Zurich solar reactor system for water and CO_2 splitting using concentrated sunlight.

Acknowledgement: EERE DE-EE0010729

References:

Tran, J., K.J. Warren, D. Mejic, R.L. Anderson, L. Jones, D.S. Hauschulz, C. Wilson and A.W. Weimer, "Pressure-enhanced performance of metal oxides for thermochemical water and carbon dioxide splitting," Joule, 7, 1759-1768 https://doi.org/10.1016/j.joule.2023.07.016 (2023)

Ascertaining the Influence of Molecular Weight on Phase Transition Temperature: Applications in Liquid Crystal Elastomer Synthesis

Summit Shrestha, David Kennedy, Timothy White

Liquid crystal elastomers (LCE) have garnered substantial attention in recent years due to their role as a stimuli-responsive soft materials with various applications in optical polymers, soft robotics, and artificial muscles. When external stimuli such as light or temperature interact with LCEs, they undergo a transition from a high order material where all liquid crystal molecules (mesogens) are aligned along an imposed nematic director, to a material with low alignment. This change in alignment initiates a transformation in shape of the material resulting in actuation or optical impedance. The merit of liquid crystal elastomers (LCEs) as programmable materials depends on the alignment of these mesogens. Greater initial alignment leads to more pronounced responses. Moreover, the temperature at which these transitions occur is crucial, as it directly determines the material's programmability. Such materials have traditionally been difficult to synthesize due to extended preparation times and elevated temperatures. Thiol-ene chemistry offers fine-tunability of synthesizing LCEs, thereby minimizing the time required to produce a monodomain LCE. Thiol-ene chemistry has also been thoroughly studied for application in polymer synthesis. In the context of liquid crystals, thiol-ene photopolymerization facilitates effective mechanical alignment of liquid crystal elastomers (LCEs) while ensuring control over the resulting material properties. Adjustments in the LCE formulation and variations in alignment levels directly influence the response's strain and temperature.

This work elucidates the influence that molecular weight of linear thiol-ene oligomers has on the temperature phase space of LCE polymers. Determining the temperature and the characteristics of how LCEs transition is critical to tuning the temperature where actuation occurs in the material. To investigate this, the monomer ThE4b4 was polymerized at varying times to increase the degree of polymerization and thus the molar mass. ThE4b4 is unique in this regard due to its thiol-ene functionalization. Possessing both a thiol and an alkene functional group enables photopolymerization in the presence of an initiator without requiring the addition of a thiol which can disguise the temperature of transition. Idealized step growth of ThE4B4 entails the production of a linear polymer. Analysis of linear polymers is desirable to decouple effects of crosslinking with the effects of increasing molecular weight as both can influence the phase transition temperature. IR spectroscopy and NMR will determine the percent conversion of thiols and alkenes and the degree of polymerization, respectively. Once the phase space is adequately defined for a linear polymer, crosslinkers will then be introduced to produce LCE films whose actuation and strain properties will then be determined utilizing dynamic mechanical analysis.

Preliminary results indicate sharp monomer peaks, as molecular weight increases these peaks begin to broaden with transition temperature increasing as well. An interesting observation is that the energy required to initiate a phase transition is concave down before reaching an inflection point where the trend reverses itself. A greater understanding of actuation and the nature of phase space allows for greater tunability and versatility when synthesizing LCEs.

Quantitative Analysis of Atomic Resolvability and Radiation Damage in Apoferritin Complexes Using Cryogenic-Electron Microscopy and Q-Score Evaluation

Sarina Hasan

Electron radiation damage is well-known to limit the resolvability in density maps reconstructed from images of single particles in cryogenic-electron microscopy (Cryo-EM). However, it is not clear whether this damage affects all side chains equally. In this study, we analyzed a set of images of vitrified Apoferritin particles that could yield a map with a final resolution of 1.69Å. To assess the effects of radiation damage at the side chain level, we computed the Q-scores of maps and models generated from later sets of frames in the movies corresponding to higher cumulative doses. Our analysis revealed that the map resolution, as computed from the Fourier Shell Correlation (FSC), decreased with increasing cumulative doses of the corresponding maps. Furthermore, the Q-scores (i.e. resolvability) of the side chain densities declined at a faster rate for negatively charged residues compared to positive and non-polar residues. We observed that aromatic residues exhibited greater structural preservation under similar cumulative dose conditions, while in the case of negative charged residues, the density resolvability could arrive from inherent scattering properties. To isolate the scattering effect from the dose effect, we modified the original map by both 1) phase randomization and 2) equalizing amplitudes at resolution lower than 1.69A in two different tests. This approach allowed us to isolate the impact of radiation damage from the scattering effects of negatively charged atoms. The observed declines in Q-score and structural visibility in the maps with higher cumulative doses were indeed attributed both to dose-induced damage and to scattering differences at low resolution, particularly involving negatively charged oxygen atoms. We demonstrated that signal falloff occurs more rapidly with increased dosage, particularly in the critical 2.1 Å and 3.2 Å resolution ranges where protein side chains, ions, and waters undergo radiolytic processes. This suggests that inherent limitations in cryo-EM's structural fidelity may exist, independent of radiation effects, potentially due to the intrinsic scattering properties of the sample or other factors such as side chain dynamic motions.

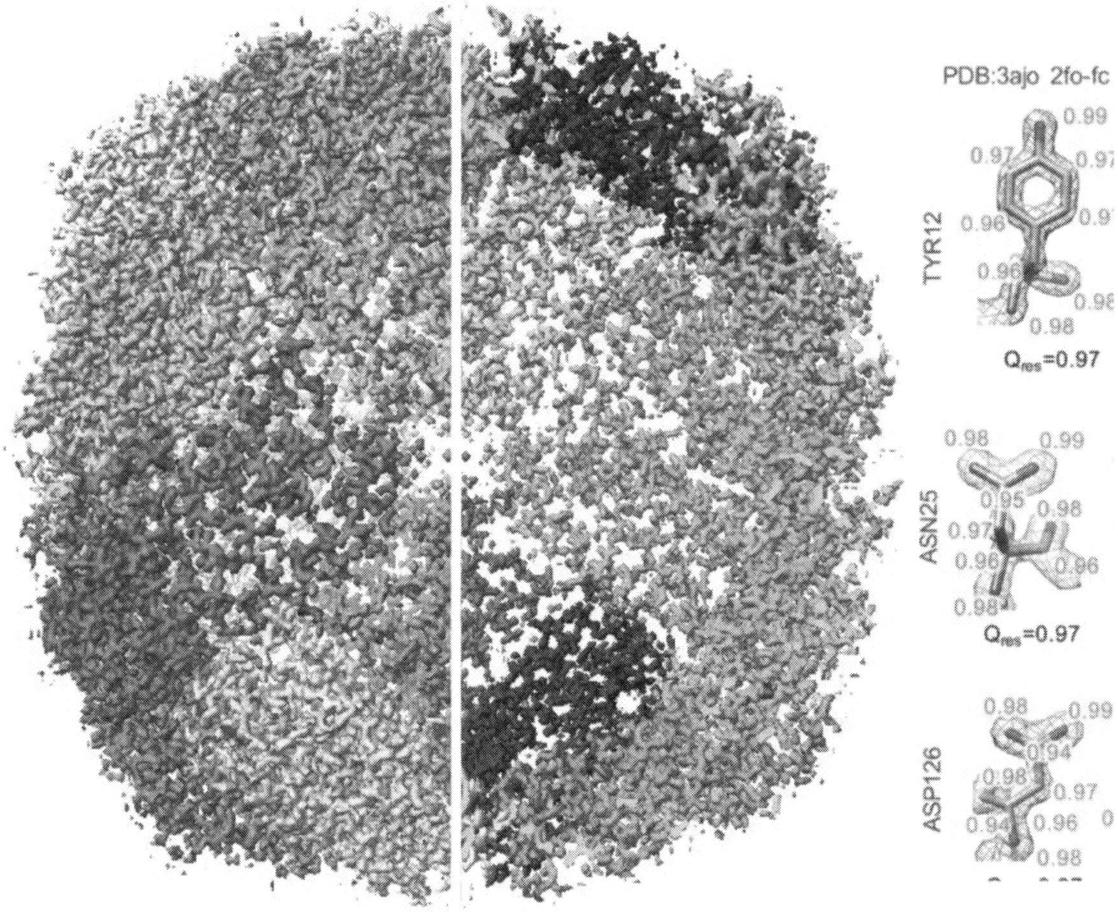
PDB:3ajo 2fo-fc
TYR12
0.99
0.97
0.97
0.96
0.96
0.98
0.98
Q_{res}=0.97
ASN25
0.98
0.99
0.95
0.98
0.97
0.96
0.96
0.98
Q_{res}=0.97
ASP126
0.98
0.99
0.94
0.98
0.97
0.94
0.96
0.98

Molding Microstructures Onto Bio-Derived Polyelectrolyte Complexes

Emily Ng, Isaac Ramírez Marrero, Sarah L. Perry

Polyelectrolyte complexation is an phase separation driven by the electrostatic forces between two oppositely-charged polyelectrolytes. This phase separation leads to a polymer-poor supernatant and a polymer-rich phase, referred to as a polyelectrolyte complex (PEC). The mechanical properties of this PEC can be tuned with the ionic strength of the solution and, given their aqueous processability, can be extracted and used as a platform for a variety of applications. These applications range from coatings and membranes to adhesives and encapsulants. Given the versatility, tunability, and aqueous processing potential of PECs, we are interested in investigating the potential for these materials as an environmentally friendly alternative to conventional plastics, which require the use of temperature and solvents to be processed. In particular, we are interested in using a liquid precursor to process solid PECs molded with different microstructures, examples including ones used for structural colors and microfluidic devices.

These microstructures often have benefits that are irreplaceable by the macro versions, such as the ability to interfere with light from the structural coloring structures, or the ability to reduce gravity effects from the microfluidic channels. From the large variety of polyelectrolytes that can be processed into films, carboxymethyl cellulose and an amino functionalized dextran are chosen as bio-derived polymers that are generally regarded as safe (GRAS) and potentially biodegradable. A food-safe amino acid-based ionic liquid is also chosen to tune the mechanical properties of both the liquid PEC precursor and the resulting solid film. We have successfully imprinted microstructures with diameters ~10 μm onto PEC films, resembling ~94% of the mold dimensions. The goal of the project is to establish a reproducible method for creating biodegradable, micromolded PEC films, as an environmentally conscious alternative in a conventional setting and laboratory setting.

Carbonated Wollastonite and Municipal Solid Waste Incinerations (MSWI) As Supplementary Cementitious Materials

Lithasha Wickramasinghe

Supplementary Cementitious Materials (SCMs) usage in concrete have been studied in relation to its benefits on the compositional properties on concrete. As the cement industry increases its emissions of CO2, the pressure to find SCM sources increases and the dependence of ordinary Portland cement (OPC) decreases. As the cement industry pursues to find ways reduce carbon dioxide emissions, multiple studies have been conducted on alternative SCMs to show their effect on reducing CO2 emissions and other compositional properties (such as strength). In this study, we studied the effects of municipal solid waste incinerations (MSWI) and wollastonite in their potential for reducing carbon emissions. We assessed MSWI and wollastonite for sequestering CO2 and evaluated the effects of carbonated wollastonite and MSWI on the cementitious matrix.

Carbonated Wollastonite and Municipal Solid Waste Incinerations (MSWI) as Supplementary Cementitious Materials

UNIVERSITY OF TEXAS ARLINGTON

Lithasha Wickramasinghe, Farzana Mustari Nishat, Warda Ashraf
Department of Civil Engineering, University of Texas at Arlington

Background & Motivation

- Ordinary Portland cement (OPC) is accountable for almost 8% CO_2 emission.
- Reducing the amount of OPC and sequestering CO_2 can contribute to the CO_2 emission reduction..
- Less reactive or waste materials can be used to store the CO_2 and used as supplementary cementitious materials (SCMs).
- The overall carbonation reaction for wollastonite and MSWI can be written as

$$Ca^{2+} \text{(from calcium bearing phases)} + CO_2 \rightarrow CaCO_3$$

Objective

- Sequestering CO_2 in less reactive materials including low calcium silicate-based materials (wollastonite) and waste-based materials (MSWI).
- Evaluating the effects of carbonated wollastonite and MSWI as supplementary cementitious materials (SCMs) on the cementitious matrix.

Materials & Methodology

- **Preparing SCM**
  - Water / powder (w/p) : 0.3
  - Carbonation time: 24 hours
- **Methods**
  - Fourier Transform Infrared Spectroscopy (FTIR)
  - X-Ray Diffraction (XRD)
  - Thermogravimetric Analysis (TGA)

Precursor SCM

- **Preparing mortar sample**
  - Water-to-binder (w/b) : 0.484
  - Sand-to-binder (s/b) : 2.75
- **Methods**
  - Compressive Strength Test

Mortar sample preparation

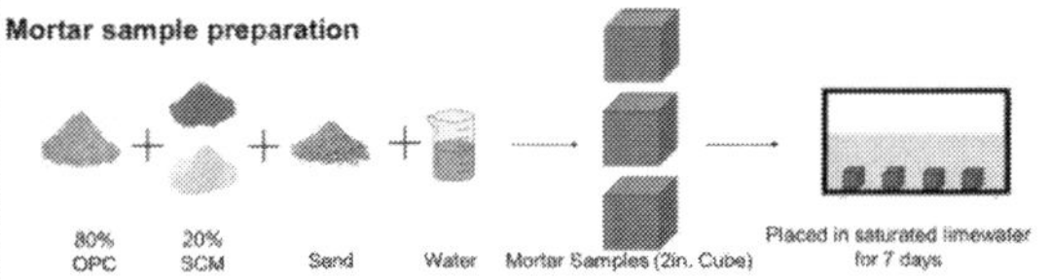

Result – FTIR

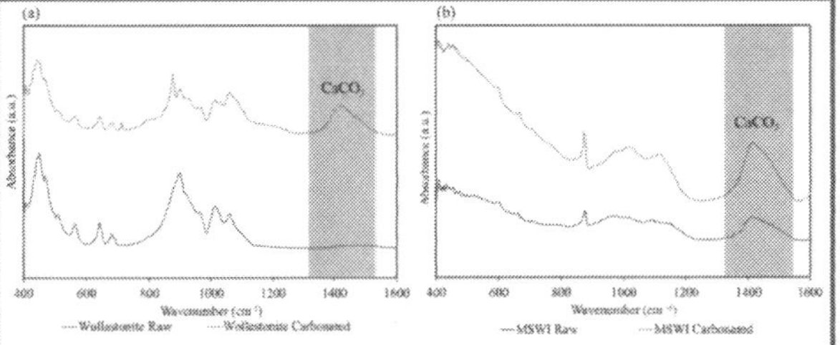

Fig. 1: FTIR pattern of (a) wollastonite and (b) MSWI before and after 24 hr. carbonation.

After 24 hour of carbonation the calcium carbonate peak increased for both samples.

Result – XRD

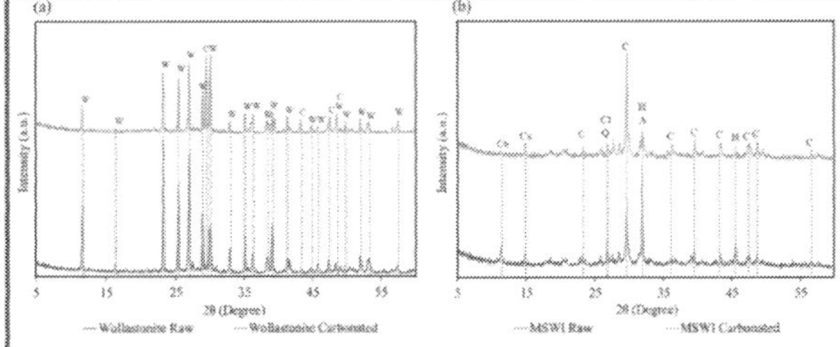

Fig. 2: XRD pattern of (a) wollastonite, and (b) MSWI before and after carbonation (C: Calcite, W: wollastonite, Q: Quartz, Cl: CaClOH, H: Halite, A: Akermanite, Ca: Calcium Sulfate Tetrahydrate, Cs: Calcium Sulfate hemihydrate (Bassanite))

The only polymorph found in the carbonated sample was calcite

Result – TGA

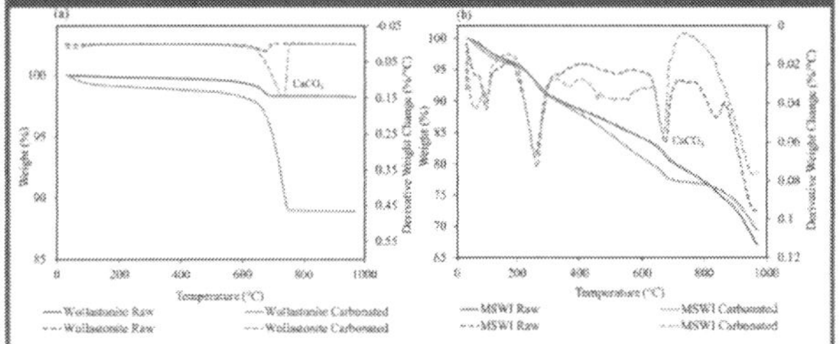

Fig. 3: TGA patterns of (a) wollastonite, and (b) MSWI.

Degree of Carbonation was calculated by the weight loss between 400°C and 800°C and was considered to calculate amount of CO_2 stored

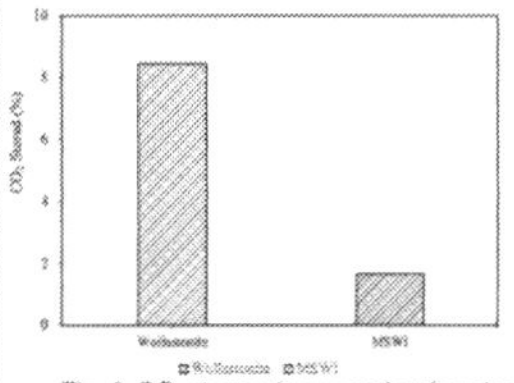

Fig. 4: CO_2 storage in raw and carbonated samples of wollastonite and MSWI

Wollastonite sequestered more CO_2 than MSWI. Wollastonite was able to sequester 8.4% and MSWI was able to sequester 1.6%.

Result – Particle Size Distribution

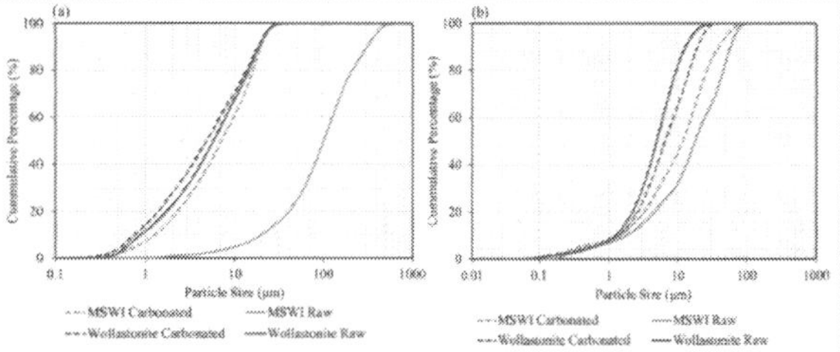

Fig. 5: Particle size distributions of (a) non-ground and (b) ground.

Particle size decreased after carbonation in non-ground. Particle size for wollastonite increased after ground but decreased for MSWI after ground.

Result – Compressive Strength

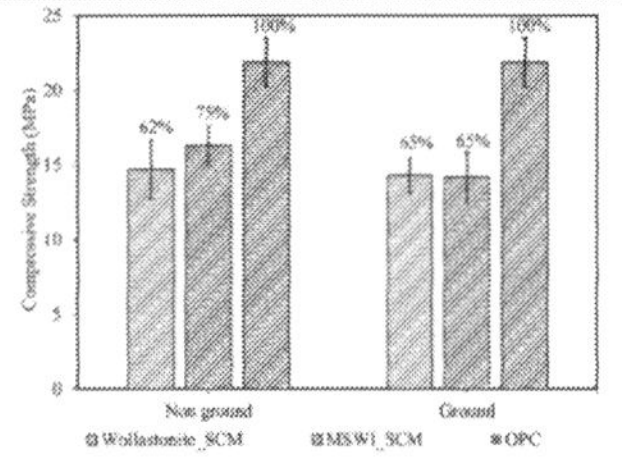

Fig. 6: Compressive strength test compared with ground wollastonite and MSWI and ground wollastonite and MSWI

As the particle size decreases, the compressive strength did not increase.

MSWI was able to achieve 75% compressive strength of OPC in non-ground batch after 7 days

Conclusion

- Wollastonite and MSWI can sequester CO_2 however the efficiency can be increased.
- Wollastonite was able to sequester 8.4% CO_2 and MSWI was able to sequester 1.6%
- Calcite is the main product formation after carbonation
- Compressive strength was lower than expected, but further treatment can help in increasing the compressive strength

Future Recommendation

- By increasing carbonation time, we might be able to increase the overall CO_2 intake
- By changing the water/powder ratio, we can increase the degree of carbonation

Acknowledgement

- National Science Foundation (NSF) funded project NSF #EFRI-2318123
- Ashraf Lab at UTA (https://www.ashraf-lab.com)

From Brown Tides to 3D Printers: Fabrication and Characterization of Novel Sargassum-Based Polymer Composite Filaments for 3D Printing

Jeziel Rodriguez Vega, Elba Herrera

As a direct effect of ocean warming and nutrient enrichment, tons of a brown macroalgae known as sargassum have been accumulating on the shores of the Caribbean region since 2011. These unprecedented annual events have been detrimental not only to some marine ecosystems but also to human health and the economy of several coastal communities. Despite the negative impact of the brown tide described above, sargassum has the potential to be a valuable resource for multiple industries, including pharmaceuticals, cosmetics, fertilizers, civil construction materials, and bioplastics. This study aimed to explore the use of this seaweed as a raw material for the fabrication of renewable micro- and nanoparticles incorporated into polymer-based filaments for fused deposition modeling (FDM) 3D printing. Specifically, the study sought to establish the process conditions for fabricating sargassum powder and polylactic acid (PLA)/sargassum composite filaments, study the effects of powder particle size and weight percentage on the microstructure, mechanical properties, thermal stability, and 3D printability of the fabricated filaments, examine the relationship between printing conditions, microstructure, and mechanical properties of the 3D printed specimens, and finally, study the biodegradability of these novel sargassum-based 3D printing materials.

In the procedure, dried sargassum previously collected from local beaches was ground in a food processor to obtain microparticles. Subsequently, a portion of the resulting material was re-ground in a planetary ball mill to obtain nanoparticles. Next, PLA pellets were coated with suitable amounts of powder (0% - 35% by weight) through mechanical mixing using small amounts of a suitable solvent as glue. The coated pellets were then fed into an extruder to fabricate the filaments. The extrusion temperatures and spooling speed were controlled to obtain the required filament thickness and quality. The sargassum-based filaments were placed in an FDM 3D printer to fabricate various dog-bone-shaped specimens and consumer products such as cell phone cases, earbud holders, and eyeglass frames.

The obtained powders were analyzed using AFM, SEM, FTIR, and Raman spectroscopy. The microstructure, thermal stability, and mechanical properties of the filaments and specimens were evaluated using SEM, TGA, and tensile tests, respectively. Preliminary biodegradability studies were conducted through soil burial tests.

The intellectual merit of this work lies in the fact that, for the first time, macroalgae powders were used in FDM 3D printing, offering alternatives to improve sustainability in 3D printing and addressing issues such as macroalgae proliferation.

Enhancing Composite Polymer Electrolytes for All-Solid-State-Batteries By $Ti_3C_2T_x$ Mxene Coupled with Anionic Covalent Organic Frameworks

Ty Leinen, Sahand Serajian, Syed Ibrahim G P, Mahmoud Shaban, Siamak Nejati, Mona Bavarian

All-solid-state batteries (ASSBs) have emerged as a possible solution to address the challenges faced by conventional batteries. Conventional batteries rely on liquid electrolytes which are susceptible to issues such as electrolyte leakage, dendrite formation, and thermal runaway. The improved safety of ASSBs, make them a suitable candidate for replacing liquid electrolyte-based batteries. However, the poor ionic conductivity of solid electrolytes hinders their deployment and commercialization. In this study, we present a composite polymer electrolyte (CPE) with a two-dimensional transition metal carbide, $Ti_3C_2T_x$ MXene, coupled with an anionic covalent organic framework (aCOF) to enhance the composite electrolyte ionic conductivity. This novel CPE was synthesized using polyethylene oxide (PEO), lithium bis(triflouromethanesulfonyl)imide (LiTFSI), freeze-dried MXene, and aCOF. This electrolyte's performance was compared against that of a control CPE consisting of PEO, LiTFSI, and freeze-dried MXene. The addition of aCOF improved the electrolyte thermal stability, and enhanced its ionic conductivity. The aCOF CPE was characterized through thermal gravimetric analysis (TGA), differential scanning calorimetry (DSC), and electrochemical impedance spectroscopy (EIS). After heating to 1000 °C, the aCOF CPE's residual mass increased by 8% through TGA. The electrolyte demonstrated an increase in conductivity by two orders of magnitude due to the decrease in its crystallinity. These results demonstrate the synergy between aCOF and MXenes through enhanced thermal properties as well as improved lithium-ion transportation.

Investigating Polymer Morphology Using Serial Block-Face Scanning Electron Microscopy

Zaw Htet Lin, Alexander Ribbe

Serial Block-Face Scanning Electron Microscopy (SBF-SEM) is a relatively recent innovation in the field of electron microscopy, offering the ability to perform 3-D reconstruction of materials with nanometer resolution. This makes SBF-SEM a valuable imaging technique for investigating the morphology found throughout a polymeric material, revealing intricate structures like phase separations commonly studied in polymer science, which are critical in determining the physical properties of materials.

Alongside minimal sample preparation, SBF-SEM's ability to automate the serial sectioning and imaging process enables the achievement of large imaging volumes that would have otherwise consisted of a long and drawn-out preparation process. Before the development of SBF-SEM, samples needed to be prepared using an Automatic Tape-collecting Ultra-Microtome (ATUM) to be then observed under an SEM. If a Transmission Electron Microscope (TEM) is employed, ultra-thin sectioned samples are also required, which calls for complex preparation protocols. On the other hand, SBF-SEM uses an in-chamber ultramicrotome equipped with a diamond knife, allowing serial sectioning and imaging to be done through automated procedures.

This poster underlines the application of SBF-SEM to polymer analysis and its interdisciplinary utility, especially in the fields of material and biological sciences. Technical aspects, such as sample preparation challenges, will also be discussed. By leveraging the high-resolution 3-D imaging power that SBF-SEM provides, a deeper understanding of the micro and nanostructures that dictate polymer properties can be achieved, paving the way for developing more advanced materials.

Partitioning Behavior of Low-Risk Drugs in Poly(ethylene-oxide)-Block-Poly(ε-caprolactone) Copolymer Films

Aidan Stewart, Ryan Van Horn

Poly(ethylene-oxide)-block-poly(ε-caprolactone) (PEO-*b*-PCL) is a copolymer that has been researched for a variety of medical applications including drug delivery. Understanding how drugs segregate within the amphiphilic PEO-*b*-PCL matrix is crucial for interpreting release profiles and developing effective drug delivery systems. This study investigated the partitioning behavior of four low-risk drugs with varying solubilities, caffeine, ibuprofen, aspirin, and acetaminophen, through analysis of changes in physical properties. Polymer films were created using the solvent-casting method, where both the polymer and drug were dissolved in chloroform, cast, and allowed to dry. The polymer films were made with varying weight percentages of the drug ranging from 0 wt% to 100 wt%. Differential Scanning Calorimetry (DSC) and Attenuated Total Reflectance Fourier Transform Infrared Spectroscopy (ATR-FTIR) were employed to evaluate the polymers' crystallinity and crystallization temperatures. Inclusion of drug molecules in the polymer matrix should decrease these properties. Evidence of lower crystallinity or crystallization temperature provides insight into drug partitioning within the hydrophilic PEO and/or hydrophobic PCL blocks of the copolymer.

A Dielectric and Calorimetric Investigation of Modified Cellulose Polymers

Kaitlyn Zuravel, Stavros Drakopoulos, Rodney Priestley

With the depletion of fossil fuels, the potential of plant-based polymers to replace traditional synthetic plastics with equivalent eco-friendly alternatives is a beacon of hope in the field of materials engineering and environmental sustainability. These polymers, abundant and usually featuring biodegradability and biocompatibility, offer a promising future for various applications ranging from biomedicine and tissue engineering to packaging. In fact, cellulose, the most abundant polysaccharide found in nature, is easily chemically modified into various types of cellulose derivatives like methylcellulose (MC) and hydroxypropylmethylcellulose (HPMC). First, we delved into the practical applications of our research by analyzing the calorimetric *Tg* and the peak water loss for MC and HPMC samples using a Differential Scanning Calorimeter (DSC) equipped with an RCS90 cooler. The samples were tested using standard heating cycles from -50.000°C to 200.000°C and then to 300.000 °C. Second, we explored the real and imaginary parts of dielectric permittivity for MC and HPMC samples using broadband dielectric spectroscopy (BSD). In a temperature-controlled chamber using an Alpha analyzer, the voltage amplitude V_{rms} of the applied field was kept constant at 1 V, from 10 mHz to 1 MHz, to perform isothermal scans ranging from -45°C to 195°C at 5°C increments. DSC analysis identified a correlation between the MC samples' molecular weight and midpoint *Tg* value. A proportional relationship was found for the compressed samples. However, the powder samples showed an inverse relationship. For HPMC samples, a correlation was found between the surface hydrophobicity (SH) and the *Tg*, in powder samples. For all compressed MC and HPMC samples, BDS analysis located segmental relaxation caused by the *Tg* at higher temperatures and lower frequencies. We also found both β and σ relaxation curves in both MC and HPMC samples. The HPMC samples showed a clear positive correlation between the SH value and the dielectric constant at both the β and σ relaxation points. A small correlation was found for the MC samples between molecular weight and the dielectric constant at low frequencies.

Waterproof Polymer Coatings for Phosphate Laser Glass-Based Actively Cooled Laser Disk Amplifiers

Justin Landis, Kenneth L. Marshall, Abigail Bonino, Prathiksha Mangalasubaskaran

All high-peak-power laser systems around the world involved in fusion research employ phosphate-based, neodymium doped laser glass (e.g., LHG-8) to amplify laser energy in the near IR wavelength region. One limitation of phosphate laser glasses is that they are attacked by water or water vapor to degrade the polished laser glass surfaces, which in turn leading to a reduction in laser energy output due to light scattering. This moisture sensitivity limits cooling methods to either "passive" gas-phase cooling (e.g., nitrogen or helium) or "active" fluid based liquid coolants such as fluorocarbons or ethylene glycol-water mixtures. Next-generation laser disk amplifiers will require active cooling in order to achieve higher energy on-target with higher shot repetition rates, but neither fluorocarbons nor ethylene glycol-water mixtures represent a viable alternative (the former is a "forever" chemical, while the latter limits laser output due to its near-IR absorption). "Heavy "water (D_2O) represents a viable alternative, but only if the LHG-8 can be coated with a protective polymer layer to eliminate or mitigate erosion of the laser glass surfaces by D_2O without compromising optical transmission or laser-induced damage. In this work, five polymer systems were evaluated as potential protective coatings for phosphate laser glass by dip-coating polished LHG-8 samples in solutions of the polymer candidates in various solvents and different % solids concentrations to produce highly transmissive coatings free of intrinsic defects ranging from tens of nanometers to tens of microns in thickness. Three of these polymer coating materials exhibited films with high surface uniformity, excellent optical transmission, and large hydrophobicity (as evidenced by contact angle measurements) on LHG-8 laser glass. These samples are currently under evaluation in a flowing coolant test bed to determine their long-term resistance to both erosion and coating defect generation induced by water.

This material is based upon work supported by the Department of Energy [National Nuclear Security Administration] University of Rochester "National Inertial Confinement Fusion Program" under Award Number(s) DE-NA0004144 and the U.S. National Science Foundation under Cooperative Agreement No. (PHY-2329970).

Impact of Degradation PEO-b-PCL on Aqueous pH

Tara Whipkey, Joanna White, Kayla Jarski, Ryan Van Horn

Polymers have unique characteristics when combined into copolymers which have characteristics of two or more polymers. Poly(ethylene oxide)-block-Polycaprolactone (PEO-b-PCL) is one such co-polyester that demonstrates both hydrophilic and hydrophobic characteristics, respectively. The amphiphilic nature of PEO-b-PCL makes it an ideal candidate for sustained drug delivery, especially with hydrophobic drugs. PCL is a class of polymers called polyesters, that are known to release carboxylic acid when they degrade in water. Through this research, it has been found that the carboxylic acid, specifically caproic acid, produced by PCL degradation in water impacts the overall pH of its environment with fast degrade and mid-range degrade PEO-b-PCL samples, but does not affect the environment of slow degrade samples. The main objective for past degradation studies was to examine the speed of degradation instead of degradation effects on its environment. This study showed how the PEO interacted with the caproic acid production of PCL. When these degradation environment studies were first conducted the pH changed similar to a weak titration curve for the slow degraded samples. New data suggests that slow degraded samples do not change the pH since it degrades over a longer period of time, but fast and mid-range degraded samples do affect their environment, confirming the previous data for fast degraded samples. As there was an initial decrease in pH for the slow degrade and mid-range degrade samples, the PEO seemed to create micelles that surrounded the caproic acid either increasing the pH or maintaining it, until the micelles eventually degraded. Micelle formation was tracked by using a UV-vis spectrophotometer that showed an increase in absorbances as the experiment progressed. Knowing how the different concentrations of PEO to PCL affect their environment ensures the safety for drug delivery. Different areas of the body maintain a constant pH for maximum efficiency, thus maintaining that pH is important. The marine environment also operates at a pH of 8.1 to 8.2 and changes can have dramatic effects. This research gave insight to the safety of using the copolymer in the body or a marine environment.

Comparative Analysis of Vacuum Filtered Vs Freeze-Dried $Ti_3C_2T_x$ Mxene in Composite Polymer Electrolytes

Micah Quirie, Sahand Serajian, Mona Bavarian

As global electrification advances, the demand for safer, energy-dense batteries have become dominant. Conventional batteries containing liquid electrolytes pose serious risks associated with thermal runaway, dendrite formation, and electrolyte leakage. To overcome these challenges, adopting solid-state batteries (SSBs), containing solid electrolytes, is one of the potential solutions due to their safety, high energy density, and prolonged lifespan. However, the low ionic conductivity of solid electrolytes remains one of the most significant challenges preventing their large-scale deployment. The focus of our research is to enhance the ionic conductivity of composite polymer electrolytes (CPEs) containing $Ti_3C_2T_x$ MXenes. We studied the effect of MXene processing used for the preparation of composite electrolyte and compared the ionic conductivity of electrolytes prepared using MXene flakes separated from suspension via freeze-drying and vacuum filtration. The films were characterized using X-ray diffraction (XRD), scanning electron microscopy (SEM), and electrochemical impedance spectroscopy (EIS) techniques. It was observed that the freeze-dried MXene-based CPE had an ionic conductivity of 1.09E-3 S/cm, presenting three orders of magnitude enhancement compared to the control sample, which was an MXene-free composite polymer electrolyte comprised of polyethylene oxide (PEO) and lithium bis(triflouromethanesulfonyl)imide (LiTFSI). Most importantly, the freeze-dried ionic conductivity exhibited one order of magnitude greater than the conductivity of the vacuum-filtered MXene-based CPE. This work has shown how significant post-processing techniques are in optimizing the performance of MXene-based solid electrolytes for solid-state battery applications.

Effect of Polymer Chemistry on the Rheological Trends of Polyelectrolyte Complexes

Yaozu Chen, Isaac Ramírez Marrero, Sarah Perry

Polyelectrolyte complexation is a phase separation phenomenon resulting from the electrostatic interactions between two oppositely charged polyelectrolytes. Polyelectrolyte complexes (PECs) are widely used in products like shampoo and produced by some organisms as bioadhesives. PECs benefit from aqueous processing and avoid high temperatures or organic solvents. Additionally, their properties can be tuned by modulating the length, charge density, and hydrophobicity of the polyelectrolyte. Furthermore, it is well understood that salt concentration strongly affects the mechanics of the PEC. Salt ions disrupt the electrostatic interactions between polyelectrolytes, which typically causes the complex to soften and swell. Thus, while electrostatics are known to be important, the effect of the identity of the ionizable group of the polyelectrolytes on the mechanics of the material is not well-understood. To bridge this knowledge gap, we are studying the effect of changing the identity of the ionizable group and its relationship with salt concentration on the linear viscoelasticity and phase behavior of these materials. Specifically, we examine the effect of anionic carboxylate and sulfonate groups, along with cationic amines with different degrees of methylation. These rheological studies give us insight into properties like viscosity and solid-to-liquid transitions. We apply time-salt superposition to combine data from individual experiments and a produce master curve for each of the systems to allow for a broader examination of how the mechanical properties can be tuned with salt concentration – from glassy complexes at low salt to viscoelastic liquids at higher salt concentration. Our preliminary data for PECs created from cationic poly(vinylamine) paired with either anionic poly(vinylsulfonic acid) or poly(acrylic acid) exhibit similar rubbery behavior from 0.05 M to 1.75 M potassium bromide. We have also determined that the different ionizable groups affect the interaction strength between the polyelectrolytes, with our preliminary data showing that cationic groups have a greater effect. By understanding how salt concentration and polymer chemistry affect PEC mechanics, we can design and aqueously process PECs for specific applications, presenting a more environmentally friendly option compared to current plastics.

Colorimetric Detection and Quantification of Low Concentrations of Hydrogen Peroxide for Applications in Stimuli-Responsive Polymers

Georgia C. Delaney, Alexa LaSasso, Katherine Pappas, Finlay J. M. Smith, Melissa B. Gordon

The detection and quantification of radicals has useful chemical, biological, and physical applications. When certain radicals are produced in aqueous solutions, they react with water in the presence of dissolved oxygen, forming hydrogen peroxide. In order to quantify the concentration of these hydrogen peroxide molecules in a given sample, we can employ the Fenton reaction in which hydrogen peroxide oxidizes ferrous ions. When this reaction occurs in the presence of xylenol orange, the resulting ferric ion complexes with xylenol orange to induce a visible color change from yellow to brown. The stability and consistency of this method make it a reliable tool for the detection and quantification of hydrogen peroxide in aqueous solutions. Though the Fenton reaction is commonly used to detect hydrogen peroxide, its conditions have not been optimized to operate over a low concentration range. In this study, we optimize the component concentrations to achieve detection and quantification of hydrogen peroxide ranging in concentration from 0.2 μM to 20 μM. For these optimized component concentrations, we determined the reaction time of hydrogen peroxide under our specific conditions, and evaluated the stability of the solution both on its own and when reacted with hydrogen peroxide. Using these conditions, we used UV-visible spectroscopy to observe changes in absorbance measured at 440 nm and 580 nm that occur with increased concentration of hydrogen peroxide and correspond to peaks associated with unbound and complexed xylenol orange, respectively. We developed a calibration curve that uses a ratio of peak heights at 440 nm and 580 nm to assess hydrogen peroxide concentration. This assay and calibration curve enable the detection and quantification of hydrogen peroxide concentration, which will be used to measure radical concentrations generated from light- or force-induced bond cleavage in polymers, and has practical applications in the testing of biological and food samples.

Engineering Solid Polymer Electrolytes for Structural Batteries with Enhanced Mechanical and Electrochemical Performances

Kyra Glassey, Paul Coman, Monirosadat Sadati

Lithium-ion batteries are widely used today due to their longevity and efficiency. They are, however, highly flammable and possess steep volumetric requirements. Structural battery composites are an emerging alternative that integrates battery and structural functions in a single multifunctional component, reducing weight and increasing safety and energy density. These composites harness carbon fibers for structural reinforcement and as electrodes, with a solid polymer electrolyte (SPE) serving as the separator and matrix. Traditionally, in SPEs mechanical and electrochemical properties are mutually exclusive. Here, we explored a bi-continuous SPE with polymer-reach and solvent-reach domains, synthesized through polymerization-induced phase separation, to decouple ionic conductivity and mechanical properties. We investigated the impact of monomer/solvent+Li salt ratio, film thickness, and the addition of glass fibers (GF) on the synthesized SPEs performance. Characterization through thermal analysis, mechanical testing, SEM and electrochemical impedance spectroscopy demonstrated that a 40/60 and 50/50 weight ratio of monomer/solvent+Li salt results in a polymer film with a bi-continuous structure leading to improved ionic conductivity. Furthermore, the incorporation of GFs in polymer film improved its mechanical strength, as confirmed by tensile testing. Surprisingly, adding GFs to the polymer films also significantly increased the ionic conductivity of the composite SPEs. SEM images revealed the formation of void spaces at the polymer/GF interface, which can potentially contribute to improved ionic conductivity via enhancing ions mobility and permeability. Developing lightweight, and compact SPEs with enhanced ionic conductivity, mechanical strength, safety and efficiency
can facilitate further advancement of structural batteries.

Novel Modeling of Highly Crosslinked Thiol-Ene Photopolymerization Reactions to Accurately Predict Diffusion Kinetics

Jamison Barcelona

A growing interest in holographic polymers requires researchers to have greater control of the polymerization process to generate improved holographic properties (increase refractive index and decrease haze). As such, a strong focus has been aimed towards improving models for holographic polymerizations. One promising approach uses thiol-ene step growth photopolymerization (also known as thiol-ene "click" chemistry) to create holographic polymers. The reaction kinetics of this "click" reaction are well understood and are traditionally controlled with temperature, concentration, light exposure, and light intensity. However, viscosity increases beyond the gel point and radical termination kinetics prevent the polymerization from reaching conversions greater than 80%, even at high temperatures. Here, experimental results are used to develop novel models for the polymerization process in an effort to predict the conditions necessary to achieve higher conversion and superior control over local conversion for improved image clarity. Previous models have effectively predicted reaction kinetics, but fail to predict diffusion kinetics due to high crosslink density and viscosity. Improvements have been made to model accuracy for highly crosslinked polymers by obtaining viscosity and reaction data from differential scanning calorimetry (DSC) and photo-DSC for high molecular weight monomers with high functionality (3 or more functional groups). Pentaerythritol triallyl ether (APE) was used as a novel high molecular weight monomer containing three alkene functional groups with many degrees of rotational freedom. It is predicted that rotational freedom will reduce glass transition temperature leading to higher conversions when reacted with PETMP. The APE PETMP polymerization was characterized by generating an empirical correlation curve between glass transition temperature and conversion. This information was used to accurately model reaction rate and conversion using reaction kinetics and free volume theory. Further investigation into "click" chemistry will focus on improving models that expand our understanding of the fundamental process of diffusion in highly crosslinked polymerizations. This has applications to other highly crosslinked step growth polymerizations (e.g. polyester and polyurethane) and will allow for greater control of local conversion; which is important in the continued development of polymer printing processes (e.g. StereoLithogrAphy) and holographic polymers for use in augmented reality, security, and information storage.

Novel Polymeric Morphologies As Positive Electrodes in Lithium-Ion Batteries

Cade Tharrington, Michael J. Petrecca, Orlin D. Velev, Peter Fedkiw

Growing demands for next-generation energy storage technology for utilization in consumer and grid energy storage applications have prompted a re-evaluation of lithium-ion batteries (LIBs), which are the preeminent electrochemical energy storage technology. Due to the cost and environmental impact of transition metals used in the positive electrode, researchers are investigating novel materials and processing techniques to introduce new chemistries and electrode architectures into LIBs. Herein, we propose a novel materials-processing platform to develop application-specific polymeric electrode architectures for next-generation LIBs. Organic radical polymers are of interest as an alternative positive electrode for LIBs. In particular, poly (2,2,6,6 tetramethyl-1-piperidinyloxy-4-yl methacrylate) (PTMA) has garnered attention for its relative ease of synthesis and its fast charge-discharge kinetics. However, before implementation as an electrode, PTMA requires a significant amount of conductive additive and binder to remedy its electronically insulating polymeric backbone and its solubility in organic electrolytes, all of which decreases the gravimetric performance of PTMA. The multiphasic processing platform presented here utilizes the precipitation of PTMA in turbulent solvent-nonsolvent induced phase separation to introduce a soft dendritic colloid (SDC) morphology. Characterized by fractal branching and nanofibrillar contact splitting, the morphology of PTMA SDCs rectifies the solubility and electronic conductivity limitations associated with PTMA electrodes by introducing improved surface area and self-adhesive properties. Rate capability and cycling stability improvements are expected from PTMA SDC electrodes compared to benchmark slurry cast PTMA electrodes due to more efficient contact with conductive additives and reduced electrolyte solubility in PTMA SDCs. This hypothesis is tested and reported upon here through electrochemical analysis via cyclic voltammetry and galvanostatic charge and discharge experiments of PTMA SDC and benchmark slurry cast PTMA positive electrodes in half-cell environments against reference lithium metal electrodes.

Conjugate Addition Polymerization of Glycal Monomers Using Dbu As Catalyst

Madelyn Funke, Mudassir Syed, Samantha Kristufek

Conjugate addition polymerization (CAP) is a form of polymerization used most in creating synthetic rubbers like polybutadiene or polychloroprene, from a small monomer that contains a double bond. The method boasts great control over crosslinking and flexibility, as well as a simple initiation step, utilizing a strong nucleophilic catalyst that is required in minimal amounts to begin the Michael Addition Reaction, which drives the polymerization. Our research group has taken this method and utilized it on a sugar-based monomer to form a renewable, sustainable polymer. Using D-glucal, a derivative of glucose, as the starting material, an oxidation and a protecting step ensures the required olefin and ketone functionalities on our monomer to polymerize with an added nucleophilic catalyst, DBU. Upon the removal of the protecting group, the structure of the material created has multiple alcohols that act as handles for further functionalization. This promises the discovery of various characteristics that imply a wide range of industrial applications. The project has seen great success in creating the polymer at a high molecular weight and now focuses on testing the material properties of the plastic and cross-linking chains to evaluate the effect on mechanical properties. However, there is extensive research that can be done with the novel polymer due to its sugar backbone, as there are many exploitable functional groups that can be changed to create new distinct polymers, each with unique material properties. Eventually, our group plans to create a library of sustainable polymers that can be developed from the base process of CAP to great yields. These fundamental studies allow for future work into scalability towards industrial applications, eventually using these materials in variety of applications, which will improve the usage of renewable and sustainable materials in plastic production.

FTIR Analysis on Solvent-Cast PCL-*B*-PEO-*B*-PCL Films

Carden Osborne, Ryan Van Horn

Biocompatible and biodegradable polymers are of interest for a variety of reasons, including their applications in drug delivery. Hydrophilic poly-ethylene oxide (PEO) and hydrophobic poly-ε-caprolactone (PCL) are two commonly used semi-crystalline polymers that can be covalently bound in a variety of hierarchical assemblies. The properties of semi-crystalline copolymers are often affected by which polymer segment crystallizes first. Crystallization order is affected by the solvent used to suspend the polymer prior to casting the polymer for its final purpose. Depending on what solvent-solute interactions are more energetically favorable, one block will precipitate out of the suspension as the solvent dries. While previous work has determined how solvents affect diblock PEO-*b*-PCL chains, other assemblies have not been thoroughly tested. One step towards a better understanding of how PEO and PCL interact is the analysis of a covalently bound tri-block configuration with both polymer types being of similar or equal molecular weight. This research shows that the crystallization of 5k10k5k PCL-*b*-PEO-*b*-PCL is affected by the solvent type using FTIR heating and cooling data. These solvent effects remain even after the polymer has been fully melted and allowed to recrystallize.

Depolymerization of Polyurethanes and Recovery of Isocyanates to Aid Plastic Circularity

Genevieve Kroll, Robert M. O'Dea, Alison J. Shapiro, Thomas H. Epps, III

Polyurethanes are a ubiquitous group of polymers used in a myriad of applications, including foams, coatings, and textiles; however, their versatility leads to a large amount of polyurethane waste with diverse chemical structures, resulting in challenges with end-of-life recycling. Polyurethanes are composed of polyols, isocyanates, and chain extenders, and current polyurethane recycling methods focus on polyol/chain extender recovery with some efforts aimed at recovering amines from the isocyanate component – yet, there have been no significant steps towards the recovery of isocyanates directly, limiting the overall circularity of polyurethane materials. This work demonstrates a first-of-its-kind depolymerization approach that enables nearly quantitative recovery of isocyanates from laboratory-made and commercially sourced polyurethanes, both thermoplastics and thermosets, using an organoboron Lewis acid under mild conditions (60-80 °C, toluene or tetrahydrofuran, ambient pressure). The recovered isocyanates were used to prepare second-generation polyurethanes that exhibited molecular weights and thermal properties similar to virgin material. Additionally, the impact of polymer additives (*e.g.*, antioxidants, dyes) was explored. Overall, this work demonstrates a significant advancement toward complete circularity of polyurethanes and could mitigate the need for the production of new polyurethane precursors (*i.e.*, new isocyanates using phosgene-intensive routes) that are toxic and harmful to the environment.

Extraction and Fractionation of PVC for Recycling and Upcycling

Logan Anguiano, Owen Lubic, Ali Alshaikh, Pravin Shinde, Jaewoo Choi, Jason Bara

Polyvinyl chloride (PVC) is a widely used polymer found in numerous products globally. Its versatility and chemical resistance make PVC essential in applications ranging from medical devices to plumbing and construction materials. However, the chemical structure of PVC which imparts these properties also creates significant challenges for end-of-life PVC. Due to its widespread use, PVC accumulates in oceans and landfills, leaching toxic additives and plasticizers into the environment. Recent research into PVC is developing methods to repurpose it, effectively diverting it from landfills and potentially reducing the need for additional PVC production.

Our research focuses on extracting relatively low MW PVC from bulk waste. By fractionating PVC, we can obtain a distribution of shorter chains that is more soluble in "green" solvents and easier to process. We have been able to obtain high yields of near-virgin quality PVC from both pristine and post-consumer sources. Additionally, the fractionated product is ideal for facilitating chemistry for modification and/or depolymerization, making it a valuable resource for new PVC applications. Ultimately, these advances could transform the perception of waste PVC from a negative-value material to a positive-value resource with significant recycling potential.

This presentation will discuss our methods and results in extraction and fractionation processes of PVC, as well as the effects of source material and solvent type on molecular weight distribution associated with each fraction.

Design of Random Heteropolymers As Next-Generation Protein Stabilizers

Akorfa Dagadu, Tianyi Jin, Alfredo Alexander-Katz

The decomposition of biodegradable plastics via nano-dispersed enzymes offers a promising solution to plastic pollution. The stabilization of enzymes using methacrylate-based random heteropolymers (RHPs) has shown potential in facilitating the degradation process. However, the mechanisms underlying RHP-facilitated enzyme stabilization remain unclear. This work focuses on understanding the morphology of the RHP-enzyme interaction, particularly with Proteinase K, known for its capability to degrade polyesters. By using biased and unbiased atomistic molecular dynamics simulations and machine learning, we explore the effect of different RHP compositions and initial conjugate structures, including globule-globule ('polymer-enzyme dimer') and coil-globule ('polymer wrapping enzyme') morphologies. The polymer globule in the dimer morphology interacts with the enzyme primarily through the side chain, and the backbone reconfiguration is hindered by the glassy nature of PMMA. The polymer coil in the wrapped morphology partially refolds, and further refolding is not favored. The preference for polymer-enzyme contacts correlates with the enzyme surface characterization and agrees with experimental results. We believe that the atomistic details of such conjugates are crucial for future material design and enzyme stabilization.

Visualizing Polymer Concentration Gradient in Organogels Using Dye

Christian Owens, Kenneth Mineart

One current goal of our research group is to create polymeric organogels with stiffness gradients by spatially varying polymer concentration within them. The present work focuses on developing a robust system for visualizing these polymer concentration gradients since they are otherwise undetectable to the human eye. This was accomplished by formulating gel samples with correlated concentrations of polymer and dye. In order to quantify dye concentration, gels were analyzed with UV-Vis spectrophotometry. We noted a considerable peak in the UV-Vis spectra (at 634 nm) when our selected dye (Blue 1) was incorporated into gels. Therefore, subsequent samples were analyzed through consideration of their absorbance value at 634 nm. A calibration curve between absorbance and dye concentration indicates that Beer's Law is obeyed up to 1 wt% dye solution and so all samples were prepared at or below this concentration. The calibration curve was then used to evaluate unknown dye concentrations in gradient samples. The resulting data demonstrate good agreement between observation of dye gradient samples and their UV-Vis spectrophotometry data. Additionally, this information shows that the applied method of gradient sample preparation is consistent and repeatable. The present work establishes a way to visually observe polymer concentration gradient, and corresponding elastic modulus gradient, in polymeric organogels.

Upcycling PVC into 3D Printing Resins

Owen Lubic, Ali Alshaikh, Logan Anguiano, Jaewoo Choi, Jason Bara

It has become evident that traditional recycling efforts are unable to handle the volume of plastic waste produced worldwide. A potential remedy to this issue may be polymer upcycling: the repurposing of waste plastic into materials with more value. Poly(vinyl chloride) (PVC) is a strong candidate for upcycling as its qualities can be leveraged in a variety of ways. One application we are pursuing is the use of PVC for 3D printing resins to produce materials with tailored properties, while simultaneously reducing the amount of "fresh" resin required for printing. We have developed methods by which PVC can be processed and then blended with commercial 3D printing resins. The mechanical properties of the resulting 3D printed objects depend on the PVC mass fraction and molecular weight; therefore, altering these variables enables the development of tunable materials to meet specific applications. This presentation will examine the relationship between the PVC source and resin type on the properties of the 3D prints, how the mass fraction of PVC affects resin performance, and how incorporating PVC influences the overall qualities of the 3D printed parts.

Phase Behavior of Precision Polyelectrolyte/Polyether Blends

Cecilia Hansen, Michael Patrick Blatt, Daniel Hallinan Jr., Victoria Horton, Rufina Alamo, Nam Nguyen, Justin Kennemur

In this work we studied the miscibility of a blend of poly(ethylene oxide) (PEO) and a precision polystyrene analogue with five-carbon phenyl spacing that is converted into a polyanion by functionalizing the phenyl groups with lithium-form (trifluoromethyl)sulfonylimide. Blends were found to be miscible at high to moderate polyanion concentration ranging from 10 to 70 wt% PEO. However, as the polyanion became more dilute (90 wt% PEO and beyond) visible, macroscopic phase separation was observed. Differential scanning calorimetry was used to study trends in the primary and secondary phase transitions of each blend to better understand the phase boundaries and interactions underpinning the phase behavior. The glass transition temperature (*Tg*) of each blend was measured as a function of composition. A single *Tg* was identified at each composition. Within the miscible range, the *Tg* values were well fit by the Gordon-Taylor model, but significant disagreement between model and experiment was observed for two-phase blends (>90 wt% PEO). Furthermore, a clear linear relationship between melting point depression and wt% PEO was visible for all miscible, semicrystalline blends apart from those in the miscibility gap. These findings give an empirical compliment to modeling efforts that show that complicated phase diagrams can arise due to ion presence and long-range electrostatic interactions in charged polymer blends.

Using Physics Based Models to Predict the Conductivity of Solid Polymer Electrolytes

David Pinegar, Paul Coman, Ralph White

This study explores the conductivity of solid polymer electrolytes (SPEs) crucial for Li-ion battery technology using computational modeling. A one-dimensional case is used on a simple electrolyte domain, writing a mass balance on each species. The individual equations for each species within the model integrate both diffusion and migration terms but are mathematically altered to obtain a singular simple diffusion equation, which becomes essential for the conductivity predictions using the Nernst-Einstein equation. The electric potential then results in an analytical solution and the domain is induced under a 1 mA cm-1 alternating current. The model is then optimized with diffusion coefficients and a rate constant being independent variables and validated against electric potential experimental data from literature concerning liquid electrolytes using a Li | Li cell. This is later generalized to SPEs with the addition of a volume fraction. Employing COMSOL Multiphysics, simulations indicate that the conductivity of a 1M LiPF6 50% SPE approximates 5.3 mS cm-1, closely aligning with other experimental measurements, with the optimized parameters reflecting similar magnitudes to experimental literature as well. Future work would be to include the SPE's ion hopping, adding to the conductivity, and continuous experimental validation. This computational approach underscores its pivotal role in advancing battery technology by enabling predictions and validation of SPE conductivities before construction. Such validation against liquid electrolyte benchmarks provides confidence in the model's ability to predict behaviors in solid electrolytes more accurately.

Miscibility of Binary Polymer Blend Confined with Asymmetrically Wetted Nanoparticle Packing

Lorenzo Galang, Robert Riggleman

Compatibilization of polymer blends can produce tunable materials with a wide range of mechanical, thermal, and transport properties, making them attractive for applications, such as advanced coatings, membranes, ion channels, and high-performance nanocomposites. However, achieving a miscible blend remains a significant challenge, as most polymer pairs are inherently immiscible. Traditional compatibilization methods, such as copolymer addition or chemical modification, often fail to provide the desired control over blend properties. Nanoparticles have emerged as promising compatibilizers due to their ability to alter interfacial interactions; however, most previous work has only been able to consider relatively low nanoparticle concentrations. This poster presents a comprehensive computational study using theoretically informed Langevin dynamics to explore the phase behavior of a polymer blend systemthat is highly loaded with nanoparticles. Specifically, we investigate thermodynamic interactions within the interstitial spaces of densely packed nanoparticle-copolymer systems, where one polymer preferentially wets the nanoparticle surface. These simulations provide detailed insight into how the morphology and phase separation of polymer blends are affected by varying degrees of asymmetric wetting and confinement in nanoparticle packings. The structure of the confined system is characterized by the structure factor computing during quenching, similar to what experiments would measure in X-ray scattering. The results of the simulation reveal that the length scale of the phase-separated structures is significantly influenced by the degree of wetting asymmetry. Systems with asymmetric polymer-nanoparticle interactions exhibit smaller phase-separated domains compared to systems with symmetric interactions. This reduction in domain size suggests that confinement by the nanoparticles can be a novel strategy to tune the phase-separation length scale. The implications of these findings are significant for advancing the design of finely tuned nanocomposite materials and may open the door to the design of novel polymer nanocomposites with unique chemical environments that could be used for membranes or other technological applications.

Investigating the Effect of Plastic Additives on Polymer-Nanoparticle Interactions

Clara Kim, Anirban Majumder, Daeyeon Lee

Polymer upcycling is a more economically favorable method to tackle global plastic pollution compared to recycling, as it chemically alters plastics to create higher-value products. Many upcycling strategies involve polymeric reactions, driven forward by heterogeneous catalysis. These catalysts often use porous support materials, which are usually dense packings of inorganic oxide nanoparticles, to maximize catalytic surface area and improve reaction yield. However, the presence of various additives in plastics can lead to unexpected issues in such reactions. Additives, such as antioxidants, plasticizers, stabilizers and more, are added to plastics to enhance performance and prevent degradation, but some might bind strongly to the catalytic sites and render them inactive. Thus, it is important to study and quantify the effect of these additives on polymer-nanoparticle interactions to identify suitable support materials for these catalysts.

One method is to measure the polymer-nanoparticle contact angle. Silica nanoparticles, a common catalyst support material, are evenly coated on hydrogenated polycyclooctene (hPCOE) films, which is a model polymer for high density polyethylene (HDPE). The sample is then heated above the polymer's melting point for a sufficiently long time to let the nanoparticles partially sink into the polymer film and equilibrate at the interface. The remaining height of a nanoparticle above the polymer interface is measured using atomic force microscopy (AFM), which is used to calculate the contact angle of the polymer on the nanoparticle. Various plastic additives are either incorporated into the hPCOE films or the silica nanoparticle suspension, and the contact angles are measured. We observe significant differences in polymer-nanoparticle interactions based on the presence of additives with different molecular structures. These findings provide insight on the effect of additives on polymer-nanoparticle interactions, helping us identify suitable support materials or surface modification techniques for polymer upcycling catalysts.

Estimation of the Mark-Houwink-Sakurada Parameters for the Characterization of the Molecular Weight of Peg/PEO

Ezekiel Negrete, Omarree Kimbrough, Orrin Shindell, Dany Munoz-Pinto

Poly(ethylene glycol)/Poly(ethylene oxide) are polymers extensively used in the biomedical field particularly in tissue engineering and drug delivery applications. The molecular weight (M) of the polymer is used to predict the diffusional properties of the material, the mechanical performance and other physical and chemical properties of the PEG/PEO based biomaterials. Viscometry, a well-established technique, offers a powerful tool for determining the molecular weight of polymers. Relative to other techniques such as gel permeation chromatography and light scattering, viscometric analysis is the most cost-efficient technique that can be easily adopted in most laboratories. This study utilizes the polymer-solvent relationship of the Mark-Houwink-Sakurada (M-H-S) equation, to correlate the M and the intrinsic viscosity of PEG/PEO aqueous-based solutions. To calculate the parameters of the M-H-S equation, seven M with estimated values of 3.4, 6.0, 10.0, 20.0, 150.0, 250.0, and 325.0 kDa were used at nine different concentrations ranging from 4.0 to 12.0 % w/w using water and a 5.0 % w/w NaCl aqueous solution as solvents. The molecular weight of each polymer was measured and confirmed using dynamic light scattering (DLS). Viscosity measurements for the two PEG/PEO aqueous-based systems were carried out at room temperature using an AMETEK Brookfield Viscometer. Using the intercept of the curve between the inherent/reduced viscosity and the polymer concentration for each M the intrinsic viscosity was estimated. The linearized form of the M-H-S equation, was used to calculate the parameters K and α. The use of 5.0 % NaCl aqueous solution allowed us to achieve viscosity measurement for low PEG/PEO M and low values of polymer concentration. The present of the ionic environment resulted in significant changes in the values of the K and α parameters of the M-H-S equation. The changes in the values of the parameters can be explained due to different polymer chain conformations in the aqueous environments given by the distinct ionic strength between the pure water and NaCl solution.

Beam Me up: Investigating Cross-Linking Behavior in Electron-Beam Polymerization of Acrylate and Methacrylate Mixtures

Garrett Simmerman, Finnis Ginder, Dr. Sage M. Schissel, Julie Jessop

Electron beams (EB) are a focused stream of high-energy electrons capable of breaking bonds, thereby creating radicals and ions. These radicals and ions can participate in cross-linking reactions, which enhance thermal and physical properties by forming bonds between polymer chains. The main benefits of using EB over methods like thermal curing are its energy efficiency and the lack of solvents required. Additionally, EB does not require costly initiators, providing a significant advantage over similar techniques like UV curing. For these reasons, EB polymerization is a promising area for further investigation. The lack of specificity in EB reactions makes polymer property prediction challenging. However, control of cross-linking reactions could provide opportunities to develop advanced engineering materials. In this study, mixtures of acrylate and methacrylate monomers were investigated since they comprise industrially relevant formulations. Gel fraction was used to quantify cross-linking by measuring the mass of polymer that is not dissolved in the solvent THF compared to the original mass of the polymer sample after EB curing. Preliminary results suggest potential relationships between acrylate structure and gel fraction, though further investigation is required to establish clear trends, particularly concerning methacrylate mixtures, which have been less responsive to EB polymerization. Future research will explore the relationship between gel fraction and monomer conversion, providing insights into how these two parameters correlate. Overall, these experiments aim to enhance our understanding of the factors influencing gel fraction, with the goal of optimizing EB polymerization processes and polymer products.

Vapor-Phase Polymer Infiltration with Al_2O_3 for Improved Nanoscale Patterning

Hamzah Farooqi, Daniel Aziz, Michael A. Filler

Many scalable nanowire patterning approaches, such as selective coaxial lithography via etching of surfaces (SCALES), rely on polymer etching for patterning. Etching soft materials, such as poly(methyl methacrylate) (PMMA), faces issues such as uneven etching (leading to imprecise patterning) and a lack of compatibility with processes involving high temperatures or rapid etching. This research aims to resolve these issues by converting soft PMMA films to hard Al_2O_3 films via Al_2O_3 infiltration and subsequent pyrolysis of the films. Spuncoat PMMA films on planar Si wafers are first infiltrated with Al_2O_3 using high temperature vapor-phase infiltration (VPI) processes. Formation of dense alumina surface layers on top of the hybrid films indicated the VPI process reached saturation. These hybrid films then undergo a 600°C pyrolysis to remove the PMMA, leaving behind amorphous Al_2O_3 films. These films can be controllably etched using an alumina etchant composed of phosphoric, acetic, and nitric acid. When compared to standard alumina films formed through thermal Atomic Layer Deposition, the converted PMMA films etched comparably suggesting a dense alumina film that can also be used in further patterning steps. Throughout the process, films were characterized through x-ray photoelectron spectroscopy (XPS) depth profiling and ellipsometry. Achieving increased control over soft materials at the nanoscale promises to facilitate more precise polymer-based patterning.

Cyanobacteria-Driven Polymerization of PEDOT:PSS

Hailey Tran, Debika Datta, Erica Hild, Nathan Soulier, Shaochen Chen, Darren Lipomi, James W. Golden, Susan S. Golden, Jonathan K. Pokorski

Combining biotic (living) and abiotic (non-living) components to engineer next-generation materials has been a growing area of interest due to its potential to create hybrid materials with enhanced functionalities from the individual components. In this work, we combine the cyanobacteria *Synechococcus elongatus* (PCC 7942) with the monomers 3,4-ethylenedioxythiophene (EDOT) and polystyrene sulfonate (PSS) to form the polymer PEDOT:PSS. PEDOT:PSS is a popular polymer choice in bioelectronics due to its conductivity, flexibility, and weight. Traditionally, the polymer can be chemically synthesized with sulfate compounds, which can induce harmful effects upon environmental exposure, such as water and soil contamination. Our hypothesis is that reactive oxygen species (ROS) released by *S.elongatus* catalyzes the PEDOT:PSS formation. *S. elongatus* are wild-type photosynthetic organisms found in abundance in nature, and they release considerable ROS when subjected to stress as part of their metabolic responses. Here, we subject the cyanobacteria cells to high-intensity light and demonstrate that the ROS generated can oxidatively polymerize PEDOT under acidic conditions within four days. Further material characterization results confirm the formation of the biosynthesized PEDOT:PSS and its conductivity and mechanical properties. We then successfully developed ink formulations and fabricated 3D-printed structures of biosynthesized PEDOT:PSS. Besides demonstrating the ability of living organisms to provide an environmentally-friendly pathway for PEDOT:PSS synthesis, the results of this project can be used to study further the fabrication of stimuli-responsive, self-healing, cyanobacteria-based PEDOT:PSS.

Polymerization of Single-Enzyme Nanogels Using Deep Eutectic Solvents

Megan Pierce, Elena Romero-Ben, Ana Beloqui, Marcelo Calderón

Enzymes are the most powerful catalytic systems in nature, but their activity is compromised under adverse conditions like high temperatures, non-physiological pH, and organic solvents. One way to increase enzyme stability is by creating single-enzyme nanogels (SENs). SENs are nanobiocatalysts where an enzyme is encapsulated by a hydrophilic polymer network. This polymeric nanostructure provides stability in conditions that typically cause enzyme denaturation and offers potential applications in drug delivery and diagnostics. However, the reaction conditions necessary for obtaining SENs could affect enzyme integrity. It was hypothesized that using sugar-based deep eutectic solvents (DESs) during SEN polymerization could improve their stability. In this project, SENs of glucose oxidase were made with and without DESs. Preliminary results show that sugar-based DESs enhance the stability and yields of SENs, improving glucose oxidase's ability to function under conditions that would typically cause denaturation.

Cycling Effects and Stability of Polythiophene-Coated Lfp Cathode Current Collectors

Maria Rizio

As global initiatives to mitigate anthropogenic carbon dioxide contributions intensify, electrochemical storage systems are poised to play an integral role in the eco-friendly metamorphosis of electricity and transportation. Lithium- ion battery (LIB) have a high energy density and low self-discharge compared to other batteries utilized within electric vehicles, making them the most common battery. While advances in LIB technologies continue, significant obstacles to LIB adoption persist in the high-demand grid and transportation sectors, such as within extending their useful life. This study explores the use of polythiophene and its derivatives (poly(3-butylthiophene) (P3BT), poly(3-hexylthiophene) (P3HT), and poly(3-octylthiophene (P3OT)) as priming materials coated on the aluminum cathode current collector surface in Lithium Iron Phosphate (LFP) LIB. The use of an electroconductive polymer would ensure good contact between the active materials and conductive additive and the current collector, which would protect the electrode from corrosion through its cycling. This method is expected to enhance battery stability by preventing electrode materials decomposition, thereby extending battery life. Additionally, the high electronic conductivity of the polymers is predicted to improve cycling ability, resulting in increased Coulombic efficiency, and rate capacity. There is also evidence for the improved stabilization of the Cathode Electrolyte Interphase (CEI) during the first several cycles that would create a protect layer outside the active material, and further limit side reactions. This preparation technique holds promise for enhancing EV battery performance and longevity, with the prospect of future research for the optimization of the technique.

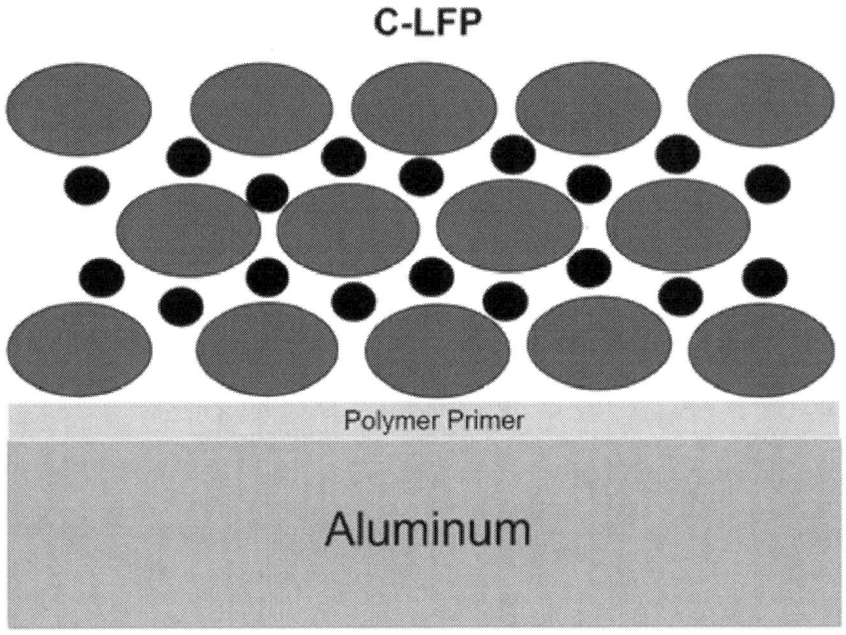
C-LFP
Polymer Primer
Aluminum

Development of Poly(ethylene glycol)-Based Hydrogels for the Enterosorptive Removal of Carcinogens and Their Precursors

Mei Weatherly, Sachin Sundar, Maria Victoria Ximenes Klaus, J. Zach Hilt

Carcinogens and their precursors, with their inherently negative reputation, are far more prevalent than many people may think. Nitrates, for one, are naturally occurring chemicals to which many people are exposed on a regular dietary basis. Typically found in leafy vegetables and as meat preservatives, multiple studies have linked them to different types of cancers (e.g. colorectal, esophageal, thyroid), despite their potential cardiovascular health benefits. When consumed, nitrates can undergo a series of reactions that result in the production of nitrosamines, a subsection of N-nitroso compounds (NOCs), many of which have been declared carcinogenic. While much of the dietary nitrate intake is often excreted, bioaccumulation may still occur; and since avoiding exposure is impractical for most people, an alternative is to develop an enterosorbent to remove the nitrates from the body before they become carcinogenic. Poly(ethylene glycol) diacrylate (PEGDA) is a monomer that when combined with cationic comonomer (3-acrylamidopropyl)trimethylammonium chloride (DMAPA-Q), formed a hydrogel composite to which powdered activated carbon (PAC) could be added as an additional adsorbent. The hydrogels were then characterized using Fourier Transform Infrared Spectroscopy (FTIR), Thermogravimetric Analysis (TGA), mass conversion, and swelling studies in DI water. Ultimately, sorption studies were conducted to determine nitrate removal efficiencies across various systems. Results demonstrated that increasing the relative amount of DMAPA-Q provided a higher swelling capacity and removal efficiency that plateaued around 75% at a 0.5 mg/mL sorbent dose due to the dominant assumed charge-based affinity. Thus, these systems have the potential for nitrate enterosorption and may be applied to other carcinogenic targets.

Photodegradable Polyacrylamide Tanglemers Enable Spatiotemporal Control over Chain Lengthening in High-Strength and Low-Hysteresis Hydrogels

Joshua Lee, Bruce Kirkpatrick, Kristi S. Anseth

ABSTRACT

Hydrogels consist of a hydrophilic network of crosslinked polymer chains with diverse applications, including medical implants, tissue engineering, environmental cleanup, and agriculture. As the field of hydrogel research continually progresses to enhance their properties, innovative strategies are being explored to improve their mechanical performance and functionality. However, conventional covalent hydrogel networks face a notable stiffness-toughness conflict: while increased crosslink density enhances the material's modulus, it simultaneously leads to embrittlement and reduced extensibility, which limits their practical use in many applications. To address these challenges, recent advancements have introduced strategies to form highly entangled poly(acrylamide) hydrogels, colloquially referred to as tanglemers, by optimizing polymerization conditions to maximize the density and length of polymer chains and minimize the concentration of crosslinker. Although entanglements are well-studied in polymer melts, assessing them in crosslinked networks is more challenging, often limited to approximating their theoretical contributions to mechanical properties. In this work, we synthesize and characterize poly(acrylamide) tanglemers using a photolabile oNB crosslinker, allowing for direct assessment of covalent trapping of entanglements under tension. These hydrogels have a relatively high modulus, which can be tuned in situ by crosslink photocleavage (from $E_T = 100$ kPa to <25 kPa). Additionally, we demonstrate that even non-degradable tanglemer formulations can be photo-softened and completely degraded through Fe^{3+}-mediated oxidation of the poly(acrylamide) backbone. While both methods of photodegradation can be used for spatial patterning and can result in softer gels with reduced fracture strength, only crosslink cleavage improves gel extensibility via light-induced chain lengthening ($\varepsilon_F = 700\%$ to >1500% following photodegradation). Moreover, crosslink photocleavage affords substantial control over localized swelling and diffusivity, opening new avenues for precise material design. In sum, we introduce a new approach for probing entanglements and asserting spatiotemporal control over stress-strain responses and small molecule diffusivity in poly(acrylamide) tanglemers, suggesting a multitude of potential soft matter applications including controlled release and tunable bioadhesive interfaces.

Fig. 1 | Highly entangled polyacrylamide hydrogels have crosslinker-independent mechanical properties before light exposure. (A) Schematic representation of the formation of entangled polyacrylamide networks using PEG-diPDA (photodegradable diacrylate, red) and PEG-diAcm (non-degradable diacrylamide, blue) crosslinkers. (B) Representative tensile stress–strain curves of tanglemers prepared with each crosslinker, strained at 10 percent/second. (C) Mechanical properties of diPDA- and diAcm-crosslinked tanglemers including (i) Young's modulus, (ii) strain to failure, (iii) strength and (iv) work of fracture. (D) Hydrogels exhibit negligible hysteresis regardless of crosslinker identity. (E) Tensile stress-strain curves under active irradiation (shaded region) beginning at 250% strain show photoresponsiveness in diPDA-crosslinked hydrogels only. n = 3-4 gels per condition, with representative data shown in panels B, D, and E.

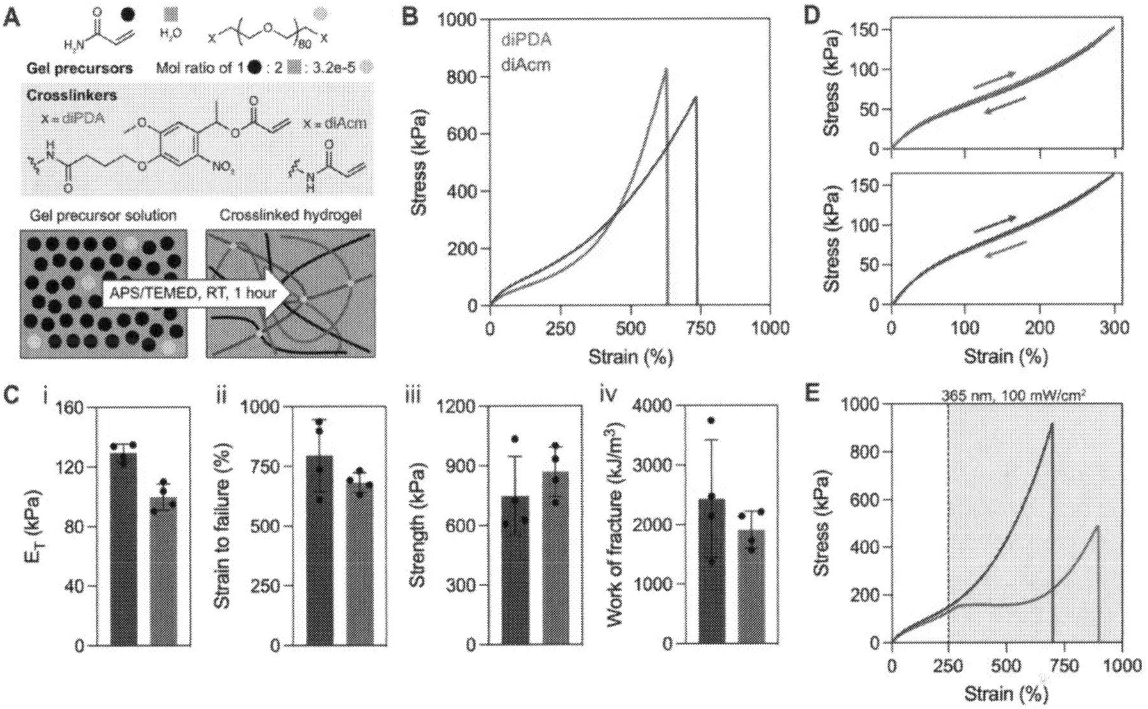

A
Gel precursors
Mol ratio of 1 : 2 : 3.2e-5
Crosslinkers
X = diPDA
X = diAcm
Gel precursor solution
Crosslinked hydrogel
APS/TEMED, RT, 1 hour
B
diPDA
diAcm
Stress (kPa)
Strain (%)
C
Strain to failure (%)
Strength (kPa)
Work of fracture (kJ/m³)
D
E
365 nm, 100 mW/cm²

Photoinitiated RAFT Copolymerization of α-Lipoic Acid and Acrylates for Degradable Polymers

Hanqing Wang, Dongjoo Lee, Rafael Verduzco

Plastic pollution has rapidly escalated into a global challenge, with one major cause being the widespread use of traditional vinyl polymers. These polymers, with their entirely carbon-carbon backbones, are resistant to degradation and continue to accumulate in the environment. Our work focuses on sustainable polymer design by utilizing photopolymerization to create copolymers of acrylate-based monomers and α-lipoic acid (LA), a naturally occurring antioxidant. While prior work has explored copolymers of LA with vinyl monomers, LA copolymers have been prepared with a limited set of cleavable comonomers. In our work, we explored photopolymerization as a more versatile route to prepare LA copolymers. Using blue light irradiation, we can simultaneously initiate the ring-opening reaction of LA and controlled polymerization of vinyl monomers via reversible addition-fragmentation chain transfer polymerization, resulting in degradable polymers with controlled molecular weight and architecture. We systematically photoinitiator concentration, monomer feed ratio, reaction concentration, and polymer chain length, to better understand the unexplored mechanisms of LA-based copolymer photopolymerization. With synthesized LA-derivatives, we also achieved complex architectures such as hyperbranched, comb, and network structures via one-step polymerization. This work provides a simple and versatile approach to preparing degradable polymers, offering significant potential to reduce the environmental impact of plastic waste.

Modeling Telechelic Triblock Polymers for Nanoemulsion Stabilization

Shane Haycock, Thomas E. Gartner III, Ryan Poling-Skutvik

Tailoring the rheological properties of emulsions is a key materials development objective for applications ranging from pharmaceuticals to 3D printing. Often, this task can be achieved by introducing telechelic triblock polymers into the emulsion, which can bridge between emulsion droplets and create a system-spanning viscoelastic network. In this work, we performed molecular dynamics simulations to investigate the interfacial behavior of model systems of telechelic triblock polymers, and evaluated the propensity to form bridges vs. loops at the liquid-liquid interface as a function of triblock polymer design. In particular, we studied coarse-grained polymer chains of varying mid-block stiffness at a flat fluid-fluid interface, mimicking the assembly of triblocks at the surface of a large emulsion droplet. We examined the polymer confirmation at the interface (i.e., the propensity to form bridges between emulsion droplets) as a function of the polymer design and the inter-droplet spacing. Our preliminary results show that, regardless of midblock stiffness, looped configurations dominate when the inter-droplet spacing is on the order of the midblock contour length. However, ongoing simulations will further evaluate how properties such as stiffness, chain length, polymer-solvent interaction strength, and inter-droplet spacing affect the interfacial behavior of the polymer, aiding in the future design of these complex fluids.

Utilizing a Robotic System to Produce Polymer Coatings for Coating Characterization

Colby Belczyk

Polymer coatings are ubiquitous in the world today, and as a result, the necessity for creating and testing new coatings is ever-growing. Formulation and drawdowns are an integral part of this testing process, though doing these steps by hand is time consuming and tedious. I explored the capabilities of a robotic system on the automation of the formulation, drawdown, and curing steps of the testing process. I used a UV curable coating system as a model for formulations. Formulations included commercial resins Ebecryl 150, 450, 8301, and 8402, reactive diluent 1,6-hexanediol diacrylate (HDDA), and UV photoinitiator Irgacure 1173. Within this robotic system, I tested parameters that influenced the accuracy of the formulations. These parameters included fine dosing amounts, fine dosing steps, and error margins. I additionally tested parameters that influenced the quality of the output polymer coatings, including initiator concentrations, drawdown mechanics, cure intensities, and cure times. Through coatings characterization, I was able to determine the ideal parameters for optimal operation. My findings indicate that with further testing and the integration of AI learning models, this robotic system can be used in the creation of polymer coatings without the need for human intervention, streamlining the process significantly.

Hydrolysis-Driven and Glutathione-Responsive Charge-Shifting Polycations for Optimized Gene Delivery

Sadie Farmer, Alex Fortenberry, Kenneth Hulugalla, Caroline Argenti, Thomas Werfel, Adam Smith

Gene therapy holds great potential for treating genetic disorders by correcting or replacing defective genes. While cationic polymers have emerged as promising carriers for nucleic acids, their high cationic charge density often results in substantial cytotoxicity and incomplete cargo dissociation within cells. To overcome these limitations, we designed cationic charge-shifting polymers by post-polymerization ring-opening reactions of poly(vinyl dimethyl azlactone) (PVDMA) with tertiary amines containing nucleophilic groups (amine, hydroxyl, thiol). This approach yielded tertiary amine-modified polymers with either hydrolytically stable amide bonds or degradable ester/thioester linkages. These modifications enabled control over the rate of cationic-to-anionic charge shifting under different environmental conditions, such as temperature, pH, and linkage type, facilitating the efficient release of nucleic acids from polyplexes. Our studies revealed that polymers with thioester linkages could undergo rapid nucleic acid release via thiol-thioester exchange reactions in response to intracellular concentrations of glutathione (GSH). This charge-shifting mechanism provides a potential trigger for intracellular nucleic acid which could thereby enhance delivery efficiency. To explore this further, we synthesized novel thioester-based cationic monomers that could be directly polymerized to produce thioester-functionalized polycations. These new materials demonstrated significantly higher in vitro mRNA transfection efficiencies compared to conventional polycations, such as PDMAEMA, due to their ability to respond to GSH and release their cargo more effectively. Our ongoing work reveals that optimizing molecular weight and monomer composition is critical for balancing polyplex stability and nucleic acid release. Higher molecular weight polymers formed more stable complexes, while lower molecular weight polymers facilitated more rapid hydrolysis. Copolymerization of thioester-functionalized monomers with hydrophobic thioester-based monomers improved hydrolytic stability without compromising GSH-responsiveness. This approach led to significantly higher transfection efficiencies than non-thioester controls. Our findings underscore the potential of thioester-based charge-shifting polymers as a new class of nucleic acid carriers, offering safer and more effective options for gene therapy applications. These materials could pave the way for advancements in therapeutic delivery systems by providing more precise control over release mechanisms in response to intracellular environments.

Recycling of Epoxy with Solar Light

Jonathan Logan

Recycling of Epoxy with Solar Light

Madeline Finale, Jonathan Logan, Arnob Saha, Matthew Durfee, Nicole Penner, John McCoy, Youngmin Lee and Sanchari Chowdhury

Epoxies are a strong polymer commonly used as structural material. However, because of the irreversible crosslinking of epoxy, it can't be reliquefied once it's cured, which makes their recycling challenging. To address this challenge, we are modifying epoxies with a thermo-reversible Diels-Alder reaction and doping with photothermal nanoparticles to provide the required heat using sunlight. We used Titanium Nitride (TiN) nanoparticles and Carbon Black nanoparticles as the photothermal nanoparticles which can absorb broad spectrum sunlight very strongly.

The photothermal nanoparticles are well dispersed in the furan precursor and then reacted with maleimide to synthesize a Diels-Alder epoxy nanoparticle composite. The composite powder was then compressed into a mold and exposed to visible light for recycling.

The epoxy could be recycled efficiently twice with 10 sun intensity light. The recycled samples were tested for their mechanical strength with a 3-point bend test. We compared he flexural modules for the light recycled and conventional heat recycled samples with the pristine samples. We found that pristine samples with or without nanoparticles, have similar flexural modulus. This is because with the loading of 0.5 weight percent, nanoparticles had no effect on the mechanical properties of the epoxy. Interestingly, the first light recycled samples had superior flexural modulus in comparison to the pristine samples. No such effect was observed for the heat recycled samples. This is attributed to the fact that in case of photothermal heating the reaction is more efficient as the nanoparticles can absorb light to generate nanoscale heating to generate volumetric heating.

In summary, our work showed that we could achieve full recycling of epoxy using sunlight without compromising it's mechanical properties, which can open up a possibility of using renewable energy for recycling epoxy.

Enhancing Proton Conductivity of Nafion Membranes with Hollow, Nanorod SiO2 Particles

Charlotte Gottilla, Mukund Kabra, Norman J. Wagner

The climate crisis motivates an increasing need to develop alternative, clean energy sources to replace fossil fuels in commercial, industrial, and residential settings. Among the technologies being considered as fossil fuel alternatives are fuel cells, which electrochemically convert chemical energy from hydrogen fuel into electrical energy. Proton exchange membrane fuel cells (PEMFC's) split hydrogen fuel into protons and electrons at the anode via a platinum catalyst, with electrons providing electricity and protons transported to the cathode through the membrane.

The most common membrane for PEMFC's is Nafion, which consists of a hydrophobic polytetrafluoroethylene (PTFE) backbone and hydrophilic sulfonic acid side chains. In Nafion, protons are conducted through the hydrated sulfonic side chains, making water management in the membrane essential; proton conductivity decreases at temperatures above 80 °C as water begins to evaporate. Improving ionic conductivity at high temperatures is important for applications in transportation (e.g., heavy vehicles), which experience a wide range of operating temperatures, and because higher temperatures (120-130 °C) otherwise improve fuel cell performance regarding carbon monoxide tolerance and catalyst performance.

As is well-established in the literature, a composite membrane of Nafion with silica nanoparticles can improve fuel cell performance at higher temperatures as pores in the nanoparticles provide physical pathways for ion transfer and may also allow for additional water storage. Previous studies have focused on spherical nanoparticles; however, the use of anisotropic particles may introduce additional benefits. Anisotropic filler particles have a larger surface area to volume ratio and may provide the same benefits as spherical particles at lower loading amounts. Furthermore, Nafion cast in aqueous alcohol solutions will aggregate into rod-like structures due to its hydrophobic backbone, offering the possibility for the silica nanoparticles to align with the Nafion aggregates and support existing proton conduction channels. Finally, the hollow tubes at the center of the nanoparticles may provide an additional avenue for proton transport.

In this work, preliminary investigation is completed into the efficacy of anisotropic silica nanoparticles as a Nafion additive. Commercial dispersions of Nafion D2821 are cast as pure films and with the addition of spherical and rod-shaped nanoparticles. The films are characterized through dynamic mechanical analysis (DMA), measurement of ionic conductivity, and water uptake measurements to evaluate the physical durability, proton conductivity, and water retention. Planned future work includes the development of a reliable method for casting equivalent films using non-commercial Nafion dispersions and investigating the effect of nanoparticle concentration.

Automation and Optimized Fabrication of Mxene-Based Textile-Based Supercapacitors

Simon Johnson, Alyssa Grube, Mona Bavarian

Power management for mobile devices is a research challenge to address global energy concerns. The development of textile-based supercapacitors (TSCs) and the incorporation of TSCs into clothing presents a promising solution. TSCs function as electrochemical double-layer capacitors, storing energy through the accumulation of electrostatic charge at the electrode-electrolyte interface. Here we explore the use of wool and casein yarns for the development of TSCs by coating them with MXene conductive material. The general formula for MXene is $M_{n+1}X_nT_x$. This compound is composed of a transition metal (M), carbon or nitrogen (X), and the surface terminations (T). With this study, the primary MXene used for coating was $Ti_3C_2T_x$ due to its conductivity, stability, and non-toxicity. Previous work demonstrated hand-knit TSCs using dip-coated wool. This process is physically tedious, time-consuming, and only allows for small-scale production. To enhance the utility of the coated material, scaling up TSCs' electrodes are crucial for increasing energy storage capacity. This project specifically focused on automating the coating and knitting process. The previous dip-coating method often resulted in uneven coating; therefore, we used roll-to-roll coating by pulling the yarn from the main spool. The apparatus consisted of a vertical stand approximately four square feet in size. Dozens of holes were drilled into the surface in an alternating pattern, each hole was threaded with a bolt and small-scale pulleys were attached to the end of each bolt. At the top of the panel, a mini scale coating bath was fastened. The bath was made of plexiglass roughly 2cm in depth, and it was equipped with a corresponding plexiglass lid with small metallic prongs protruding out of the bottom. The dimensions of the bath were designed specifically to allow the yarn to be submerged in the solution and drawn continuously through the bath. At the base of the apparatus was a small 4" fan, as well as a larger bolt connected to a small motor. The mounting of the lower bolt allowed for it to be continuously rotated clockwise, away from the end of the pulley system. To coat the thread, first a sufficient length of leader yarn was threaded throughout the pulley system; afterwards, the coating bath was filled with prepared MXene solution, and the yarn was continuously drawn through the solution and air dried using the fan until it reached the spool. After the desired length of yarn was coated and wound onto the spool, the yarn was removed from the bath and allowed to proceed through the pulley system until the uncoated portion had reached the collection spool as well. With this process, single coated samples were prepared as well as quad-coated samples. This automated method resolved the time-limiting factor for the production of the TSCs which was the issue with the previous dip-coating method.

Co-Products from Guayule Resin

Justice Armijo, Eric Palmer

AICHE Annual Meeting, San Diego, 2024, October 24th

Title: Co-products from Guayule Resin

Authors: Justice Armijo, Mostafa Dehghanizadeh, Catherine Brewer

Division: Materials Engineering & Sciences

Guayule (*Parthenium argentatum*) is a desert shrub of the Asteraceae family that thrives in arid and semi-arid regions with low water demand. This crop is native to the southwestern U.S. and northern Mexico and is being used to produce high-quality, hypoallergenic natural rubber. Industries utilizing natural rubber include pharmaceuticals, tire production, and food industries. Due to the economic feasibility of guayule-derived rubber production, sale of co-products are required for sustainable and commercial viability of guayule as an industrial crop. During rubber extraction, two byproducts are produced: a woody bagasse and a complex resin. Both byproducts were characterized using various methods including Fourier-transform infrared spectrometry (FT-IR), gas chromatography–mass spectrometry (GC–MS), and high-resolution Fourier transform ion cyclotron resonance mass spectroscopy (FT-ICR MS). Whole guayule resin, along with vacuum-distillation fractions, were identified to have significant value in monoterpene chemical, insect repellant, and adhesive applications. Specific insect repellency was tested against Turkestan cockroaches and found a significant number of compounds present in the resin distillation fractions as in commercially used active ingredients for repellents, and a repellency action time to at least seven days for some fractions. An economic review of the co-product applications was conducted and found to support the development of guayule natural rubber production.

Increasing the Biodegradability of Thermally Reprocessable Vanillin-Based Thermosets

Owen Sabolcik

Thermosets are a widely used group of plastics that have strong thermal and chemical resistance. Because of these beneficial properties, thermosets are used widely in cookware, electrical insulation, and lamination. Unfortunately, due to the significant degree of crosslinking and high thermal resistance that they exhibit, recycling and reprocessing prove to be challenging. This leads to a large amount of plastic waste that does not degrade in a reasonable amount of time. Vitrimers are a versatile group of polymers that exhibit both thermoset and thermoplastic attributes as a result of their covalent adaptive network (CAN). Although they are processable, these polymers will quickly lose their stability after a few cycles, once again resulting in unwanted plastic waste. This work attempts to overcome this barrier by selectively doping a vitrimer with a cellulose derivative to enhance its biodegradability while not compromising its desirable characteristics. A vanillin-based vitrimer was selected due to its sustainable source material, high thermal resistance, and processability. A copolymer was synthesized using the cellulose derivative and the vitrimer in conjunction. Afterward, the copolymer's physical properties and biodegradability were observed. The biodegradability was ascertained via biochemical oxygen demand (BOD) and there was a noticeable increase in the BOD of the doped sample. Differential scanning calorimetry (DSC) was employed to further support the indication of concordant thermal stability. Finally, solvent resistance tests across different mediums were taken of the cellulose-doped vitrimer and the results indicated similar stability to the base vitrimer. These results highlight the potential of cellulose doping as an inexpensive and effective way to facilitate the degradability of vitrimers and other thermosets.

Studies in Powder Segregation — a Computational and Experimental Approach

Victor Wu, Miao Na, Rakesh Gummalla, John P. Hecht

Segregation is the physical phenomenon in which a multicomponent powder mixture separates into its components due to differences in size, shape and density. Preventing segregation is of special interest to the Procter & Gamble Company in applications where a powder with a specific composition is required. To understand how segregation occurs in an industrial hopper on the production lines of certain P&G products involving powders, an existing numerical simulation process, based on solving an advection-diffusion-segregation PDE using gPROMS software, has been used with literature-validated results. However, due to necessary manual calculation steps, work was required to reduce manual errors and enhance useability. Additionally, a potential new powder binary mixture for a P&G product was identified, requiring an experimental assessment of segregation tendency of these materials on the production line. On the computational side, work was carried out regarding the segregation simulations to make it more usable, and develop capabilities for automation of gPROMS simulations in Python and numerical simulation implementations in Python and MATLAB. Subsequently, a Python-based tool which can interface with any gPROMS-based simulation was developed then applied to the powder segregation simulations, plus additional code eliminating manual calculations and implementing automatic postprocessing of simulation results. On the experimental side, the segregation tendency of the new proposed binary mixture was assessed in pilot scale experiments: a specialized "modified heap" segregation testing apparatus was used to assess sifting segregation, a pilot-scale loss in weight feeder was used to simulate real production line conditions, and a twin-shaft paddle mixer was used to assess mixing difficulty and segregation in handling. Results suggest a weak tendency for this binary mixture to segregate in real manufacturing conditions. However, achieving a well-mixed sample of this specific mixture is very difficult in the first place, with an abnormally long time required in the mixer to reach a well-mixed steady state, and segregation readily occurs in post-mixing handling. These experiments also served as method development for future experiments with other mixtures. Future research on the computational side will largely focus on other applications of numerical PDE simulations, including alternate uses of the Python-gPROMS interfacing code as well as applications of simpler systems in Python or MATLAB. On the experimental side, rheology work will focus on determining the impact of certain amounts of segregation of this proposed mixture on the consumer appeal of this P&G product.

Investigating Red Phosphorous and Liquid Metal As Next Generation Active Materials for Monovalent Cation Energy Storage

Rex Colvard III, Zach Park, Peter Fedkiw, Michael Dickey

The demand for energy storage in applications such as electric vehicles, consumer electronics and grid storage is ever increasing. Lithium-ion batteries (LIBs) are currently the state-of-art in electrochemical energy storage and are extensively used in electric vehicles (EV's) and consumer electronics. The annual EV Li-ion battery demand for sustainable development is projected to grow to up to .7 TWh per year by 2030, a 7x increase compared to the current demand projection of .1 TWh per year for 2025.[1] In order to meet this energy demand, we propose to use EGaIn, a eutectic alloy of gallium and indium that is liquid at room temperature, as a "self-healing" agent for red phosphorous in anodes to achieve longer cycle lives and higher capacities. Red phosphorous (RP) has been the subject of extensive research due to its relatively low cost compared to other commonly used anode materials in LIB's, stability in air, low toxicity, and extremely high theoretical capacity for Li^+ ions at 2596 mAh g^{-1}. Graphite, a currently commercially used anode material, has roughly a 7x lower capacity of 372 mAh g^{-1} compared to RP for Li^+ ions. RP's main drawback is that it has a volume expansion of up to 300% during charging. This can cause delamination of the RP from the anode, resulting in capacity fade due to the RP becoming electrically disconnected from the anode.[2] To mitigate this delamination, we propose to use EGaIn to maintain electrical contact with delaminated red P particles and fill in defects formed from red P's large volume expansion. EGaIn has been used as a "self-healing" agent for silicon, another high-capacity anode material in LIB's, that also suffers from high volume expansion.[3] In our work, we fabricate a composite film electrode comprising EGaIn and RP nanoparticles in a carbon polymer matrix. We perform material characterization of the electrode before and after electrochemical characterization of the electrode in a Li half-cell. This strategy of using EGaIn in combination with RP may help develop effective approaches for the integration of other high volume expansion active materials in anodes.

1. IEA (2021), Global EV Outlook 2021, IEA, Paris https://www.iea.org/reports/global-ev-outlook-2021, Licence: CC BY 4.0
2. Zhou, J. *et al.* Phosphorus-Based Composites as Anode Materials for Advanced Alkali Metal Ion Batteries. *Adv. Funct. Mater.* **30**, 2004648 (2020).
3. Yang, J. *et al.* Self-Healing Silicon Anode via the Addition of GaInSn-Encapsulated Microcapsules. *ACS Appl. Energy Mater.* **5**, 12945–12952 (2022).

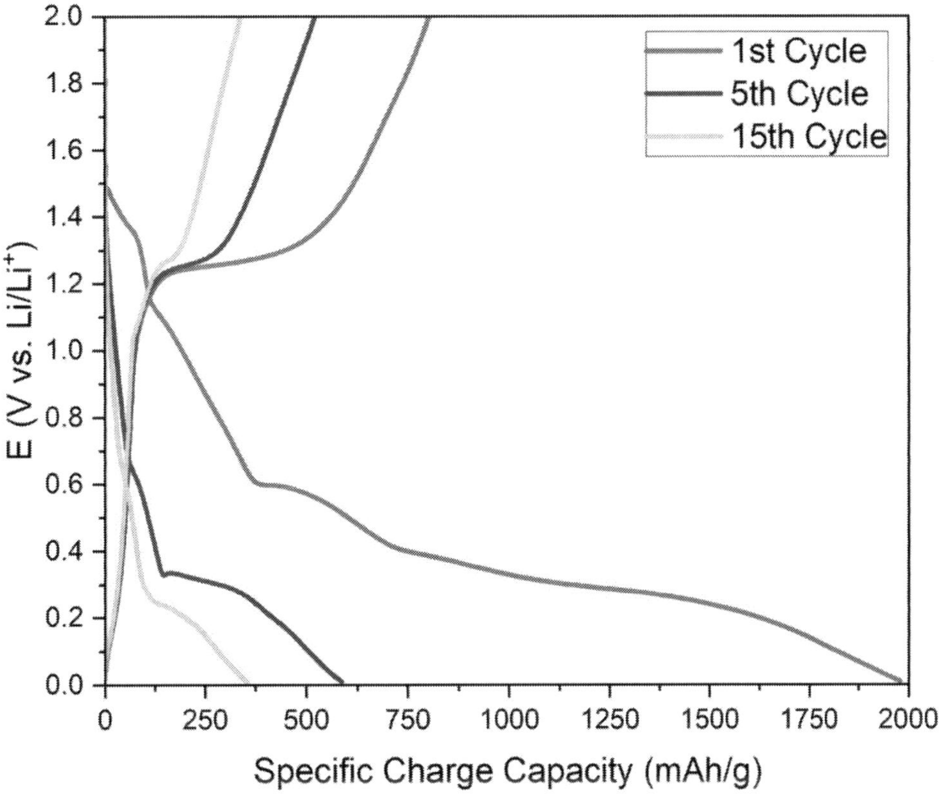
1st Cycle
5th Cycle
15th Cycle
E (V vs. Li/Li+)
2.0
1.8
1.6
1.4
1.2
1.0
0.8
0.6
0.4
0.2
0.0
0
250
500
750
1000
1250
1500
1750
2000
Specific Charge Capacity (mAh/g)

Characterization of Chiral Self-Assembly in Sequence-Defined Mesogenic Dimers

Eva Reed, Clement Chan, Emily Ostermann, Emily Davidson

Liquid crystals (LCs) are phases of matter that possess properties in between those of liquids and crystalline solids. Due to the intrinsic anisotropy of calamitic LC mesogens, they frequently adopt LC and crystalline structures with useful optical, magnetic, and electronic properties. Recently, the Davidson research group has developed a general approach to using orthogonal protection/deprotection reactions to synthesize discrete and sequence defined liquid crystalline oligomers. This work examines the self-assembly of mesogenic dimers synthesized with two different mesogens; a phenyl benzoate core both with (***M***) and without (***N***) a methyl-substituted aromatic ring. The two dimers with alternating sequence (***MN*** and ***NM***) and their mixtures are both examined. Both pure oligomers exhibit chiral superstructures composed of ~100 nm nanocylinders; however, their crystalline morphology and phase behavior are divergent as observed via optical, thermal, and X-ray characterization. We find that blends of the two materials assemble to form banded spherulites. Scanning electron microscopy of the alternating dimers grown in the confined melt or in solvent advances our understanding of their chiral self-assemblies and motivates possible applications in chiroptical materials or catalysis via chemical functionalization of the nanostructures.

Effect of Fused Silica Deposition Method on a Laser Performance of Multilayered Dielectric Gratings

Lena Lederman, Leonid Solodov, Brittany Hoffman, Alexei Kozlov, Stavros Demos, Alexander Shestopalov

Chirped pulse amplification (CPA), a method for producing ultrashort, high peak intensity laser pulses, is used in multiple applications including medical treatments, material processing, ultra-fast spectroscopy, and high-energy-density physics. An arrangement of four optical gratings stretches our laser pulses temporally and spectrally, amplifies them, and compresses them again, producing shorter pulses with a much higher peak intensity. One current limitation to CPA is the optical gratings damage at lower laser fluences than the intrinsic damage threshold of pure unprocessed silica coating that is used to make the grating lines, suggesting that the grating manufacturing process induces defects and contaminants that absorb laser light and cause premature damage. Of the possible sources of damage, the effect of the silica coating method on the laser damage performance of the CPA gratings is not well evaluated. To better understand how different types of silica coatings change during the grating manufacturing, we developed a protocol to evaluate physical, chemical, and laser damage properties of the silica coatings before and after the grating manufacturing cycle. After performing a reactive ion etch process, etch selectivity, surface chemical composition, surface roughness, and damage test performance were evaluated for a silica coating deposited via electron beam evaporation. Future experiments will compare these metrics with other SiO2 coating methods to determine the optimal coating process for maximizing laser damage resistance.

This material is based upon work supported by the Department of Energy [National Nuclear Security Administration] University of Rochester "National Inertial Confinement Fusion Program" under Award Number(s) DE-NA0004144.

Colloidal Diffusion on Curvature Landscapes

Adam Tobin-Williams, John E. Bond, Michael A. Bevan

Ordered structures of self-assembled colloidal particles can form metasurfaces with exceptional properties (optical, electromagnetic, thermal, etc). Collective particle dynamics govern the transition between disordered and ordered states during self-assembly, and so are crucial for quickly achieving target structures. However, significant characterization has only occurred on planar and constant curvature surfaces, avoiding spatially varying curvature. In contrast, flexible and/or curved surfaces dominate potential use-cases including protein assembly on cell membranes, flexible quantum dot displays, and plasmonic particle assembly. Here, we study how curvature affects colloidal dynamics on surfaces of varying curvature up to the concentrated fluid regime. We simulate spherical colloids with short-range repulsive potentials using Brownian Dynamics from infinite dilution to a dense fluid and measure self-diffusion. We find that the apparently curvature-dependent mean-squared displacement can be accounted for using only local planar approximations, leaving the diffusivity itself independent of curvature up to freezing conditions. Our results demonstrate that curvature does not intrinsically affect dynamics in fluid systems. Understanding these fluid-state dynamics is critically important for guiding any self-assembly process which involves them and so we expect these findings to undergird further simulations of colloidal assemblies which transition between disordered and ordered states.

Sensational Mxenes: Evolving the Manufacturing of Textile-Based Supercapacitors

Abaigeal Aydt, Mona Bavarian, Alyssa Grube

The widespread use of portable electronics in everyday activities encourages interest in the research of wearable supercapacitors. Current methods to create textile-based supercapacitors (TSCs) involve the synthesis of conductive materials to coat fiber such as cotton or wool. Among various materials, MXenes, a family of two-dimensional transition metal carbides or nitrides, have gained significant attention. Methods of coating and different types of yarn are of particular interest for optimization to address limiting factors in current techniques. In comparing two yarns, wool and cotton, the wool yarn in comparison to the cotton yarn has higher MXene content on the yarn's surface when examined with X-ray Photoelectron Spectroscopy (XPS). Additionally, the wool yarn exhibited more pseudocapacitive behavior when compared to the cotton yarn due to its fiber's increased aqueous electrolyte uptake. Yarn coating was previously done by simply dipping the yarn into the MXene colloidal solution and drying it under a continuous vacuum. This coating and drying method is a two-step process. To optimize the method, an auto-coater was developed to streamline yarn coating into a one-step process where the yarn is simultaneously coated and dried. Although this method is still undergoing refinement, the current auto-coater employs a fan and motor system. The motor, in conjunction with a shaft, spins the newly coated and dried yarn onto a spindle. Once the yarn is coated and run through a series of tests, the yarn is knitted into a TSC. Currently, the TSC is knitted by hand, but with the help of a Kniterate knitting machine, this process can be improved. The knitting machine can be programmed to knit electronic components that can be used in this research. In order to utilize the Kniterate machine, fundamentals of operation and coding knowledge are needed. Using a knitting machine offers several advantages; TSCs can be knitted with more consistent sizing than what can be done via hand knitting, mass production is more efficient, and the knitting environment is more controlled. Using wool fibers, the auto-coater, and the new Kniterate machine, the research of developing wearable MXene TSCs is expanding to include studying the effect of yarn substrate and processing on performance.

Synthesis of High Entropy Alloys through Ultra Fast Laser Pulses for Catalytic Applications

Hamood Qureshi, Khadijatu N. Alhassan, Kevin McPeak

Research into high entropy alloy nanoparticles (HEA NPs) has the potential to progress catalysis research. HEA NPs have been shown to enhance product selectivity, catalytic efficiency, and catalyst stability of processes ranging from ammonia oxidation to carbon monoxide oxidation. Our project is focused on developing HEA NPs that can be used in hydrogen generation. To do this, we have developed a robot that can vary the fluence and pulsewidth of an ultrafast 1064 nm laser to deliver high bursts of energy to metal salt coated supports to convert the metal salts into HEA NPs. These nanoparticles are then examined underneath SEM imaging and XRD to determine the size and structure of the nanoparticles. Future tests will involve synthesizing larger arrays and varied compositions of HEA NPs to determine which nanoparticles are viable for testing. Then, we will test the catalytic performance of these nanoparticles to determine whether they are viable for use in hydrogen generation in rechargeable hydrogen batteries.

Topographic Analysis of Niobium during Incremental Electropolishing

Madison DiGuilio

Superconducting radio frequency (SRF) niobium cavities are pivotal for experimentation at Thomas Jefferson National Accelerator Facility (TJNAF). The peak electromagnetic fields achievable in these cavities depend on the smoothness of the surface. At high accelerating fields, topographic defects may create opportunities for excessive dissipation and thermal runaway turning the cavity normal conducting. Scientists at TJNAF place immense emphasis on preparing and analyzing niobium samples for optimization, as well as improving methods of analysis. The effect of electropolishing on surface defects was investigated via atomic force microscopy. Relevant parameters were analyzed from the topographic data for the maximum accelerating field in Nb cavities. Magnetic field enhancement (MFE) and superheating field suppression (SFS) factors of the surface were calculated to develop a deeper understanding of the effect of electropolishing on irregular defects in polished Nb. Via incremental EP of Nb, there was an overall decrease in the depth of the groove and MFE. It would be of interest to develop a model for estimating the change in shape of irregular topographic defects with EP. Nb_3Sn hosts areas of high slope angles and groove depths, and next steps involve applying a polishing technique to ameliorate these locations. This comparative analysis of Nb will assist in estimating the amount of material needed to be removed from the sample surface of cavity materials to eliminate negative impact on high field performance. Such a model may be useful for next generation SRF cavity material.

Flow Visualization of a Marangoni Surfer: Relating Flow to Surfer Motion

Paloma Suarez Davila, Arkava Ganguly, Ritu Raj, Ankur Gupta

Marangoni surfers are self-propelling particles that move due to surface tension gradients. Understanding their behavior in the presence of surfactant gradients is crucial for advancing microfluidic systems and for controlling particle dynamics in both scientific and industrial applications. The Marangoni effect, responsible for these surfers' motion, plays an important role in enhanced oil recovery by releasing trapped oil from rock pores and in drug delivery by improving the distribution of therapeutic agents.

In this study, we developed an experimental setup to visualize and quantify fluid flow around objects at an air-water interface under surfactant gradients. Using Particle Image Velocimetry (PIV), we characterized the fluid flow around 5 mm Marangoni surfers. Our setup, simpler than traditional PIV systems, consisted of a tank, tracer particles, and a camera. We initially investigated how surfactant concentration and the method of application affect the swimmer's maximum velocity. By analyzing the velocity field and determining the time constant, we derived a Gaussian fit equation with adjustable parameters, A and b. Fitting this equation to experimental data revealed that the values of A and b varied depending on the surfactant concentration and application method, suggesting that convective forces, rather than purely diffusive ones, influence the transport dynamics due to high particle velocities.

Additionally, we optimized tracer particle concentration by testing various levels to balance resolution, variability, and efficiency of tracer usage. We conducted experiments to identify the ideal concentration that provided the best combination of these factors. Finally, we studied the flow around free swimmers with different surfactant application methods, integrating results from tracer concentration, surfactant effects, and application methods. This allowed us to visualize the flow field around Marangoni surfers and compare it to free particle velocity. Future work will focus on stabilizing tracer solutions and achieving a precise quantitative match between tracer PIV experiments and free particle tracking to further investigate the dynamics of these systems.

Additive Manufacturing (DIW) of Energetic Inks

Faris Wald, Caleb Kwi, Ernest McVey, Chloe Sample

Additive manufacturing (AM), known as 3D printing, is widely used today to produce complex geometries quickly and cheaply. The ability to produce exact, repeatable, designs has intrigued the energetics community who would like the ability to rapidly produce explosive designs for wave shaping. Two main problems that occur when AM energetics, heat and density. Energetic performance is exponentially related to the density of material; discovering an AM technique that maintains high solids loading and density is required to be useful for military applications. Based on Pluronic F-127, we developed a stackable, high solids loading, high-density direct ink writing (DIW) ink that can be printed without heat or UV-curing. Additionally, the DIW ink developed utilizes a water-based solvent which greatly increases safety during the processing and printing of energetic materials. Currently, microgram and nanogram copper particles have been developed into printable inks which will be used to form the shape charge portion of our 3D model. Initial work incorporating explosive pentaerythritol tetranitrate (PETN) into the DIW ink has been successful with the ink printed into basic geometric shapes. Future work will be conducted to optimize the print settings for the PETN DIW ink.

Layer-By-Layer Nanofilm Coated Hollow Glass Microspheres for Rapid Isolation and Recovery of Salmonella in Resource-Limited Settings

Timothy McGinnis, Wei Li

Efficient and rapid detection of salmonella and other bacterial infections is needed to reduce infections in low-income settings. The current detection methods are not viable due to high costs and long timeframes. Conventional approaches require 4 to 5 days for results, and methods such as Surface-Enhanced Raman Scattering (SERS) remain inaccessible due to their complex nature and high expense.

This study provides an alternative bacterial isolation system using self-floating hollow glass microspheres (HGMS) coated with layer-by-layer (LbL) nanostructured polymetric films. Using anti-salmonella antibodies, these microspheres target and rapidly capture salmonella cells. HGMS facilitates bacteria cell isolation within 3 hours, confirming Salmonella presence within 24 hours. This systematic approach reduces detection time while reducing cost, making this a more ideal approach for resource-limited areas.

Characterizing Hydration of Ultrathin Film Anion Exchange Membranes Via in Situ Spectroscopic Ellipsometry

Ubanni Opashi, Wyatt Tenhaeff

Anion exchange membranes (AEMs) play a critical role in optimizing the performance of various applications, including redox flow batteries and other ion separation technologies. This study investigates the balance between membrane selectivity and anion transport in ultrathin AEMs. Poly(4-vinylpyridine) was synthesized via initiated chemical vapor deposition (iCVD) and subsequently quaternized with bromoethane. Spectroscopic ellipsometry, combined with an in situ environmental cell, was employed to analyze film thickness, refractive index and volume % of water as a function of relative humidity. Results indicate that humidity results in increased membrane thickness (swelling) and a decrease in refractive index due to water absorption.

The extent of P4VP quaternization was found to increase water uptake, as the introduction of more fixed charges increases the thermodynamic driving force for swelling. Water content in the quaternized P4VP films was determined by modeling the optical properties of the hydrated films with the Bruggeman effective medium approximation, showing the direct correlation between water content and extent of quaternization. Additionally, experiments revealed that over prolonged exposure to high humidity, the maximum membrane thickness stabilized at approximately the same value, remaining consistent across subsequent tests. These findings highlight the important interactions between quaternization and hydration, providing insights essential for optimizing AEM performance in various ion separation applications.

Enhanced Glucose Sensing Via Metal-Phenolic Networks Modified Gold Leaf Electrode: A Novel Approach for Improved Hydrogen Peroxide Oxidation and Detection

Xinyi Wu, Song Yi Yeon, Sunanda Dey, Ariel Furst

In 2021, 38.4 million people in the U.S. alone had diabetes, a chronic disease caused by an inability in regulating blood glucose. Glucose monitoring is crucial for them to avoid complications and prevent disease progression. Though glucose monitoring is well-established, there is always the need for improved sensing. To that end, we have developed a gold leaf electrode that possesses increased glucose detection sensitivity through surface modification with metal-phenolic-networks (MPNs). MPNs are simple, inexpensive, and sustainably-made, and they have the ability to serve as electrocatalysts for hydrogen peroxide (H_2O_2) oxidation. The gold leaf glucose sensor involves a coupled reaction between glucose and H_2O_2 oxidation. Glucose is oxidized by a glucose oxidase coating on the electrode, generating H_2O_2, which is then oxidized through the functional electrode. The current produced by H_2O_2 oxidation is proportional to glucose levels, enabling quantification. After simple, manual electrode fabrication, the system sensitivity was evaluated through titration of H_2O_2 solutions onto the surface. However, the gold leaf electrode was insufficient for quantitative peroxide oxidation. We therefore incorporated an electropolymerized MPN layer on the electrode, yielded currents proportional to peroxide concentration, based on catalysis at the MPN metal centers. To stabilize the MPN and increase the difference in electrochemical current between each H_2O_2 concentration, chitosan was drop coated onto the working electrode. The gold leaf electrode, modified with MPNs and chitosan, showed significant improvement in the sensitivity of glucose sensing, and future tests with urine and serum will be conducted to validate its performance in more complex biological environments.

Studying the Effects of Voltage on Degradation Pathways in Perovskite Solar Cells

Laurel Wasiniak, Peter Lillich, Jeffrey Christians

Silicon solar cells currently dominate the photovoltaic market; however, they are already approaching their thermodynamic performance limit. By pairing halide perovskites with silicon in a tandem solar cell, or using halide perovskite tandems, the thermodynamic limit can be raised from 33% up to 45%. The key technological hurdle to realize these higher performance solar cells is material stability in real-world solar cell operation. Halide perovskites consist of an ABX_3 crystal structure, where the A (1^+ cation), B (2^+ metal cation), and X-site (halide) components are manipulated to obtain desired properties. In this work, $CsPbI_3$ was used as a model halide perovskite for a variety of experiments to better understand the phase change that are common across many halide perovskite compositions. Material stability is challenging, as there are many degradation pathways, including pathways involving multiple stressors. Due to the ionic nature of the constituent species, voltage was used to promote ion migration and observe the phase change from $CsPbI_3$ perovskite to the nonperovskite phase. These experiments were run in a controlled environment to isolate the materials from external stressors, such as humidity, and the phase transition was imaged in real-time using optical microscopy. Additional factors, such as heat and voltage were added to the system to track how their presence influenced the rate at which the samples changed from the dark perovskite phase to the light non-perovskite phase. This work provides a step toward a better understanding of voltage-driven and mixed degradation modes in halide perovskites that are critical to real-world stability.

Fighting Freezing Rain: Effects of Surface Texturing and Temperature on Droplet Impact

Bruce Baker, Katie Kersten, Nick Wayman, Syed Ibrahim Gnani Peer Mohamed, Siamak Nejati

Reduction of droplet-to-surface contact time and area has a wide variety of industrial benefits and is a primary focus of anti-icing technology. It is known that contact dynamics may be affected to some degree by surface texture, droplet radius, variations in the hydrophobic coating, and temperature variation of the droplet or surface, among other factors. The team is dedicated to quantifying the influence of these factors and assessing how strongly they each affect droplet impact. In this study, poly(1H,1H,2H,2H-perfluorodecyl acrylate) (pPFDA) is deposited via initiated chemical vapor deposition (iCVD) on a smooth silicon surface as well as femtosecond lasers roughened aluminum surfaces. The surfaces are each imaged at the millimeter scale to quantify roughness. The coated substrates are placed on a peltier device at varying temperatures, and droplet impacts are then measured and evaluated at a high framerate to yield kinetic information. The roughened surfaces saw a large reduction in contact time and a much more elastic collision between the droplet and the surface. Droplets adhered to chilled surfaces more strongly, and this effect was enhanced on rougher surfaces.

Phase Separation of Liquid Crystals Produces Filaments

Aaron Boyd, Christopher Browne, Chinedum Osuji

Liquid-liquid phase separation has received growing interest due to the newly recognized role it plays in biological systems in the formation of condensates. Some biological fluids can, under certain conditions, exhibit liquid crystallinity. It is unknown how this liquid crystallinity may impact condensate formation and the shape of condensate domains. This work examines phase separation in synthetic systems that are a mixture of a liquid crystal mesogen and a simple solvent. While interfacial forces generally drive the formation of discrete spherical condensates that coalesce into macroscopic domains, our work has shown that liquid-liquid crystal systems can form sample-spanning filamentous networks. The formation of these networks is difficult to study due to highly dynamic interaction of many growing filaments. By confining this phase separation to microdroplets, we are better able to observe filament dynamics and we find that filament morphology is determined by an interplay of interfacial forces, distortion of the smectic layers, and geometric confinement. We observe a characteristic relationship between drop and filament radius. This relationship appears to be mediated by the evolution of filaments by intermittent collapse, where some critical threshold length is reached, and the filament coalesces before regrowing with a thicker fixed radius. High-speed imaging suggests that the collapse mechanism itself may be mediated by filament-filament contact which propagates by zipping and twisting. Further exploration of this system may yield insights into biological condensates, smectic behavior under non-equilibrium dynamics, or new materials.

Localization of Microparticles in Solution By Microfluidic Interactions on a Patterned Substrate

Ian Forrer, Yian Cheng, L. Jay Guo

Accurate and efficient fabrication is critical to the progress of nanotechnology. Most existing nanofabrication techniques use top-down methods, although some bottom-up methods are used. By comparison, in macroscopic fabrication processes modular prefabrication and subsequent assembly allows for high efficiency and reduced costs. However, the assembly of prefabricated micron-scale particles with high accuracy and throughput remains challenging. One example is the assembly of Micro-LED (LED) arrays. LED displays offer superior brightness, stability, and power consumption[1] but the individual LEDs must be grown on separate wafers then transferred to form a display. Traditional pick and place methods are accurate but limited in throughput. Existing fluidic assembly methods offer a higher throughput, but accurate techniques exist only for the confinement of particles at over 10 m[3] or submicron[2] size ranges.

This work develops a fluidic assembly technique to localize microparticles onto a patterned wafer. By patterning a silicon wafer with regions of gold and silicon oxide then thiolating the gold to increase hydrophobicity, microfluidic interactions can localize colloidal particles. A microfluidic channel is formed to press a droplet of nanoparticle solution into a thin film, then the cover is removed causing the bulk fluid to recede and leave behind microdroplets that localize the particles as they evaporate. Localization can also be controlled by altering the solvent properties. For example, adding isopropanol to the primarily aqueous solvent increases surface energy but also decreases the cohesion in the bulk solvent to improve droplet formation. On a thiol-treated gold substrate adding 30% isopropanol by volume gave the highest localization density as shown in Figure 1. This study presents preliminary results and proof of concept for fluidic assembly of microparticles with potential applications in microLED technologies and other microelectronics.

(1) Scholz, S.; Kondakov, D.; Lüssem, B.; Leo, K. Degradation Mechanisms and Reactions in Organic Light-Emitting Devices. Chem. Rev. 2015, 115 (16), 8449–8503. https://doi.org/10.1021/cr400704v.

(2) Chen, L.; Panday, A.; Park, J.; Kim, M.; Oh, D. K.; Ok, J. G.; Guo, L. J. Size-Selective Sub-Micrometer-Particle Confinement Utilizing Ionic Entropy-Directed Trapping in Inscribed Nanovoid Patterns. ACS Nano 2021, 15 (9), 14185–14192. https://doi.org/10.1021/acsnano.1c00014.

(3) Chang, W.; Kim, J.; Kim, M.; Lee, M. W.; Lim, C. H.; Kim, G.; Hwang, S.; Chang, J.; Min, Y. H.; Jeon, K.; Kim, S.; Choi, Y.-H.; Lee, J. S. Concurrent Self-Assembly of RGB microLEDs for next-Generation Displays. Nature 2023, 617 (7960), 287–291. https://doi.org/10.1038/s41586-023-05889-w.

Figure 1: Microscope image of particles localized on a hexagonally patterned substrate.

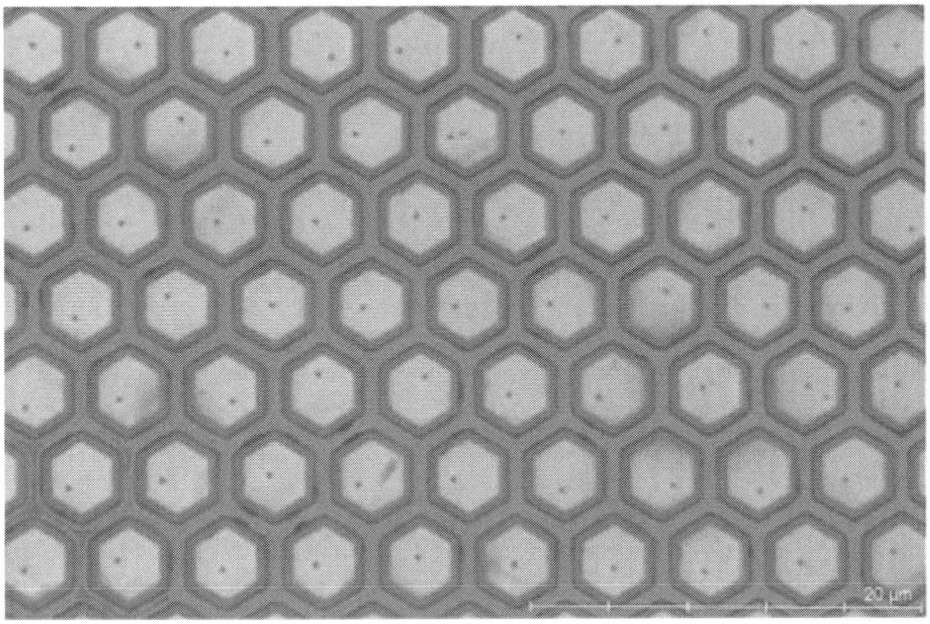
20 µm

Adhesive and Physical Characterization for Adfors Wall Finishing

Szymon Kuzniar

ADFORS Wall Finishing (AWF) is a novel woven material with enhanced properties compared to a traditional painted finish used to reduce interior finishing times in off-site home manufacturing (OSM) by up to 50% and realize a >10% overall productivity gain. However, no commercially available adhesives have been recommended for use resembling that in OSM, as the finished modules can be stored in harsh environments for up to a year. Without a prescribed adhesive and standard procedures, adverse effects on AWF adhesion have been observed after as little as three months. Using a full factorial experimental design – varying environmental aging, adhesive type, and application methods – the recommendation was narrowed down to two adhesive candidates for the system. Larger-scale samples have also been used in a screening study to find possible delamination causes, determining the probable cause of system failure as adhesive misapplication. Furthermore, AWF properties were characterized using permeability, impact resistance, washability, and infrared thermography using chemical engineering fundamentals such as heat and mass transfer, in addition to quantifying aesthetic qualities including color, gloss, and roughness. Having confirmed the crucial points of AWF's value proposition and aesthetic suite, it was soft launched into OSM with strongly positive feedback and substantial time savings.

Computational Modeling of Organic Vapor Phase Infiltration

Eitan Feldman, Brian Welch, Tamar Segal-Peretz

Vapor Phase Infiltration (VPI) is a sequential, repeated process where precursors embed themselves into a substrate, altering material properties such as UV resistance or conductivity. VPI has previously been used to make organic-inorganic hybrid materials. However, VPI with organic precursors and an organic substrate is not well studied and could potentially result in significant material enhancements. Here, the thermodynamics of potentially significant VPI reactions between common Molecular Layer Deposition organic precursors and a common Atomic Layer Deposition organic substrate are discussed. Three reactions were examined: polyvinyl alcohol (PVA) with isophthaloyl chloride (IC), PVA with 1,4-Diisocyanatobenzene (CB), and PVA with terephthalaldehyde (TA). Each of these reactions was first tested experimentally and then analyzed using computational methods. While the reaction of IC and PVA was observed experimentally, the other two reactions did not occur to a significant extent, despite the considerable reactivity of CB. Through the use of ARC, an open-source software, the thermodynamics of these reactions was determined (with PVA approximated by smaller oligomers). The computational results confirmed that the reaction between IC and PVA is much more favorable, than the other two reactions. However, the reaction between CB and PVA is still somewhat favorable. While the reaction between alcohols (PVA) and isocyanates (CB) is generally quite favorable, the rigidity of the PVA and the lack of solvent in this gas phase reaction could explain why there is a significant kinetic barrier in this particular reaction. Both the experimental and computational results indicate the viability of the IC PVA reaction, and the lack of viability for the TA PVA and CB PVA reactions. Further research can build on this proof of concept to develop organic VPI.

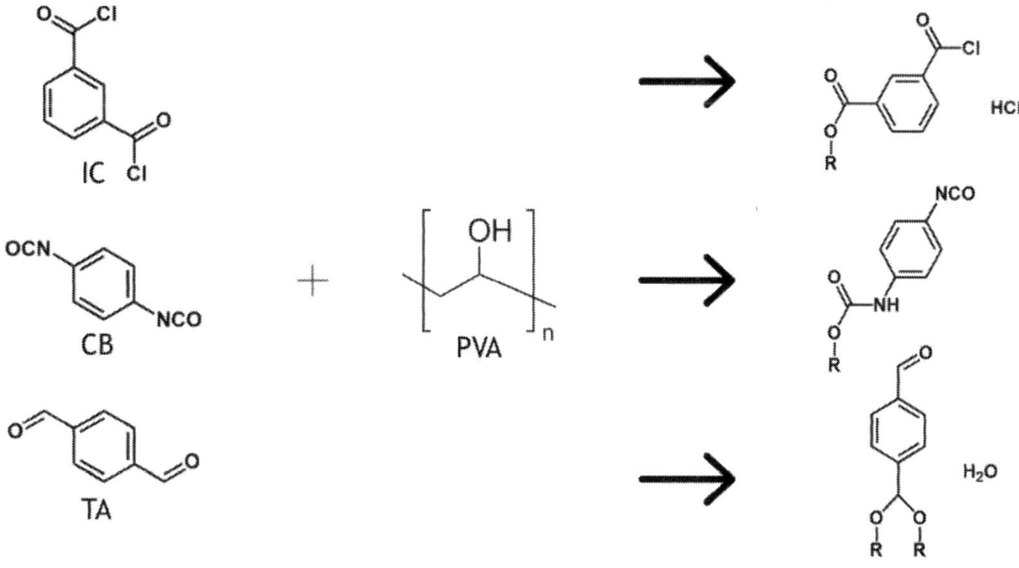

Effect of Dipole Asymmetry on the Interfacial Behavior of Polar Fluids

Ananya Venkatachalam, Samuel Varner, Pierre Walker, Bilin Zhuang, Zhen-Gang Wang

Recent research using first-principles molecular dynamics has shown that water, despite being a neutral molecule, can generate a strong electric field at the air-water interface. We postulate that this anomalous interfacial behavior, which can catalyze atmospheric reactions, arises from dipole asymmetry, where the molecular center of mass is offset from its net dipole vector. Our study generalizes this idea, focusing on the underlying mechanisms for strong polarization and ion adsorption at the vapor--liquid interfaces of polar fluids. We find that asymmetry in molecular dipoles results in preferred orientations at the interface, leading to a finite electrostatic potential and stronger electric fields. Higher dipole moments increase fluid density and structural organization, while greater dipole offset asymmetry produces more distinct alignment and stronger electric fields at the interface. These results contribute to an improved understanding of interfacial polarization and offer potential applications in predictive climate change models.

Stretchable, Skin-Conforming Laser-Induced Graphene Sensor Array for Non-Invasive Stress Monitoring in Mental Health Applications

Ruotong Gao, Muhammad Khatib

About one in four American adults suffer from mental illness with over half of them not receiving any treatment or diagnosis. The average time between initial symptoms and treatment sits at around eleven years. Roughly ten percent of Americans adults suffer from serious mental health issues with half reporting serious consideration of suicide. This number is only rising with the mental health crisis at it's all time high, hitting teens and young adults the hardest. People afflicted by mental illness suffering severely with severe risks of cardiovascular and metabolic diseases, substance abuse, unemployment, dropout rate, and more. Stress is one of, if not the leading cause of mental health disorders, and requires significantly more awareness on it's negative ramifications. Initial diagnosis and monitoring of mental illness is especially difficult due to the financial and social repercussions of such actions. Therefore, there is an urgent need to develop a new device, one that is highly precise in it's measurement, affordable, portable, and multifaceted. Laser-Induced Graphene is at the forefront of material science in its excellent conductivity and properties. Due to the porous nature of graphene, sensor-liquid interfacing is facilitated through the high amount of contact the liquid has with the graphene. These properties allow for graphene to be an excellent electrode for a electrodermal skin-conductance/impedence sensor, the premier method for stress detection and monitoring for psychologists. Anxiety and extreme levels of stress rise the body temperature, and increases perspiration significantly, both of which can be tested using similar resistance sensor technology in a compact format. These sensors are created through elastomeric encapsulation of Laser-Induced Graphene (LIG) and other polymers such as styrene-isoprene-styrene (SIS) to optimize their specific sensing properties. A skin-impedance sensor for electrodermal activity measurement, a temperature sensor for detecting stress-induced thermal changes, and a humidity sensor for monitoring perspiration levels was created in order to monitor stress through a holistic metric of the three factors. Testing for the skin impedance sensor was conducted with comparison of a electrolyte-water solution and a DI water control to establish the efficacy of the sensor. Afterwards, the skin impedance sensor was tested on myself using the cold pressor test to induce a sympathetic autonomic stress response that was recorded in comparison to a control standard state. The results received for the temperature sensor demonstrated a linear response ($r^2 = 0.9465$) across a range of 0-50°C. The humidity sensor registered spikes of 8*10-4 and the skin-impedence sensor registered a significantly sharper curve, taking 6.75 seconds to reach 0.0683 microamps within the simulated stress compared to 60 seconds for 0.0682 microamps for the control. This developed sensor array provides immediate applications for medical usage for mental health monitoring in a precise, skin-comforming, cost-effective, portable device that is easily accessible. Integration of these sensors was achieved through a custom-designed printed circuit board assembly (PCBA), utilizing a parallel circuit design for multiple channel measurements. The system operates at low voltage (< 0.5 V) and employs a Wheatstone Bridge configuration to minimize noise and inter-sensor interference. Initial results indicate promising performance, but requires further developments for consistency. While refinement is needed, this technology shows potential for immediate applications in medical and mental health monitoring, offering a precise, skin-conforming, and cost-effective solution for stress assessment. This research contributes to the growing field of wearable bioelectronics and demonstrates the potential of advanced materials in addressing critical healthcare challenges. Future work will focus on large-scale validation studies and the development of robust data analysis algorithms to enhance the system's diagnostic capabilities.

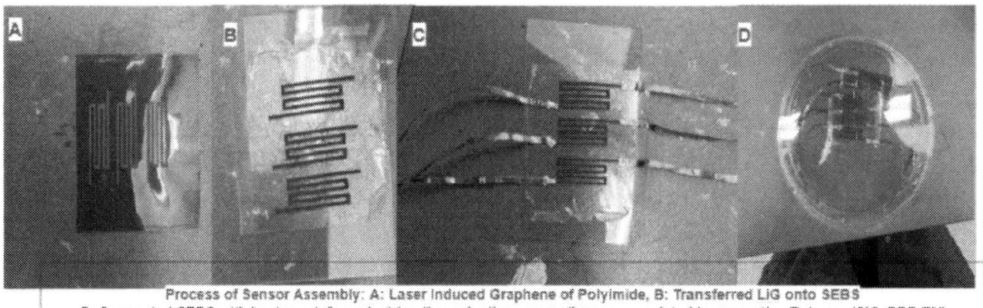

Process of Sensor Assembly: A: Laser Induced Graphene of Polyimide, B: Transferred LIG onto SEBS
C: Connected SEBS with kapton-reinforced wiring through silver epoxy then encapsulated in respective Polymer (SIS, PPG PU)

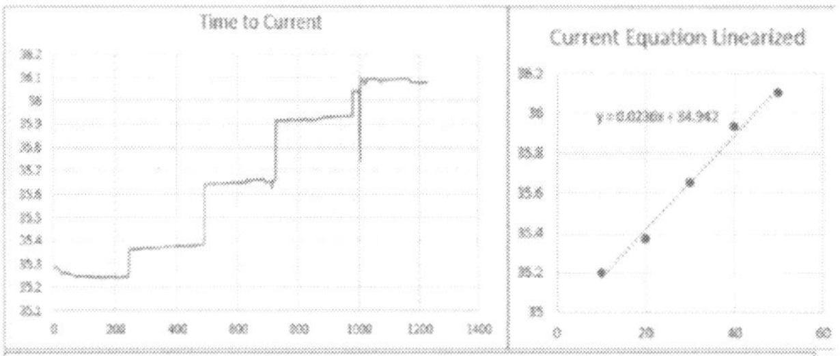

Graph 3: Temperature Sensor surface testing on 500 mL beaker in increments of 10°C using chronoamperometry on a Palmsens 4 Potentiostat data in microamps plotted over 300 seconds in intervals of 60 seconds per 10 degrees.

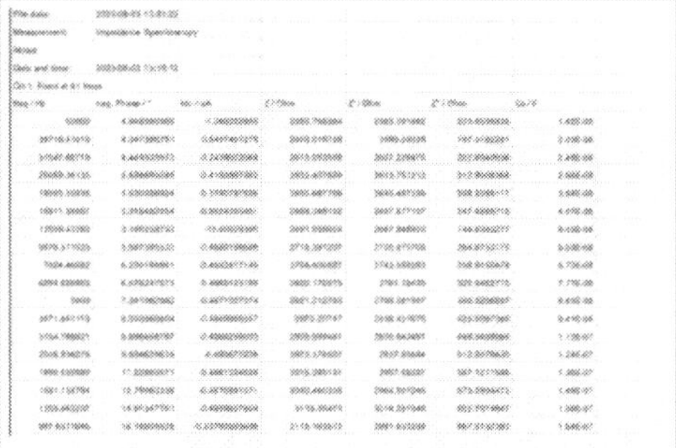

Graph 2: Electrolyte testing of Skin-Impedence Sensor to test for Efficacy using NaCl compared to DI Water Impedence Spectroscopy Data using Palmsens

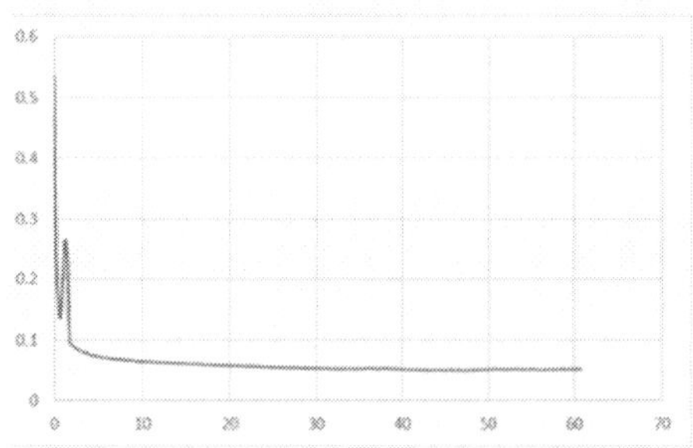

Figure 5: Skin-impedance human data over 60 seconds with the cold pressor test to induce autonomic sympathetic stress, High stress.

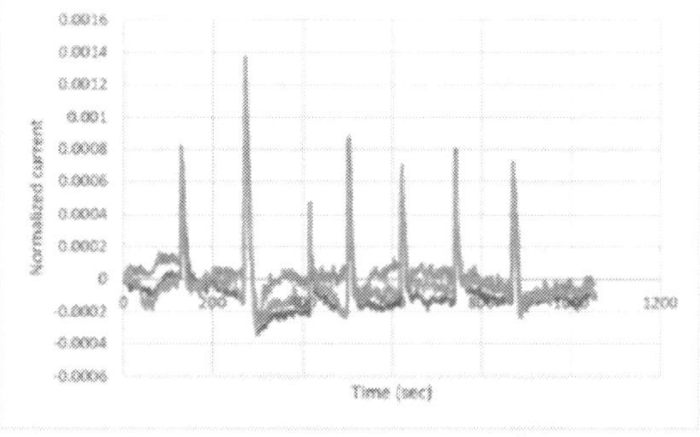

Figure 4: Humidity Sensor data over 1000 deciseconds (1/10 a second) tested with hot breaths on the surface of the electrode head on the sensor

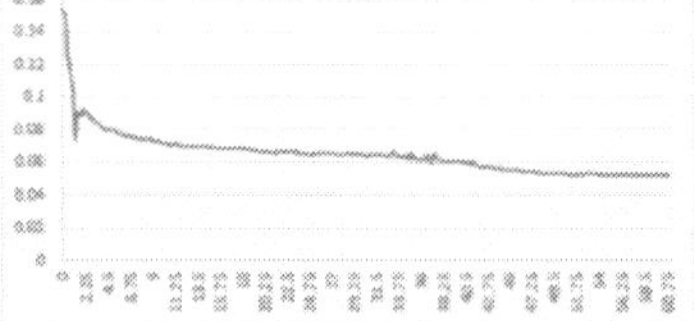

Figure 6: Skin-impedance human data over 60 seconds with the cold pressor test to induce autonomic sympathetic stress, No Stress

```
$ python oneclasssvm.py
Number of outliers: 561
Number of inliers: 525
```

Data received when inputted a high-stress dataset into the One-class svm type machine learning algorithm

Solvent and Concentration Effects on Polymorphic Forms of Glutaric Acid

Victoria Burek, Ryan Snyder

Glutaric acid, a five-carbon dicarboxylic acid, is known for its polymorphic properties, characterized by the ability to crystallize in multiple distinct forms. Polymorphism, the phenomenon where a single chemical compound can adopt different crystalline structures, plays a significant role in determining the physicochemical properties of substances. A notable example is carbon, which can manifest as either graphite or diamond. The polymorphic behavior of glutaric acid is particularly relevant in pharmaceutical applications, where the specific polymorph can profoundly impact a drug's bioavailability, stability, and therapeutic efficacy. Glutaric acid predominantly exhibits two polymorphs: the metastable alpha form and the stable beta form.

In this study, we investigated how using different solvents at varying concentrations affects the polymorphic outcome of glutaric acid. Particles of glutaric acid were generated by using an atomizer and dried using two different drying media: silica gel and zeolite when water was employed as the solvent, and charcoal and zeolite when alcohol was used. The resulting particles were collected on an X-ray diffraction (XRD) disk in a PIXE Cascade Impactor, with any residual particles being captured on a NeoPore filter. The collected samples were subsequently analyzed using powder X-ray diffraction (pXRD). Data analysis was performed using MATLAB and Excel to determine the polymorphic composition of the samples. The results show that the choice of solvent impacted the polymorph that was present, and that the water based experiments displayed polymorphism that isn't found through other methods of water crystallization.

Influence of Surfactants on Electrode Slurries Used for Flow Battery Application

Samhita Vasudevan, KangJin Lee, Christopher L. Wirth

Flow batteries are an emerging technology used for energy storage on the grid-scale level. The performance of a flow battery greatly depends on the efficiency of the internal slurry electrode, which affects the conductivity, capacity, and stability of the slurry during inactive periods. To improve on slurry stability, surfactants have been added into the carbon black slurry electrode. Previously, one non-ionic surfactant, Triton-X 100, has shown a relationship between concentration of surfactant and slurry stability. It was shown that surfactant concentration affected slurry stability at a critical mass ratio of surfactant to carbon black ($\alpha = c_{surf}/c_{cb}$) which for Triton-X 100 was 0.7. At this point, the surface area of the carbon black particles are saturated with surfactants which inhibit the interparticle interactions, leading to an unstable slurry. This study was extended with two more surfactants, Triton-X 45 and Triton-X 405, to investigate the relationship between surfactant molecular weight and slurry stability. The adsorption studies concerned finding critical α for both Triton-X 45 and 405. Adsorption studies were completed using UV-Vis spectroscopy for α values ranging from 0 – 1.5. Results show that there is no significant difference in critical α between Triton-X 100, Triton X-45, Triton X-405, showing that the molecular weight of surfactant does not affect surfactant saturation at a CB particle surface. In addition, sedimentation studies have been conducted in order to measure the settling behavior of carbon black particles with the addition of surfactants. Previous studies have demonstrated that carbon black slurries behave like a colloidal gel interparticle attraction, but collapse under gravitational stress, leading to phase separation. The sedimentation behavior of gels can be influenced by various factors such as particle concentration, gel strength/elasticity, and the presence of surfactants. Thus, as was done previously with Triton-X 100, sedimentation experiments with Triton-X 45 and Triton-X 405 were conducted. This is crucial to understanding how concentration of surfactant and surfactant size affects gel collapse behavior over time. Time studies are ongoing for various carbon black weight percentages and α values.

Alternative Sintering Processes and Their Influence on the Electrical Resistance of Printed Ag Nanoparticle Conductive Networks

Gray Podolak, Naimul Arefin, Moni Hur-E-Jannat, Minxiang (Glenn) Zeng

While post-printing treatment is essential for printed electronics to achieve high device performance, the evaluation of different sintering techniques has been fragmented and mostly qualitative, leaving their effects on device transport properties unclear. This study systematically examines four distinct sintering methods, thermal, vacuum, chemical, and photonic (ultraviolet light), to quantitatively evaluate their performance and feasibility in large-scale fabrication of printed electronics. In particular, we systematically studied chemical sintering by quantifying the role of chemical agents in the device's conductivity, which led to the identification of HCl vapor in optimizing the electrical conductivity of printed silver devices. As a proof of concept, a flexible strain sensor was successfully printed and sintered chemically. While this work does not intend to cover all existing sintering methods for printed electronics, our findings do offer a detailed evaluation, shedding light on how different sintering approaches can concretely affect the performance and quality of printed electronic components and architecture.

Exploring Janus Particle Microroller Swarms in Mazes

Tyler Richardson, James Gilchrist

Title: Exploring Janus Particle Microroller Swarms in Mazes

Student: Tyler Richardson (collaborating with Junqi Cui, Mahir Enan, Olivia Percaccio, Genevieve Powell, John Riffle, and Noah Tobin)

Advisor: Professor James Gilchrist

This study aims to understand Janus particle microroller behavior under constraints by investigating the dynamic navigation of Janus particles through 3D printed mazes. The particles are 44 mm PMMA beads that are coated on one half with iron oxide and they are activated using an oscillating magnetic field. The mazes are made of transparent materials for clear visualization and efficient lighting, and they have complex patterns with branching pathways and dead ends. To understand their motion, particles are investigated under more simple systems where they swarm and move unidirectionally and their velocity and microroller flux is quantified. Using these rules, better design systems to actuate swarms of rollers and navigate them through complex geometries.

Lifetimes of Excited States in Ge-72 from Inelastic Neutron Scattering

Blake Tomas Lopez, Logan Martin, Elizabeth A Chouinard, Sarah E Evans, Sally F Hicks, Jeffrey R Vanhoy, Yongchi Xiao, Steven W Yates, Erin E Peters

Nuclear structure data are essential for a better understanding of the atomic nucleus and its properties. The nucleus of an atom is generally described as possessing a 'shape' that can be spherical or deformed, which arises from the interactions of nucleons. For some nuclei, 'shape coexistence' has been observed, which is a phenomenon that describes different deformations of the same nucleus with increasing excitation energy. In addition, nuclear structure data are necessary for constraining calculations of the nuclear matrix element for neutrinoless double-beta decay ($0\nu\beta\beta$). If this theoretical process is observed, the matrix element is necessary for determining the mass of the neutrino.

Currently, the study of nuclear structure data for the isotopic chain of Ge is being conducted at the University of Kentucky Accelerator Laboratory (UKAL). Ge nuclei have been shown to exhibit shape coexistence and ^{76}Ge is a candidate for $0\nu\beta\beta$, thus studying these nuclei are of multiple interests. For this project, data for ^{72}Ge were collected using the UK 7 MV Van de Graaff particle accelerator to produce a 5 MeV proton beam that impinged upon a tritium gas target. The subsequent reaction produced monoenergetic neutrons at an energy of 4.0 MeV. The neutrons were inelastically scattering from an elemental target enriched to 97.85% ^{72}Ge. The resulting gamma rays were detected at various angles utilizing a high-purity germanium detector. Lifetimes of the excited states were obtained from the angular distribution data making use of the Doppler effect. This analysis utilized the Doppler-shift attenuation method following inelastic neutron scattering. A linear function relating gamma ray energy as a function of the cosine of the detection angle was used to express the slowing down of the recoiling process through the material, and the resulting attenuation factor was compared to a theoretical curve to extract the lifetime value. New lifetime measurements were found in the femtosecond region for gamma rays of excited states in ^{72}Ge, which have not been previously recorded in the Evaluated Nuclear Structure Data File (ENSDF) database maintained by the National Nuclear Data Center (NNDC). In addition, lifetime values that were previously measured agree well with those found in ENSDF. The lifetimes obtained were found within the range of 35 to 900 fs with their appropriate errors. The analysis of these data for the ^{72}Ge isotope at UKAL will be used with other data to determine if the isotope exhibits shape coexistence and for a better understanding of its structure.

This material is based upon work supported by the U.S. National Science Foundation under Grant No. PHY-2209178. Travel funding was provided in part by the University of Kentucky's Chellgren Center for Undergraduate Excellence and donor Terry Strange (CME Alumni). Additional funding was provided by the AIChE Chapter at the University of Kentucky and the Department of Materials and Chemical Engineering.

Analysis of Phase Transitions in Lead Halide Perovskite Cesium Lead Chloride

Joseph Fink, Sammy Fieser, Joshua Choi

Understanding the reaction kinetics of a material is an integral part of its characterization. The focus of this project is on the formation of the perovskite cesium lead chloride (CsPbCl3). Materials containing a perovskite structure, such as CsPbCl3, have defect tolerance and tunable electronic properties, which make them a topic of intense research. For instance, ytterbium-doped cesium lead chloride (Yb:CsPbCl3) is a promising scintillator material for radiation conversion in detectors. The solid-state synthesis allows for easy scale-up to produce large amounts of polycrystalline scintillators. Developing an understanding for the formation allows us to better refine the synthesis as well as predict the phase transitions of the material. Interdiffusion of constituent elements is the limiting step in the conversion to the perovskite phase. Increased heating temperature and time is expected to increase the rate of diffusion and will result in a higher conversion. X-ray diffraction (XRD) and differential scanning calorimetry (DSC) were the main methods of quantifying the amount of each phase in a sample after heating. A solid state synthesis, beginning with the same composition of reactants, was utilized to create samples. Samples varied in both temperature and reaction duration in an ambient atmosphere controlled furnace. Diffraction patterns from the XRD were analyzed to determine the percent composition of the crystalline phases present in the sample by comparing relative peak intensities for the perovskite and precursors. Furthermore, an unreacted sample was measured through heating on a hot-stage in the XRD to track the phase development in-situ. Measuring the heat flow into the sample by DSC also provided an indication of the conversion of the material from its orthorhombic to cubic phase. Initial results show that as temperature and time increase in the furnace, a pure 3D phase composition encompasses the material. Additionally, the interdiffusion coefficient was determined for the precursors, CsCl and PbCl2, to understand the rate of interdiffusion of the precursor elements. To accomplish this, pellets of CsCl and PbCl2, together, were heated and measured using EDS linescans and fitted to characteristic equations. Hot-stage heating and calculating the interdiffusion coefficient are on-going. Future work looks to add the dopant Yb and quantify its diffusion characteristics in the perovskite.

Investigation of Fabrication Methods to Enhance Proton Conductivity in $Ba(Zr_{0.4}Ce_{0.4}Y_{0.1}Yb_{0.1})O_3$ electrolytes for Solid Oxide Electrolysis Cells Using Sintering Aids

Ha Tran, Siyuan Liu, Awa Kalu, Xingbo Liu, Wenyuan Li

High-temperature solid oxide cells (SOCs) can be reversibly operated as fuel cells (SOFCs) or as electrolyzer cells (SOECs) for applications like water splitting or syngas production. SOECs are particularly attractive due to their high electrical-to-chemical energy conversion efficiencies, making them ideal for hydrogen or syngas production in renewable energy contexts, such as 'power2gas.' Among ABO_3 perovskite-type proton conductors, $Ba(Zr_{0.4}Ce_{0.4}Y_{0.1}Yb_{0.1})O_3$ exhibits a relatively high conductivity of 0.013 S cm^{-1} at 600°C. However, its densification at low temperatures is challenging, requiring sintering at 1500–1700°C to achieve a high relative density. Such high sintering temperatures can lead to problems like BaO evaporation, phase segregation, and Ni diffusion, all of which reduce proton conductivity and affect the cell's performance. This project aims to develop a $Ba(Zr_{0.4}Ce_{0.4}Y_{0.1}Yb_{0.1})O_3$ electrolyte with high protonic conductivity for proton-conducting solid oxide electrolysis cells. By using sintering aids like ZnO, Y_2O_3, and BaO, we propose different fabrication methods to transfer from $Ba(Zr_{0.4}Ce_{0.4}Y_{0.1}Yb_{0.1})O_3$ electrolyte pellets to apply on NiO anode in order to enhance the densification process and reduce grain boundary resistance. In this research, different wt% of ZnO, Y_2O_3, and BaO were added to $Ba(Zr_{0.4}Ce_{0.4}Y_{0.1}Yb_{0.1})O_3$, leading to improvements in conductivity and grain size. The effects of these additives on the relative density, microstructure, and electrochemical performance of the electrolyte for steam electrolysis were investigated, showing increased conductivity and enhanced electrochemical performance. The phase purity and proton conductivity of the electrolyte were also measured and analyzed.

Quantifying Colloid Sizes in Purified Tin(II) Iodide Precursors for Superior Tin-Halide Perovskite Films

Aditya Pradhan, Jessica Persaud, Aditya Mohite

Perovskites are notable semiconductors for photovoltaic devices, achieving power conversion efficiencies (PCEs) exceeding 26%. However, concerns over commercialization persist due to lead-based precursors. Tin-based perovskites present a promising alternative, sharing similar band structures, band gaps, and lattice properties, but they suffer from lower stability and reduced PCEs due to the oxidation of tin(II) to tin(IV) and rapid crystallization during casting. Literature suggests that uncontrolled crystallization results from the poor solubility of tin(II) iodide (SnI_2), a key precursor in tin perovskite synthesis, in aprotic solvents. This leads to defect trap states, iodide vacancies, and cascading surface oxidation of uncoordinated tin(II). Our study aimed to quantify SnI_2 solubility by measuring colloidal size distribution at varying temperatures and concentrations to optimize crystallization and improve stability.

Initially, to account for the spontaneous oxidation of SnI_2, we performed sublimation to separate tin(II) iodide from tin(IV) impurities by leveraging the differences in their sublimation temperatures. Differential scanning calorimetry (DSC) indicated tin(IV) iodide sublimes at 150°C, while tin(II) iodide sublimes at 330°C. X-ray photoelectron spectroscopy (XPS) confirmed a significant reduction in tin(IV) impurities after separation, validating sublimation as an effective purification method.

Following purification, we used dynamic light scattering (DLS) to study the effect of varying SnI_2 concentrations in N,N-Dimethylformamide (DMF) solvent. Increasing the concentration from 0.8M to 0.9M led to a 0.357% increase in particle hydrodynamic radius, from 0.546nm to 0.548nm. A further increase to 1M resulted in a 1.30% rise, reaching 0.555nm, thus indicating a positive correlation between concentration and particle size. We also examined temperature effects on colloidal size distribution. At 25°C, 0.9M SnI_2 solutions showed two particle size groups with radii of 0.364nm and 46,500nm. After placing 0.9M SnI_2 initially at room temperature in a 50°C bath, the larger particles shrank to 10,300nm in 3 minutes and disappeared completely after 10 minutes. We hypothesize that these large particles are colloidal agglomerates formed at room temperature, which gradually dissipate as temperature increases, leading to their disappearance from DLS measurements. These agglomerates could result in defects and poor crystallization of the film during casting.

In the future, we plan to synthesize tin perovskite films using SnI_2 precursors prepared at different temperatures and concentrations. Through X-ray diffraction and time-resolved photoluminescence characterization, we aim to identify the optimal SnI_2 precursor conditions to control crystallization and maximize film quality.

Depositing Graphene Oxide Composite Films Onto Cotton Substrates for a Topical Controlled Release Drug Delivery System

Andrew Zerby, Zachary Saleeba, Robert Hurt

Our group aims to leverage Graphene Oxide (GO) film properties to sustain controlled antibiotic release from cotton wound dressings for biomedical applications. In order to accomplish this, we must create a GO film–cotton fiber interface which can exhibit controlled release of intercalates. As it has recently been reported that multilayered 2-D GO films exhibit controlled basal plane release of intercalates trapped within the layers of the film, this makes our objective more feasible. As for the creation of the interface, our work demonstrates that GO composites can be deposited as uniform films onto complex cotton substrates. The GO composite solution consists of Graphene Oxide, a crosslinker, and a polymer. To deposit the film onto the cotton substrate, the cotton dressings were saturated with stable composite solutions (with and without antibiotic) and air dried at elevated temperatures (~60°C) for 30 minutes. Scanning electron microscopy (SEM) and optical microscopy (OM) were used to characterize the films, and biological assays were used to confirm non-toxicity to human tissue and antibacterial. SEM images show textural differences between non-treated and treated cotton fibers. Fibers treated with the GO composites show smooth exteriors while fibers not treated with the composites exhibit textured exterior surfaces. OM results display color changes in the treated cotton dressings in contrast with the untreated dressings. Initial biological assay results indicate no hemolytic activity. The SEM results suggest GO composite film deposition onto the fibers. GO films, if mechanically undisturbed after deposition, are typically very smooth. The transition of textured, untreated cotton fibers to smooth fibers after treatment indicates film deposition onto the entire fiber. This is advantageous because it will maximize the area of film deposition and, therefore, will maximize the amount of antibiotic loading. The change in color of the cotton weaves ultimately indicates retention of GO on the fibers, as GO has a brown-like hue.

Modification of Low-Cost 3D Printed Electrodes with Noble Metals

Hao Dang

The ability to use fused fluid fabrication (FFF) 3D printing to generate electrochemical cells and electrodes has increased electrochemistry's accessibility of electrochemistry. Much of the research on FFF 3D printed electrodes has utilized conductive carbon-based composite materials. While these materials have shown tremendous potential in a variety of applications, the conductivity and composition of these composite electrodes is still a limiting factor. In some cases, modification of FFF 3D printed parts with nickel has dramatically improved the conductivity, though nickel has limited stability in some chemical environments. Here we demonstrate our recent progress in modifying nickel-based electrodes with noble metals using both electroless deposition and electrodeposition. The noble metal thin films were evaluated using a combination of gravimetric, microscopic, spectroscopic, and electrochemical methods. Results demonstrate that the conditions during the noble metal deposition have a profound effect on the properties of the resulting electrode, likely due to the stability issues of the underlying 3D-printed electrode. Further optimization will result in the ability to generate custom electrode geometries coated with thin films of noble metals at low cost using widely available 3D printing techniques. Such flexibility in fabrication methods will continue to advance the ways electrochemistry is employed in academic and research environments.

Slowing the Phase Transition of a Suite of Perovskites Via Passivation

Brian Tran, Ian McGovern, Jeffrey Christians

The global drive for expanding renewable energy production to combat global warming demands for increasing solar energy production. Metal halide perovskite solar cells (PSCs) have gained notoriety in recent years due to their high efficiency, comparable to silicon photovoltaic (PV) cells, and low manufacturing costs. They realize their full potential in tandem solar cells with silicon which have now reached over 34% power conversion efficiency, compared with 27.3% for silicon alone. A major barrier preventing widespread commercial implementation is the instability of halide perovskites when exposed to ambient conditions. One such critical factor in MHP stability are surface defects (halide vacancies) during fabrication generating under coordinated Pb^{2+} ions at the surface and grain boundaries. These sites act as nonradiative recombination sites which reduce the PSC performance and, more importantly, can act as sites for degradation processes to begin. Therefore, surface defect engineering is a crucial component in improving the stability of PSCs and is done via surface passivation. The ligands acting as lewis bases will bind to the under coordinated Pb^{2+} ions reducing the number of surface defects. A class of passivator ligands have shown to improve both the phase and thermal stability of the perovskite with the chemical composition of $Cs_{0.05}(FA_{0.85}MA_{0.15})_{0.95}Pb(I_{0.85}Br_{0.15})_3$. In this work, we use this suite of surface passivators in a dip coating process which is broadly compatible with a suite of halide perovskite films. To determine which ligands improve thermal and phase stability, initial explorations of thermal stability and phase stability have been performed. This work extends the applicability of surface passivation to a wide range of halide perovskite films and film formation methods, and is a stepping stone towards a better understanding of surface defects in halide perovskite degradation and to improve perovskite stability in the real world environment.

Quantifying Oxidation of Hafnium, Niobium, Tantalum, and Zirconium High Entropy Alloy

Alexander Lascheid

Newer generation heat engines and exchangers – such as hydrogen internal combustion engines – achieve high efficiencies because of the high temperatures that are reached inside of the combustion chambers. These temperatures can reach as high as 1500 °C, which many current steels and alloys cannot withstand. As temperatures and efficiencies continue to rise, new materials are needed to be developed that are able to withstand these high temperatures. High entropy alloys (HEAs) are a class of materials composed of 4 or more elements mixed in nearly equiatomic compositions and have shown the potential for improved material performance in high temperature, corrosive, and irradiation environments. This work explores the oxidation and irradiation performance of a HfNbTaZr. Isochronal annealing at 600, 700, 800, and 900 °C coupled with mass change measurements enable calculation of the oxide growth rate, while in-situ Thermogravimetric Analysis (TGA) and in-situ X-ray Diffraction (XRD) up to 900 °C decipher the real-time phase and oxide growth changes. Post-exposure analysis with XRD and SEM offer information on the development of new phases and the spatial distribution of oxide formation. XRD analysis provided valuable information about the likelihood of each element oxidizing, with Ta and Zr being much more likely to oxidize as compared to their Hf and Nb counterparts. This was in contrast to the in-situ SEM annealing experiments, highlighting the differences between testing techniques. Finally, initial results on the irradiation stability of the HfNbTaZr HEA under 2.8 MeV Au ion irradiation offers insights on the performance of this HEA composition in multiple extreme environment conditions. This work can inform future alloy design to develop optimized HEAs and materials to better withstand higher temperatures, which is a key cornerstone to next generation Carnot heat engines – and the next step towards a zero-emission world.

Click Chemistry of Organic Color Center-Tailored Carbon Nanotubes

Anna Huszar, Brandon Heppe, Geyou Ao

Single-wall carbon nanotubes (SWCNTs) can be used as versatile near-infrared (NIR) fluorescent probes for biological applications, such as biosensing and imaging. The use of click and bioorthogonal chemistry can open many possibilities for tailoring the SWCNT surface with a diverse range of biological functionalities for targeted biomedical applications. This work focuses on using organic color center (OCC)-tailored SWCNTs with specific functional groups as a material foundation for click chemistry. Specifically, we performed the isolation of pure-chirality (6,5) SWCNTs, the creation of fluorescent quantum defects (i.e., OCC) on (6,5) SWCNTs using aryl diazonium chemistry, and the copper-catalyzed azide-alkyne cycloaddition click chemistry. The newly formed defect-state emission peak, that is red-shifted compared to the pristine emission peak, confirmed the creation of fluorescent quantum defects on SWCNTs. Following the membrane filtration to remove the excess diazonium, $CuSO_4$, THPTA and varying [aryl azide]: [defect] molar ratios were mixed for reaction at various time periods. The vis-NIR absorption and fluorescence of samples were characterized using the NS3 NanoSpectralyzer. Our preliminary findings show a possible azide-alkyne cycloaddition reaction, providing potential for versatile surface functionalization of SWCNTs for biosensing, early diagnostics, and bioimaging applications.

Predicting the Nonfouling Properties of Peptides Using Model Ensemble

Shea Bailey, Ibrahim Imam, Qing Shao

Nonfouling peptides, which bind specifically to the body, are invaluable in biomedical applications such as drug delivery. However, recognizing nonfouling peptides poses a significant challenge due to the limited number of known peptides and their complex synthesis. The vast array of untested peptides makes the experimental process both costly and time-consuming. To address this, we employ protein language models (PLMs) to facilitate the recognition of nonfouling peptides. This study investigates the impact of a model ensemble on the accuracy of nonfouling peptide predictions and leverages this ensemble to test peptides from natural sources, aiming to discover more biomedically useful peptides.

Pre and Post-Processing Effects of Protein Structure in Dynamic Thermosets

Yansi Foong, Shelby Surprenant, Wui Yarn Daphne Chan

Petroleum-based polymers are not biodegradable but are commonly used in various applications, including medical supplies, food packaging, and for convenience. To lessen the environmental impact, polymers composed of proteins have been investigated to replace petroleum-based ones. In this study, we measured effects of processing on protein structure and mechanical properties on copolymers prepared using globular and fibrillized proteins. The copolymer was produced through a series of processes, including fibrillating the protein using an HCl solution, performing protein methacrylation to enable polymerization, and polymerizing with an acrylate comonomer. After the protein-copolymer was made, the polymers were remolded under heat and pressure and later analyzed for their properties. We characterized the pre and post processed copolymers using Fourier Transform Infrared (FTIR) spectroscopy to compare the ratio of the α helices to β sheets. The polymers were also subjected to a swell test to measure their crosslink densities. In addition, the mechanical properties of the polymers were also recorded through a tensile test. Through these tests, we found that for proteins that were both fibrillized and methacrylated, the ratios of alpha helices to beta sheets were more consistent, their polymer strengths was higher, and their polymer crosslink density was higher compared to polymers that were just methacrylated.

Applying High Throughput Experimentation Techniques to Assemble Particles of Arbitrary Shapes Using DNA Bridges

Naomi Kern, Huat Thart-Chiang, Lilo Pozzo

Future technological developments in fields including alternative energy and medicine require next-generation materials. Synthesizing each new material requires exploring a multi-dimensional parameter space. Developing laboratory automation tools for automating lab procedures and data analysis will be key to efficient discovery of optimal, novel materials. Some automation tools utilized in this work include automated sample loading and analysis for both Small Angle X-ray Scattering (SAXS) and Dynamic Light Scattering (DLS), and a newly constructed sonication robot.

The goals of this project are to apply these lab automation tools to construct and characterize crystalline structures of particles of arbitrary shapes encapsulated in a lipid membrane and connected with DNA bridges. With high throughput methods, the impact of design parameters on the crystal structure can also be determined. Parameters of interest in the self-assembly of particles of any geometry include the ratio of lipid types in the membrane and the nanoparticle surface area to membrane surface area ratio.

The first step in constructing a new crystalline material is embedding the nanoparticles in a lipid membrane of optimal composition. Next, the cholesterol end of synthesized DNA-cholesterol fragments embeds in the membrane and complementary DNA fragments are added to connect the nanoparticles. Heating the solution above the melting temperature of the complementary regions of the DNA fragments and dropping the temperature slowly encourages crystal formation. The aggregates formed are analyzed via SAXS before and after DNA bridge addition and heating to determine if crystals are formed.

This technique has been used on spherical silica particles previously, and it is hypothesized that it can be applied to self-assemble crystals comprised of other particle geometries. Preliminary results from this project are presented here.

Engineering Stable Protein Nanosheets for *In Vivo* Applications

Raka Chakraborty, Ishmamul Sadab, Julie Champion

Protein nanosheets (pNS) are two-dimensional, self-assembling protein biomaterials with a thickness less than 100 nm. These pNS are made in an aqueous environment under end-over-end rotation using two fusion proteins, a hydrophobic elastin-like protein (ELP) fused between two arginine-rich zippers (Z_R) and a hydrophilic functional globular protein attached to a leucine-rich zipper (Z_E). The mechanism behind the assembly of these pNS is the hydrophobic interaction between the ELPs and the alignment of the hydrophobic and hydrophilic moieties in the assembled structure along the continually expanding and contracting air-water interface due end-to-end rotation. Since the ultimate goal is to use these pNS as scaffolds for surface display of therapeutic proteins for *in vivo* studies, assessing the stability of the pNS are vital in this respect.

With this in mind, the motivation for this research was engineering the pNS with ELP variants of different lengths and hydrophobicity to determine how they affect the stability of pNS in human pooled serum (HPS). Six different ELP variants were tested: Z_R-ELP, Z_R-ELP_{35}, ZR-ELP_{50}, I_5-Z_R-ELP, Y_5-Z_R-ELP. The more hydrophobic ELPs (I_5 and Y_5) were hypothesized to result in greater hydrophobic interactions between ELPs, the longer ELPs (ELP_{35} and ELP_{50}) were theorized to maximize inter-ELP interactions, and the removal of one ZR was thought to increase ELP interactions at the air-water interface by reducing steric hinderance.

All pNS were made with a Z_E/Z_R ratio of 1 where the globular protein utilized was green fluorescent protein, sfGFP. Notably, for all hybrid pNS 80% of all ELP population were contributed by Z_R-ELP-Z_R and the remaining 20% by the ELP variants of interest. An initial week-long stability test was conducted on the five hybrid pNS and two pure pNS (Z_R-ELP-Z_R and Z_R-ELP respectively) where they were added to HPS and stored at 37°C. Flow cytometry was run at predetermined time intervals. The forward-scattering was observed to establish the size of pNS after incubation with HPS. The side-scattering did not indicate much variability meaning the pNS were being oriented consistently without altering their configurations or crumpling. The hybrid pNS using Z_R-ELP-Z_R and Z_R-ELP was determined to be the most stable as it was the only sample that maintained a fairly constant size and concentration over the duration.

A subsequent week-long stability test was run to determine if the ratio of Z_R contributors made a difference on the stability of these hybrid pNS. Samples where Z_R-ELP contributed 20%, 50%, and 80% of the Z_R population (with the requisite from Z_R-ELP-Z_R) were incubated with HPS. Irrespective of the ratio, the pNS remained stable for five days after which point they appear to disassociate since the mean forward scattering decreases.

Future scope of this research includes substituting a different fluorescent protein in place of sfGFP and varying the ZR contributors after doing so. Upon determining which configuration of hybrid pNS is most stable, the functional protein can be substituted for a biologically active or specific ligand and ultimately be used as active cell engagers.

Degradation of Particle Laden Silk Fibroin Sponges for Controlled Payload Release

Jack Girton, Elizabeth Aikman, Marisa O. Pacheco, Nisha Kotta, Whitney Stoppel

Silk Fibroin (SF), a protein derived from *Bombyx mori* silk fibers, has demonstrated significant potential as a versatile biomaterial for medical applications. SF can be processed into biocompatible, biodegradable, and highly customizable forms such as gels, sponges, and particles, making it ideal for tissue engineering [1, 2]. A key challenge in these applications is the controlled delivery of small molecules in specific quantities and sequences to guide tissue regeneration. To address this, we propose a hybrid material consisting of SF particles [3] embedded in an SF sponge [4]. The sponge structure provides a scaffold for cell infiltration and molecule retention, while the particles enable controlled release of small molecules.

We hypothesize that this particle-laden sponge will retain the mechanical properties of the bulk sponge while adding an enhanced level of controlled release. The sponges were prepared via lyophilization of frozen SF solution, followed by water annealing, while the particles were formed through liquid-liquid phase separation of SF and PVA solutions [1, 3]. To create the hybrid material, SF particles were incorporated into the SF solution used to form sponges [4, 5], followed by freezing, lyophilizing, and water annealing, which induced beta-sheet formation. The beta-sheet content is known to influence degradation rate, providing a tunable property based on annealing conditions (temperature, time, and pressure) [5-7].

To study the degradation kinetics of the hybrid material, we conducted a comparative analysis of pure SF sponges and 50/50 particle-laden sponges, degrading the samples with Protease XIV or Collagenase and visualizing changes using scanning electron microscopy. Time-dependent degradation data was fitted to a Michaelis-Menten kinetic model, yielding a modified first-order rate constant to describe the material's degradation and the associated release rate of an encapsulated payload [7]. Our results offer insights into how material composition and preparation can be optimized for tissue engineering applications.

References:

[1] D. N. Rockwood, R. C. Preda, T. Yucel, X. Wang, M. L. Lovett, and D. L. Kaplan, 'Materials fabrication from Bombyx mori silk fibroin,' (in English), *Nat Protoc,* Article vol. 6, no. 10, pp. 1612-31, Sep 22 2011, doi: 10.1038/nprot.2011.379.

[2] M. O. Pacheco *et al.*, 'Progress in silk and silk fiber-inspired polymeric nanomaterials for drug delivery,' *Front Chem Eng,* Mini Review vol. 4, 2022, doi: 10.3389/fceng.2022.1044431.

[3] M. O. Pacheco *et al.*, 'Silk Fibroin Particles as Carriers in the Development of Hemoglobin-Based Oxygen Carriers,' (in English), *Adv Nanobiomed Res,* vol. 3, no. 9, p. 2300019, Sep 2023, doi: 10.1002/anbr.202300019.

[4] E. L. Aikman *et al.*, 'Impact of crystalline domains on long-term stability and mechanical performance of anisotropic silk fibroin sponges,' (in eng), *J Biomed Mater Res A,* vol. 112, no. 9, pp. 1451-1471, Mar 12 2024, doi: 10.1002/jbm.a.37703.

[5] J. Rnjak-Kovacina *et al.*, 'Lyophilized Silk Sponges: A Versatile Biomaterial Platform for Soft Tissue Engineering,' (in eng), *ACS Biomater Sci Eng,* vol. 1, no. 4, pp. 260-270, Apr 13 2015, doi:

10.1021/ab500149p.

[6] Y. Kambe, Y. Mizoguchi, K. Kuwahara, T. Nakaoki, Y. Hirano, and T. Yamaoka, 'Beta-sheet content significantly correlates with the biodegradation time of silk fibroin hydrogels showing a wide range of compressive modulus,' *Polym Degrad Stabil,* vol. 179, p. 109240, 2020/09/01/ 2020, doi: https://doi.org/10.1016/j.polymdegradstab.2020.109240.

[7] J. F. Jameson, M. O. Pacheco, J. E. Butler, and W. L. Stoppel, 'Estimating Kinetic Rate Parameters for Enzymatic Degradation of Lyophilized Silk Fibroin Sponges,' (in English), *Front Bioeng Biotech,* Original Research vol. 9, no. 537, p. 664306, 2021-July-06 2021, doi: 10.3389/fbioe.2021.664306.

Thiol-Functionalized Silk Biomaterials, a Study of Two Different Synthetic Routes

Sanjana Srinivas, Sanyukta Patil, Cassidy Soard, Kelly A. Burke

Biomaterials are extremely useful for local and controlled delivery of therapeutics as well as for tissue engineering purposes. In this work, brush-like polymers are generated on the surface of biomaterials, and two different synthetic approaches are explored to enable the generation of brush-like polymers containing thiol groups. The thiol groups are compatible with a wide range of molecules, including bioactive molecules and various drugs, to enable the use of the materials in drug delivery and tissue engineering applications. We use *Bombyx mori* silk fibroin, as it has been widely researched and used for its biocompatibility, controlled degradation, and drug release properties.

We are comparing two methods of forming thiol-functionalized silk biomaterials. Both methods result in the formation of brush-like polymers on the surface of silk biomaterials. Method A began with hydroxyl-enriched silk films, which were subsequently modified to attach a chain transfer agent (RAFT agent). The films were then polymerized with AEMA (amino ethyl methacrylate) using reversible addition-fragmentation chain transfer polymerization, to result in the formation of brush-like polymers. Traut's Reagent was subsequently utilized to convert the primary amines present on the brush side chains to thiol groups.

Method B begins by forming a thiol-containing monomer via a two-step synthesis reaction. The films were then modified to enrich hydroxyl groups and attach the chain transfer agent (RAFT agent). The brush-like polymer was then synthesized from the thiol-containing monomer. It is hypothesized that forming brushes from a thiol-containing monomer directly on the silk films (Method B) will lead to brushes with greater thiol content than converting primary amines to thiols on existing brushes (Method A). The enhanced thiol content is expected to effectively allow a greater amount of drug product to attach to the biomaterial.

To quantify the difference between Method A and Method B, a variety of characterizations and assays were conducted. ATR-FTIR and Raman spectra were used to characterize the films before and after the attachment of the RAFT agent. Contact angle studies using sessile drop method showed increased angles after polymerizing the films with a hydrophobic monomer. Ellman's Assay was used to quantify the amount of thiol contained in the molecule and a ninhydrin assay was used to quantify the amount of primary amine present. The ability to form a thiol functionalized biomaterial with a greater thiol content is expected to enable increased drug attachment that is suitable for various biomedical applications.

Ultrasound-Responsive Phospholipid-Coated Microbubbles for Controlled Drug Delivery across Mucosa

Evalynn Ellison, Aastha Shah, Canan Dagdeviren

Precise and targeted drug delivery is a major area of interest in the field of biomedical engineering. A popular approach to this goal is ultrasound-responsive materials which have shown to increase the precision and targeting in drug delivery across the skin. However, many areas of the body have a mucosal layer that impedes the effectiveness of passive drug diffusion. This includes areas such as the stomach, intestines, and bladder that have great potential as drug delivery sites for the ability to by-pass first pass metabolism. A wide range of ultrasound responsive drug carriers have been used to mitigate these challenges from the mucosal layer by enhancing passive diffusion. Microbubbles have been highly effective as a drug delivery vehicle due to the adaptability of their coatings. Phospholipid-coated microbubbles increase the stability and lend amphiphilic properties to the drug carrier. Our focus is to develop a proof of concept for an ultrasound-responsive phospholipid-coated microbubble drug carrier for a new drug of interest. In order to accomplish this, we imaged the phospholipid-coated microbubbles through microscopy to characterize their response to ultrasound waves. To achieve the best enhancement in passive diffusion we wanted to observe both acoustic streaming and cavitation. Electricity was run through a piezoelectric device (PZT) at different frequencies and the microbubble behavior was recorded over a period of three seconds. From these results, we were able to conclude that 217 KHz was the resonant frequency of the device leading to the highest amount of acoustic streaming, measured by the mass flux value. However, no cavitation was observed from the imaging. We hypothesize that this was caused by the additional stability added by the phospholipid layer on the microbubbles. Therefore, a different microbubble coating would be better suited for the desired ultrasound response behavior. This further research will hopefully advance ultrasound responsive drug delivery devices for faster, controlled drug release in order to increase patient comfort and compliance.

Biodegradable and Renewable Protein-Based Elastomers

Irene Cui, Wui Yarn Daphne Chan

Currently, agricultural industries generate large quantities of byproducts which can be utilized as feedstocks for biomass-based polymers rather than discarded as waste. Proteins are unique in their ability to form stiff structures through extensive hydrogen-bonding. Some of the amino acids in proteins, such as lysine, serine, and threonine, contain functional groups that can initiate the growth of polymer chains from the protein's surface. We hypothesize that proteins can be copolymerized with soft polyester segments to create elastomers with both high mechanical strength and high elongation at break, combining the stiffness of the proteins with the flexibility of the polymer chains. Our system uses a common dairy byproduct whey protein isolate as the hard domains and δ-lactones, which naturally occur in plants and animal products, as the soft segments. Whey protein is not soluble in the hydrophobic δ-lactones, so we first introduce a surfactant benzalkonium chloride to enable miscibility between the two components. Next, we show that whey protein and δ-lactones can be copolymerized under solvent-free conditions in the presence of the surfactant. Further investigation is needed to identify the copolymerization mechanism as well as determine the effects of protein content and modifications on the mechanical properties of the resulting material. Ultimately, this research contributes to the design of mechanically robust, renewable materials to support the shift toward sustainability.

Establishing Enzyme-Loaded Eutectogels

Milena Mesfun, Matias Picchio, Elena Romero-Ben, Ana Beloqui, Marcelo Calderón

Enzymes are natural catalysts with remarkable selectivity and specificity in biological reactions, but their use outside biological systems is limited. Utilizing an appropriate supporting medium for immobilization presents an appealing approach which allows the confinement of the enzymes with cofactors or chemical catalysts, enhancing the stability of enzymes in harsh conditions and offers synergistic catalytic activity from natural (enzymes) and artificial (inorganic) components. Eutectogels, recently gaining attention due to their unique properties, therapeutic activity, ease of preparation, low cost, low volatility, and environmental friendliness, serve as an excellent support medium. My project involves encapsulating Horseradish peroxidase (HRP) in eutectogels made from therapeutic deep eutectic solvents Menthol: Lactic Acid and sugar-based eutectic solvents Xylitol: Glycerol and Sorbitol: Choline Chloride with low molecular weight gelator 1,3:2,4-Dibenzylidene-d-sorbitol (DBS). These enzyme/eutectogel hybrids catalyze peroxidation reactions using hydrogen peroxide as a substrate, showcasing their effectiveness in non-biological reactions.

Novel Fouling Release Coating Based-Dynamic Surface Instability

Morgan Galvao, Manoj K. Chaudhury

This research aims to develop environmentally friendly coatings for ship hulls that effectively deter biofouling by weakening any adhesion to the point where hydrodynamic forces from the ship's motion can easily dislodge them. The elastomer utilized contains hygroscopic inclusions that absorb moisture, causing it to swell upon contact [*Figure 1*]. When this elastomer is bonded to a rigid substrate and absorbs water, it undergoes surface instability. The wavelength of this instability increases proportionally with the thickness of the coating. Increasing the amount of hydrophilic inclusion enhances moisture absorption, leading to greater swelling and increased surface instability. The research involves varying the levels of hydrophilic inclusion in the coatings and observing their effects. It's observed that higher hydrophilic content not only increases swelling but also alters the pattern of water absorption in the coatings. Below, is an image of the distinct pattern exhibited by one film with 40% hydrophilic inclusion as it absorbs water as a function of time [*Figure 2*]. This swelling in the film occurs due to differences in the chemical potential of water between the external source and the film. The process slows down as the osmolarity difference between internal and external environments decreases towards equilibrium. Furthermore, the research demonstrates that attaching a dome-shaped adherent to these coatings leverages the elastic force generated during surface instability to effectively dislodge it. This property makes these coatings promising for applications on ship hulls to prevent fouling accumulation during extended voyages. *Figure 1.* Schematic representation of a foulant in contact with a hygroscopic elastomer bonded to a rigid substrate (A), which swells by absorbing water from the surrounding and undergoes a morphological instability (B) thereby pushing the foulant out of the surface. *Figure 2.* A) An example of the surface morphology observed with 40% hydrophilic inclusion in the coating. B) Wavelengths of the different coatings with varying amounts of hydrophilic inclusion plotted vs time.

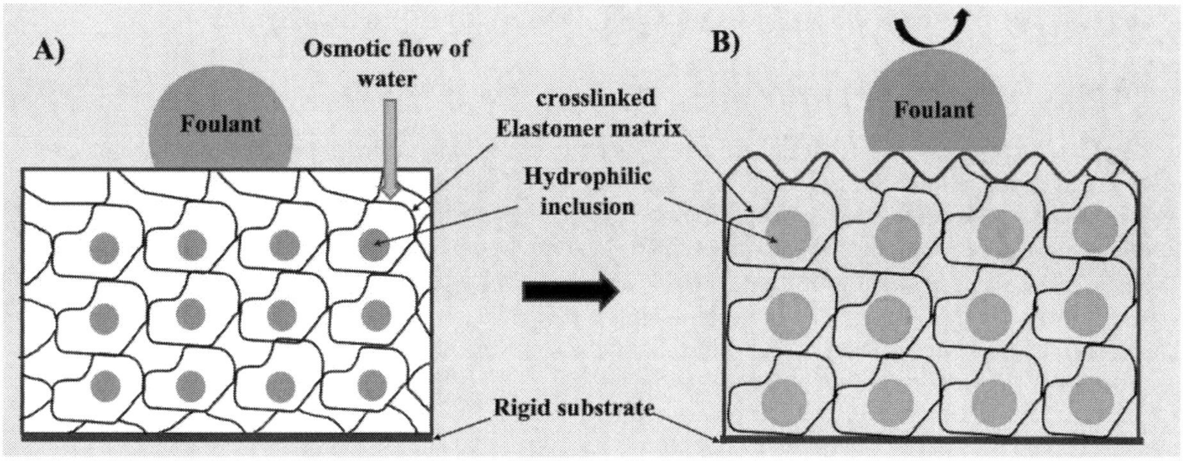
A)
Osmotic flow of water
Foulant
crosslinked Elastomer matrix
Hydrophilic inclusion
Rigid substrate
B)
Foulant

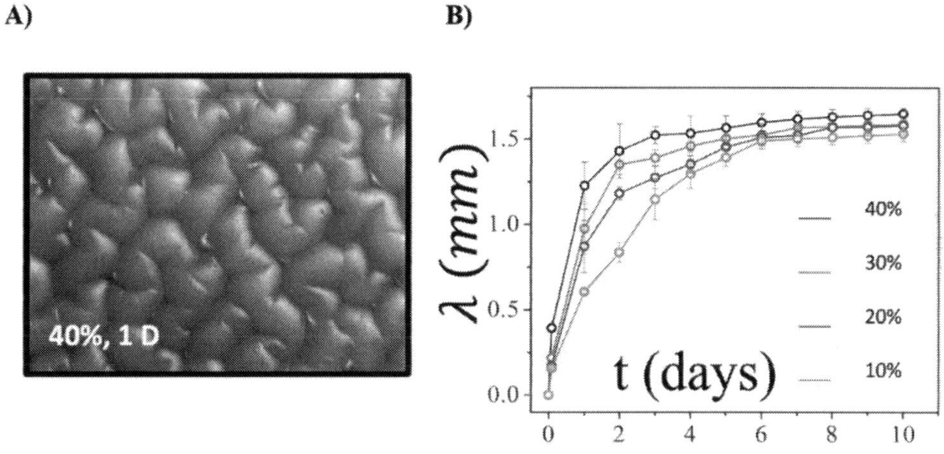
A)
40%, 1 D
B)
λ (mm)
t (days)
40%
30%
20%
10%
0.0
0.5
1.0
1.5
0
2
4
6
8
10

Porous Copper Nanowire Synthesis on CNT-DNA Thin Film Substrates

James Badger, Samuel Lowell, Mason Remondelli, Enoch Nagelli, F. John Burpo

Biological hydrogels can serve as 3-dimensional bio-templates for tunable nano-porous materials that serve as both electrochemically active high surface area and structural material. We present a general approach to 1) form a composite biopolymer-carbon nanotube hydrogel using covalently crosslinked DNA-CNT, 2) equilibrate the hydrogel bio-template using Na_2PdCl_4 solution to mediate direct metal deposition, and 3) utilize an electroless copper deposition solution to form a porous copper nanowire array on the DNA-CNT substrate. Nanostructure is dependent on the biotemplate molecule, crosslinking mechanism, and drying method. Such multi-functional electro-mechanical materials are envisioned to enable a broad range of applications such as sensors, photovoltaics, catalytic systems, fuel cells, and energy absorption. Previous work allowed for the synthesis of metal aerogels and biotemplated metal films using a drop casting method where the gel was cast directly onto a substrate from a micropipette. The cast samples were then processed using three solutions to prepare and deposit metal directly onto the DNA-CNT substrate forming a biotemplated copper nanowire array. This method limited the bulk size of the cast sample to an approximate surface area of $1cm^2$. To increase the overall surface area and volume of the nanowire arrays, we sought an alternate casting method. The new method cast heated DNA-CNT solutions into a parafilm mold on a glass microscope slide. This molded sample was then processed in a similar method to the drop cast samples. This method is preferable to a drop casting method as it will enable the future manufacturing of larger gels in terms of both surface area and volume. Nanostructured porous metals offer a wide range of applications including catalysis, energy storage and conversion, and sensing. The formation of free-standing metal aerogels with high specific surface area and hierarchical porosity confers significant advantages for conductivity, mass transfer properties, reaction specificity, and strength. This has previously been achieved with a direct reduction method,[1] as well as on gelatin,[2] and cellulose nanofiber templates.[3] To further demonstrate multi-functional nanostructured electrodes, hierarchical DNA-CNT nano-composite aerogels will continue to be explored.

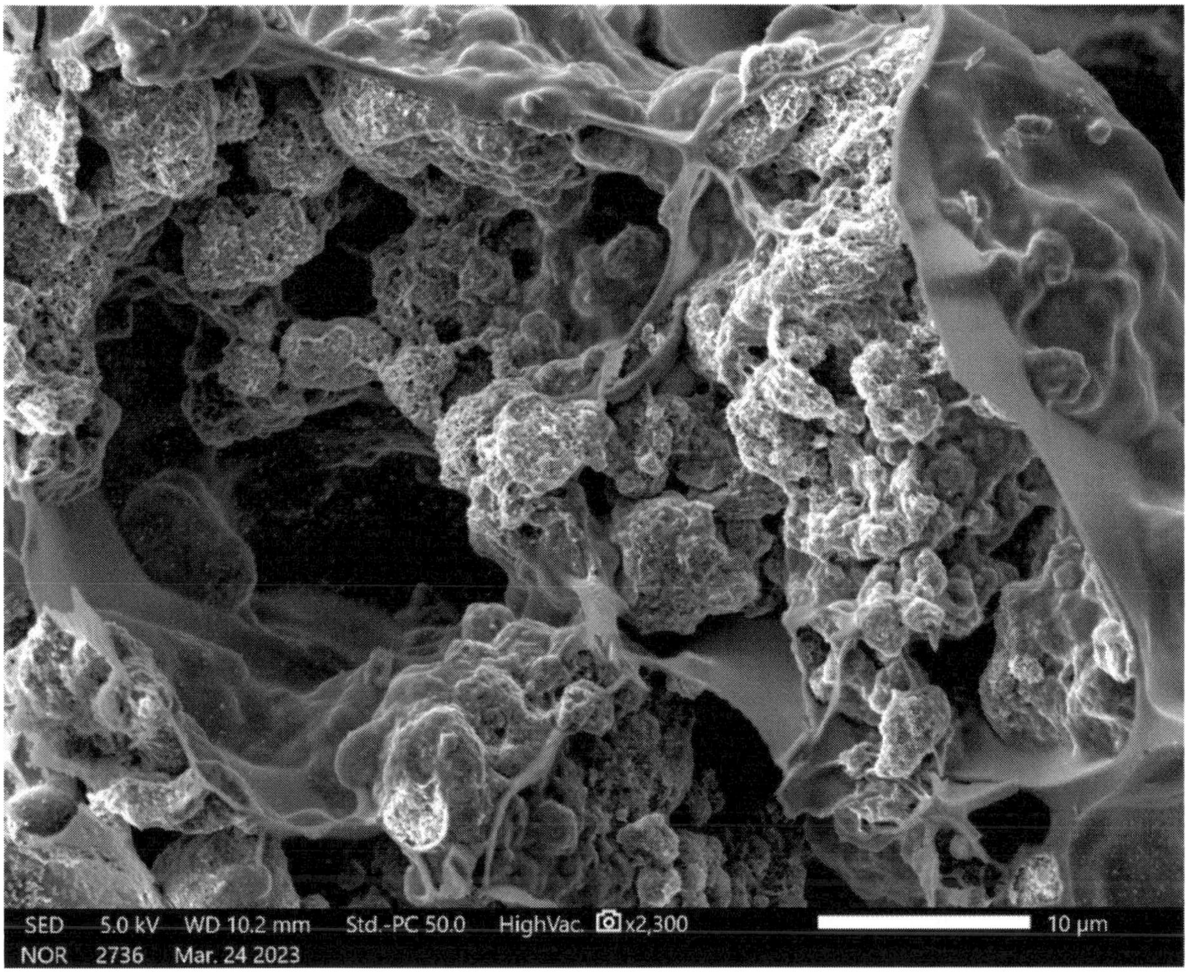
SED 5.0 kV WD 10.2 mm Std.-PC 50.0 HighVac. x2,300
10 µm
NOR 2736 Mar. 24 2023

Enhancing Diffusion with Active Particles in a Fluctuating Environment

Katherine Ellis, Tobias Dwyer, Philipp Schönhöfer, Ian Wyllie, Anni Shi, Daniel K. Schwartz, Sharon C. Glotzer

Mass transport and diffusion in porous environments has been widely studied and characterized. In many studies, the porous environment is assumed to be a static structure, however, in nature, these environments are dynamic, having pores of constantly fluctuating size. Together with experimentalists who are able to isolate the effect of environmental fluctuations on passive particles, we aim to model active particles to further study the effect of a dynamic environment on diffusion. We employ Brownian Dynamics in the HOOMD-blue particle simulation toolkit to model active particles diffusing through a dynamic porous environment. By tuning the forces at work in the model, we are able to recover many of the remarkable dynamics observed in experiments. Further, we can explore questions about the system, such as the effects of pore geometry and active force magnitude, which are difficult to assess in experiment alone. Knowledge of such effects lay the groundwork for engineering the diffusion of a particle by controlling its environment.

Development of a Wearable Fabric Sensor for Noninvasive Lactate Monitoring during Exercise

Lisa Chambers

This work focuses on the development of a wearable fabric sensor using novel nanocomposite materials. Currently, there is no established method for noninvasive and continuous monitoring of lactate levels during exercise. This sensor will facilitate the gathering and analysis of sweat biomarkers, specifically lactate, furnishing insights into human health and performance monitoring. Lactate was selected as the analyte of interest to measure the exertion of the participant, and the material to be developed will provide real-time health assessment and performance optimization. Excess lactate build-up has been implicated in causing cramping, fatigue, and diminished performance in athletes. To fabricate such a sensor, an electrospun nylon fabric is treated via vacuum filtration with a solution. The solution utilizes carboxylic acid functionalized multi-walled carbon nanotubes (MWCNTs) that are dispersed in water with a surfactant using a probe sonicator. These lactate sensing materials were characterized via SEM and FTIR to determine the morphology of the nanocomposite sensor material. The sensor's active layer is then synthesized by electropolymerization using pyrrole, lactate oxidase, and a second aqueous dispersion of carboxylic acid functionalized MWCNTs. Characterization of MWCNT dispersion quality and particle size were completed through UV-vis spectroscopy and Zetasizer analysis. Electrochemical properties were evaluated through cyclic voltammetry of the sensor's response to synthetic sweat solutions containing lactate. Further testing involved analyzing sensor responses to spiked and synthetic sweat solutions containing common biomarkers within sweat such as glucose, potassium, calcium, and sodium. Initial benchtop tests have confirmed the sensor's ability to detect physiologically relevant levels of sweat lactate. The nanocomposite sensors are to be validated thoroughly via an IRB-approved on-body lab testing, which will provide feedback to the design to be used for further development. Participants will cycle for 60 minutes and sweat samples will be collected with sterile gauze adhered to the skin using Tegaderm. We aim to test 30 subjects for validation of the wearable fabric sensor. After testing, the gauze is centrifuged and analyzed for the lactate concentration using the gold standard laboratory technique, HPLC.

High-Throughput Analysis of Recycled Battery Graphite Using Electron Microscopy

Katelyn Lyle, John Mangum, Kae Fink

With the push towards an electrified grid, there has been a sharp increase in the number spent lithium-ion batteries. Specifically, earlier generation electric vehicles (EVs) are being improved upon and replaced by newer generation EVs with higher performing batteries. This highlights the need to develop environmentally friendly and low-cost methods to recycle battery materials and feed them back into a circular energy economy. Among these materials, graphite is a key component for anodes. Although it has been historically viewed as a waste product for end-of-life batteries due to its low cost, the massive scale increase in energy storage requirements has created a need to circularize every component of battery materials. Due to leaps in graphite optimization since the dawn of batteries, the morphology of graphite in early generation batteries does not resemble state of the art anodes, as spheroidized particles are replacing early-generation flake-like particles.

This project seeks to optimize the graphite upcycling process by developing methods to convert spent flake-like graphite particles into high performing spheroidized graphite anodes at scale. Advanced microscopic techniques are used to analyze reclaimed battery anode graphite that has been through a series of mechanical tuning processes. High throughput scanning electron microscopy (SEM) imaging is used to capture large sample sizes of particles and extract data about their external properties, while focused-ion beam milling is used to capture data about internal particle properties. These properties are assessed for thousands of particles per sample using a machine-learning program that segments and analyzes SEM images. It is found that a broad range of recycling conditions were able to produce graphite particles with variable size distributions, circularity distributions, and morphological properties that can be compared to optimized spheroidized graphite to further tune recycling processes. The study of internal particle characteristics further revealed the mechanism by which graphite is spheroidized to be the flakes folding in on themselves to form a larger ball. Finally, characteristics within each sample were compared to reveal that particle size is linked to particle circularity, with smaller particles skewing towards greater circularity. In future work, properties of particle size, circularity, and porosity in upcycled graphite can then be linked to cell performance to better inform recycling processes.

An Inorganic/Organic Molecular Hybrid Material for Fully Solution-Processed Photonic Structures

Mehul Dhoot, Natalie Stingelin, Victoria Quirós Cordero

Photonic crystals are nanostructures that efficiently control light with negligible losses by creating a photonic band gap, a range of wavelengths that are reflected rather than transmitted. This effect arises from the periodic variation in refractive indices within the structure. With sufficient contrast, light can be reflected, guided, or confined, a mechanism which can be applied to nanoscale technologies in optics and photonics. Here, we show that films of a titanium oxide hydrate (TiOH) / poly(vinyl alcohol) (PVA) hybrid material can be fully processed from solution under ambient conditions, and can be precisely designed to reach target properties for applications in a variety of photonic structures. The refractive index n of hybrid thin films can be tuned between 1.4 and 1.8 by adjusting the material's composition and through post-deposition techniques like photonic curing. Higher inorganic (TiOH) content and photonic curing in the presence of ultraviolet light was shown to correlate with a higher hybrid refractive index. Meanwhile, thickness of the films could be increased with increased inorganic content, but was shown to decrease after photonic curing. These properties could be leveraged to generate multilayer photonic structures with high reflectance, low optical losses, and a precise photonic band gap. Distributed Bragg reflectors (DBRs) could be reliably produced by layering the high-n inorganic/organic hybrid with a low-n polymer, with solution-processed structures of up to 21 layers yielding efficient performance. Solution-processed optical microcavities with up to 12 layers on either side of the cavity were also shown to exhibit strong light-matter coupling with low optical losses, with performance on par with structures fabricated from more resource-intensive processes. Placing an active material at the center of a microcavity was also shown using photoluminescence measurements to generate coherent condensates of light-matter quasiparticles known as exciton-polaritons. This research will be a major step in unlocking the countless applications of photonic crystals in nanotechnology, including optical communication, sensing, diodes, photovoltaics, and quantum information systems.

Beta-Sheet Co-Assembly Interactions Guide Organization of Charged Peptide-Polydiacetylene Conjugates

Jeffrey Li, Tarunya Sudarshan, Sujeung Lim, Alicia Robang, Leel Liberty, Herdeline Ardoña, Anant Paravastu

Thermochromic polymers, such as polydiacetylene (PDA), have shown promise as molecular sensors due to their environmental stimuli-dependent chromogenic properties. These polymer materials have potential for applications such as food quality indicators, chemical detectors, and temperature sensors. The color-changing responses arise from structural changes and are dependent on polymer sidechain characteristics; however, the effect of sidechain chemistries on thermochromic responses remains poorly understood. Previously, Lim et al. demonstrated that amphiphilic peptide-PDA co-assemblies form 1D nanostructures that exhibit thermochromic properties unique from self-assemblies of the positive and negative peptide-PDA monomers. However, the structure and order of these materials has not been extensively studied. Here, we expand upon prior work by using 2D solid-state NMR to characterize the self-assembled and co-assembled structures of charged peptide-PDA materials. We find that selectively ^{13}C- and ^{15}N-labeled self-assemblies of KK_2KGV_5-PDA form anti-parallel β-sheet structures, while DD_2DG_4V-PDA forms semi-ordered random coil structures, consistent with previous CD and FTIR results. Furthermore, we demonstrate that a 1:1 mixture of KK_2KGV_5-PDA and DD_2DG_4V-PDA exhibits cooperative co-assembly interactions between oppositely charged peptide sidechains that suppress self-assembly. The resulting ordered parallel β-sheets are structurally distinct from the peptide-PDA monomers. Our results confirm that elements of peptide co-assembly design can indeed be applied to polymer sidechain engineering, enabling the study of sidechain chemistries using the language of protein design. This expands future possibilities for synthesis of tunable, bioinspired polymer sensors.

Cholesteric Liquid Crystalline Elastomers: Tunable Light Filters for Infrared Applications

Judy Chen, Alexis Phillips, Tim White

Within various industries such as automotive, aerospace, and wearable technology, the demand for soft optical materials continues to rise for functions like tunable filters, sensing, and light modulation. Cholesteric liquid crystalline elastomers (CLCEs) are a unique class of soft optical materials that self organize into a helicoidal structure. The periodicity of the refractive index across the thickness of the material causes the elastomer to exhibit selective reflection. This study focuses on fabricating stretchable, adjustable light filters that selectively reflect light within the mid-wave and short-wave infrared spectrum. We will delve into how the elastomer's optical and material characteristics enable control over reflection tunability and polarization, along with the ability to switch optical properties in response to mechanical and electrical stimuli.

Additive Manufacturing (FDM) of Viton-Ammonium Perchlorate-Aluminum (VAPAL)

Valenica Ramirez, Faris Wald, Matthew Ellsworth

The purpose of this study is to investigate the compatibility of ammonium perchlorate and aluminum (AP/Al) with Viton (VAPAL) for use in additive manufacturing (AM). After compatibility testing to ensure safety, VAPAL of various solids loading was successfully formed into filaments using a Filabot extruder. The procedure developed for suspending AP/Al in Viton was effective, with a theoretical maximum density of over 90%. The results from both sensitivity testing and analysis showed that Viton is compatible with AP/Al as the stability of the product increased when compared to ammonium perchlorate by itself. Additionally, results from a differential scanning calorimeter (DSC) show that the decomposition of VAPAL occurs near the same temperature as ammonium perchlorate, indicating that VAPAL is safe to print at fused deposition modeling (FDM) temperatures. Lastly, samples taken intermittently along an extruded sample show uniform density of AP/Al, validating our filament extruding process. Burn rates of the extruded filament were conducted for various solids loadings. Further research will be done to explore the feasibility of VAPAL within 3D printing. Challenges we expect with the next phase of research include how VAPAL will behave in the extruder head and what definition we can print with this filament.

Leveraging Supercritical Fluids for the Functionalization of Cellulose Packaging Substrates

David Ball, W. Keshani R. Perera, Obiora Muojama, James D. Sheehan, Brenda Hutton-Prager

In recent years, plastic packaging materials have generated environmental concern due to a lack of recyclability, with tens of millions of tons directed to landfills annually in the United States. Cellulose, abundant in nature, is a biopolymer with emerging potential as a renewable alternative to plastic packaging material. Though cellulose-based materials are gaining traction in food packaging, cellulose is naturally hydrophilic and requires further treatment to enhance viability. Cardboard packaging materials are often treated with synthetic waxes to improve hydrophobicity, which makes recyclability difficult. Novel methods using greener technology and sustainable materials are becoming popular to mitigate environmental harm and prolong the life of expirable packaged goods. Supercritical fluids (SCFs) are compressible fluids comprising molecular species exceeding their critical pressure and critical temperature, combining the properties of gases and liquids and becoming an effective solvent. This project evaluates the feasibility of supercritical impregnation (SCI) processes of biobased additives into cellulose-based packaging materials for improved functionality.

Previous research has established the viability of Alkyl Ketene Dimer SCI, a surface sizing agent, into filter paper to improve hydrophobicity. The goal of the present investigation is to develop vanillin solute impregnation processes using supercritical carbon dioxide ($scCO_2$) for increased UV resistance of cellulose substrates as packaging material. Vanillin is initially dissolved in polar cosolvents and introduced into $scCO_2$ at different concentrations. SCI of cellulose is performed at varied pressures and temperatures to optimize process conditions. Preliminary investigations demonstrate SCI of vanillin into cellulosic substrates successfully increased sample UV absorbance as measured by UV-Visible spectroscopy. FTIR spectroscopy of SCI-modified cellulose identified IR bands associated with vanillin. Quartz crystal microbalance testing determined that vanillin is recoverable from cellulose with solvent flushing, indicating the potential recyclability of vanillin-impregnated packaging material. This study will impact cellulosic packaging material properties and limit plastic pollution if the proposed SCI process reaches a manufacturing scale.

Effects of Intrinsic Impurities on Hydrothermal Relithiation for NCM811 Regeneration

Nuchcha Manaanuntakul

The increasing use of lithium-ion (Li-ion) batteries necessitates efficient recycling methods to mitigate environmental impact and resource depletion. While hydrothermal relithiation, a method of direct recycling, has shown promise in laboratory settings for regenerating NCM811 ($LiNi_{0.8}Co_{0.1}Mn_{0.1}O_2$) cathodes, scaling this method to industrial levels remains challenging due to the need to address and remove impurities, such as binder residues and conductive additives in spent materials. These impurities, which vary in concentration across different spent cathode materials, have not yet been systematically studied. Our research focuses on understanding how different levels of inherent carbon binder impurities affect the hydrothermal relithiation process. This was accomplished through analysis of NCM811 materials with varying percentages of Polyvinylidene Fluoride (PVDF) after hydrothermal relithiation. Increasing percent of impurities lead to a less fully lithiated product. Temperature can also be optimized to aid the relithiation and limit effects of PVDF impurities. By systematically investigating the impact of these impurities, we aim to develop scalable relithiation parameters that maintain efficiency and effectiveness despite the variability in industrial-grade materials.

Kinetics Studies of Anatase-Rutile Transition

William Steere, Seth Drahusz, Nan Yi

Titanium dioxide (TiO_2) is a material that has drawn attention for its chemical stability and versatile catalytic properties. The transitions among three crystalline structures, namely anatase, rutile, and brookite, are important for building and understanding the structure-property relationship of TiO_2. Meanwhile, chemical kinetics studies provide an efficient and powerful methodology to identify mechanisms behind chemical reactions. Differential thermal analysis (DTA) was employed to study the underlying kinetics of the TiO_2 phase transition. Various heating rates (2.5-20°C/min) were investigated and, using the Ozawa-Flynn-Wall (OFW) method, values of the activation energy (E_a) and pre-exponential factor (A) were determined. Lastly, the DTA data was fitted to a group of common kinetic models. My results suggested the anatase-rutile transition is an exothermic process and, importantly, the transition follows a nucleation and nuclei growth mechanism. This work provides the groundwork for understanding the structure transformation of TiO_2, leading to insight on its catalytic performances.

Manipulation of Cohesive States in Colloidal Chiral Fluids

Allison Cornelius

Chiral fluids composed of spinning colloidal particles are an emerging topic in active matter systems, exhibiting distinct self-organizing and emergent behaviors. A typical synthetic chiral fluid involves rotating colloidal magnets, where magnetic and hydrodynamic forces between particles give rise to circulating clusters with unidirectional edge flows. In this work, we incorporate additional diffusiophoretic interactions to externally influence the collective behavior of spinning colloids. The system is composed of ellipsoidal hematite microparticles suspended in H2O2. Under an external rotating magnetic field, phase separation into circulating clusters, a characteristic of chiral fluids, is observed. However, UV light exposure triggers the decomposition of H2O2 on the hematite surface, introducing diffusiophoretic forces that disrupt the structural cohesion of the clusters. This chemical interaction expands the circulating clusters to a limited degree while preserving their overall interconnectivity. The expansion observed is fully reversible, with clusters reverting to their original cohesive state once UV illumination ceases. Incorporating chemically-driven interactions into chiral fluids opens up new possibilities for self-organization, both in these systems and in broader active colloidal systems.

Distillers' Spent Grains Used Towards Degradable Plant Pots

Annie Miloser, Jason Stallings Jr., Malgorzata (Gosia) Chwatko

The distillery industry has grown tremendously over the past decade, related to the rise of ethanol for biofuels and alcoholic beverages such as bourbon. At the same time, this growth has led to many positives, such as job creation or tourism in the case of bourbon. One of the industry's biggest challenges is the production of distillers' spent grains (DGS). The distillers' spent grains are the dried bottom products related to distillation. Due to its chemical composition, it cannot be simply disposed of, and in bourbon production, most of the DGS is donated to local farmers or even encouraged with cash rewards. While successful thus far, other alternatives need to be developed for this product. In the circular economy, this product poses many benefits for plant growth. As such, this project aims to develop plant pots made up predominantly of DGS for growing seedlings. Various concentrations of polymeric binder and three drying methods: room temperature, oven drying at 80°C, and freeze-drying were tested. Mechanical properties were assessed using the three-point bend test. Results demonstrated that DSG-based pots can be effectively synthesized, highlighting their potential as a sustainable alternative to plastic pots and a practical approach to upcycling distillery waste.

Towards Understanding Ionic Liquid Activity in Aqueous Electrolytes from Infinite Dilution to Water-in-Salt Electrolytes

Justin Paredes Catalan, Oscar Nordness

Ionic Liquids (ILs) are organic salts characterized by low melting temperatures below 100°C. These materials offer a vast design space that is readily assessable through judicious anion and cation selection, thus enabling their thermophysical properties — including density, viscosity, conductivity, and surface tension — to be easily tuned. In recent years, aqueous IL systems have becoming increasingly important in fields such as biotechnology, fuel cell development, working fluids, pharmaceutical solvents, and protein stabilization. Accordingly, developing a comprehensive understanding of how aqueous IL systems behave across the entire IL-water composition range, i.e. from infinite dilution to pure IL is paramount for their design and implementation.

To date, various models have been proposed for calculating the activity coefficients of aqueous electrolytes with the vast majority based on the Debye-Huckel model for inorganic electrolytes in which ion-ion interactions are dominated by electrostatics. However, significantly less consideration has given to more complex organic ions (e.g., ILs) in which various intermolecular interactions including ion pairing, van der Waals forces, and hydrophobic self-assembly are known to significantly contribute to electrolyte behavior. Alternatively, the Non-Random Two Liquid (NRTL) model, is designed to capture IL behavior at higher ionic concentrations but is limited in its ability to capture the behavior of dilute IL systems.

In the present work, we leverage data obtained from the NIST ILThermo database in order to investigate water and IL activity coefficients in aqueous IL electrolytes. In addition to water activity coefficients directly obtained from the reported data, we calculate the ion activity coefficients of various ILs. In particular, we focus on the effectiveness of dilute electrolyte (i.e., Debye-Huckel) and concentrated electrolyte models (i.e. the NRTL model) in three concentration regimes: near infinite dilution, intermediate (i.e. bi-percolated concentrations), and high IL concentrations similar to those found in water in salt electrolytes in order to access the limitations and efficacy of these models and develop a more robust understanding of the intermolecular interactions underpinning these systems.

CO_2-Controlled Swelling in Hydrogels

Finlay J. M. Smith, Sarah Sergi, Melissa B. Gordon

Swelling is a key behavior exhibited in hydrogels. Despite their widespread use, few hydrogels offer control over the swelling process via application of an external stimulus. Poly(N-isopropylacrylamide) (pNIPAM) hydrogels exhibit a volume phase transition temperature (VPTT), in which the gel reversibly transitions from an unswollen, hydrophobic state to a swollen, hydrophilic state by altering the temperature. Here, pNIPAM hydrogels are engineered to respond to CO_2 which imparts enhanced hydrophilicity to the gel, thus elevating its VPTT. Under isothermal conditions, CO_2 is used as a trigger to increase the VPTT from below to above the swelling temperature of the gel which enables control over the onset and extent of swelling. We characterize the CO_2-controlled swelling behavior by measuring the uptake of water every ten minutes for three hours. Specifically, the swelling kinetics and swelling extent are studied as a function of hydrogel composition. To further understand and predict swelling behavior, a Voigt model was fitted to the experimental swelling data for each composition to determine the maximum swelling capacity and time constant that describes the swelling behavior. Across the range of compositions tested, gels swelled up to four times their original mass when exposed to CO_2. Furthermore, argon was sparged into the system to displace the CO_2 and reverse the swelling. We show that this swell-deswell behavior is repeatable at least three times via cyclical sparging of CO_2 and argon, and observed longer deswell times compared to swelling times. This CO_2-controlled system enables easily adjustable swelling characteristics via small modifications to composition, which may lend itself to applications in the medical field particularly for wound treatment in diabetes patients where CO_2 therapy is used to improve vascularization.

CO_2-Modulated Volume Phase Transition in Pnipam Hydrogels

Sarah R. Sergi, Finlay J. M. Smith, Melissa B. Gordon

Poly(*N*-isopropylacrylamide) (pNIPAM) is a well-studied thermoresponsive polymer exhibiting a lower critical solution temperature (LCST) in aqueous solutions or a volume phase transition temperature (VPTT) in crosslinked hydrogels. The volume phase transition is characterized by a transition from a shrunken, hydrophobic state to a swollen, hydrophilic state. For a specific hydrogel composition, this transition is typically modulated by a temperature change. In this study, we conferred CO_2 responsivity to pNIPAM hydrogels and show that these gels display a CO_2-switchable volume phase transition to enable isothermal and reversible gas-triggered swelling. We systematically measured the effect of hydrogel composition on its VPTT before and after exposure to CO_2 using UV-Visible spectroscopy. Over the compositional range presented, we observed shifts of up to 4.2ºC upon CO_2 exposure. The reversible nature of the CO_2-induced VPTT shift was demonstrated via cyclical addition and removal of CO_2, and the morphology throughout this cycle was observed using scanning electron microscopy (SEM). Through analysis of the SEM images, we found that the average pore size of the hydrogels more than doubled after exposure to CO_2 and that upon removal of CO_2, the pores returned to their initial size. The CO_2-responsive system described in this study can be tuned and applied to induce pronounced, isothermal swelling in hydrogels for use in a variety of biomedical applications, including drug delivery and wound dressings.

Photodegradation Kinetics of Trithiocarbonates

Alexa T. LaSasso, Katherine M. Pappas, Georgia C. Delaney, Melissa B. Gordon

Aqueous reversible addition fragmentation chain transfer (RAFT) polymerization is a technique used to synthesize polymers with precise control over molecular weight and structure while offering a biocompatible and environmentally friendly alternative to traditional organic solvents. Light-mediated aqueous RAFT polymerizations in which the polymerization is initiated using UV or visible light enables spatiotemporal control over the polymerization process and offers benefits as light is safe, abundant, and inexpensive. However, use of water and UV light can lead to the hydrolysis and photodegradation of chain transfer agents, respectively. Therefore, consideration into polymerization conditions must be taken. This study investigates the degradation of a trithiocarbonate-containing RAFT agent in water when exposed to UV light at its peak absorbance of 310 nm, as a function of concentration, pH, and time. The photodegradation kinetics were monitored using UV-Vis spectroscopy to gain insights into the effects of UV exposure on these compounds. For the concentrations studied, first order reaction kinetics were observed, and rate constants and half-lives were determined. These findings enhance our understanding when using trithiocarbonate RAFT agents in photopolymerizations.

Quantifying Radical Production from UV-Irradiation of a Trithiocarbonate-Containing RAFT Agent

Katherine M. Pappas, Alexa T. LaSasso, Georgia C. Delaney, Melissa B. Gordon

In the past few years, advancements have established reversible addition-fragmentation chain transfer (RAFT) polymerization as a favorable approach for the synthesis of polymers with a diverse range of applications such as drug delivery and smart materials. Trithiocarbonate (TTC) groups, widely used in RAFT polymerizations, are susceptible to light, heat, and force due to their labile carbon-sulfur bonds, resulting in radical formation via homolytic cleavage. Conducting these polymerizations in aqueous solution provides a cost-effective, biologically safe, and environmentally friendly alternative to traditional organic solvents. In this work, we quantify radical production generated from the photolytic covalent bond scission of a TTC-containing RAFT agent in water. The method relies on the Fenton reaction which, when activated in the presence of the colorimetric indicator xylenol orange (XO), leads to an observable color change from yellow-to-brown, quantified using UV-Vis spectroscopy. Specifically, aqueous solutions of a TTC-containing RAFT agent were irradiated at its peak absorbance of 310 nm, and resulting color changes were measured as a function of concentration. For each concentration, the color changes were quantified by comparing the ratio of the peak absorbances at 580 nm to 440 nm, corresponding to complexed XO and unbound XO, respectively. Using known concentrations of radical generators, a calibration curve relating the extent of color change to the radical concentration was developed. This spectroscopic analysis enables the quantification of radicals produced from the stimuli-induced cleavage of TTC-containing polymer solutions and hydrogels, contributing to the use and further understanding of TTC-containing RAFT agents, and can be adapted for use with other radical generators.

Investigating Antioxidative Coatings on Carbon/Carbon Composites for High Temperature Applications

Jacob McCormick, Mosiur Rahaman, Hema Ramsurn

Carbon/carbon (C/C) composites have been extensively researched due to their increasing demand in the aerospace industry for applications such as rocket nozzles, heat shields of space vehicles, and disk brakes of airplanes. Advantages of C/C composites include their high tensile strength, high heat conductivity, low coefficient of thermal expansion (CTE), and ability to maintain these properties under high-temperature environments. A challenge inherent with C/C composites is their susceptibility to mass loss when exposed to oxygen in environments that exceed 500°C. To protect the C/C composites in these moderate to high-temperature environments, high-temperature ceramic coatings are used as anti-oxidative coatings onto the C/C composite.

Several variables can affect the efficiency of these coatings such as adhesion to the substrate, similar CTE to that of the substrate, high melting point, low reactivity with oxygen, and ability to prevent diffusion of oxygen. The formation of oxides can be beneficial by providing a thin layer that prevents the diffusion of oxidative gases and, in some cases, acts as a self-healing agent which fills the pores and eventual cracks. However, some coatings rely on oxides that can only form at high temperatures (around 1500°C), making them not as effective at lower temperatures. Eventual cracks in the coating can also be closed due to the expansion of the coating at high temperatures (> 1000°C), preventing the oxidative gases from reaching the C/C composite substrate. With a high melting temperature (2,730°C) and a CTE close to carbon, SiC (silicon carbide) has been widely used as HTC coating for C/C composites. In addition, the formation of a glassy SiO_2 film when SiC oxidizes can prevent the diffusion of oxidative species through the coating.

It is common for C/C composite coatings to be multilayered with a combination of different methods and chemicals to optimize performance. In this work, some commonly explored methods for obtaining SiC coatings namely pack cementation (PC) and slurry-dipping/CVD will be explored as the first layers of protection. To this effect, six C/C composite samples will be used in this study: uncoated, PC, PC with boron, slurry/CVD, PC + slurry/CVD, and PC with boron+ slurry/CVD. The hypothesis is that the first SiC buffer layer should be from PC as the conversion of the carbon from the C/C composite to SiC will allow high adhesion of the coating. Since the viscosity of SiO_2 glass is too high to flow into and seal the microcracks at intermediate temperatures, boron will be added as B_2O_3 has lower viscosity and thus can flow into and seal coating cracks. The second SiC layer will be applied through slurry coating with ZrB_2 and a polymeric SiC precursor (SMP-10, Starfire Systems) in an endeavor to improve the self-healing properties of this coating. To understand the effectiveness of preventing mass loss, each sample will be tested using thermal gravimetric analysis in an air environment at 850°C. The corresponding data will then be compared to understand the advantages of different methods, chemicals, and multilayered coatings.

Encapsulation of Phase Change Materials By Microemulsion Processes

Arren Mallott, Moonchul Park, Jaewon Lee

Phase change materials (PCMs) play a crucial role in thermal energy storage systems by utilizing the latent heat of fusion to store and release energy. $KF\cdot 4H_2O$ is a particularly promising PCM due to its phase change temperature of 18°C, making it well-suited for enhancing the energy density of air conditioning systems. However, its low thermal conductivity, a common drawback of salt-based materials, limits its power density and hinders its potential for commercialization. Another challenge with PCMs is the risk of leakage, which can compromise the reliability of the storage system. Encapsulation has been proposed as a solution to prevent leakage, but scalable encapsulation methods face difficulties. For instance, the impregnation method, which involves embedding liquid PCM into micro-sized matrices, requires expensive solid adsorbents. While effective in incorporating PCMs, this approach necessitates careful consideration of cost and commercialization viability.

To overcome these challenges, we investigated $KF\cdot 4H_2O$-based microemulsion techniques to develop innovative nano- and micro-sized core-shell structured PCMs, designed to deliver high power density and leak-free thermal energy storage systems. This size scale was chosen to optimize the available heat transfer through the coating layer. Using a microemulsion process involving oil, water, and surfactants, we successfully synthesized TiO_2-encapsulated $KF\cdot 4H_2O$ core-shell structures. TiO_2 was chosen as the coating material due to its larger thermal conductivity compared to $KF\cdot 4H_2O$ and high strength. As the TiO_2 precursor diffused to the oil-water interfaces, it reacted with water, resulting in the formation of TiO_2, which fully encapsulated the $KF\cdot 4H_2O$. After purifying the $TiO_2/KF\cdot 4H_2O$ core-shell structures, no residual liquids were detected. Transmission electron microscopy (TEM) confirmed the complete encapsulation of $KF\cdot 4H_2O$ by TiO_2. Particle sizes ranged from 60 nm to 120 nm and had coating thicknesses of 6 nm to 12 nm. Furthermore, differential scanning calorimetry (DSC) studies demonstrated the phase change behavior of $KF\cdot 4H_2O$ (solid to liquid) without any evidence of water leakage during thermal cycling, validating the effectiveness of this encapsulation method.

Gradient Gel Inserts: Innovative Solutions for Knee Pain Relief

Antoinne Robinson, Kenneth Mineart

The present research focuses on developing a novel shoe insert made of a gel material featuring a controllable stiffness gradient to improve biomechanical foot alignment, reduce knee joint loading, and alleviate knee pain. The material's stiffness was adjusted by varying polymer concentration, and compression and indentation testing provided insights into the optimal range of these concentrations. A specialized template was designed to ensure consistent gradient sample production, facilitating repeatable results. Fatigue tests simulated real-life conditions, demonstrating the material's durability under repeated stress. The findings evidence the feasibility of creating a customizable gel material for shoe inserts, offering a promising solution for managing knee pain.

Statistical Analysis of Environmental Factors on SAM Systems

Ryan Zmarzlak

By 2036, continuous resolution increase in CMOS chips in computers will become technologically demanding, leading to an invalidation of Moore's Law. As an alternative, Organic Molecular Electronic Devices could potentially achieve Angstrom-scale resolution. Self-Assembled Monolayers (SAMs) derived from organic molecules, coupled with liquid metal EGaIn electrodes, provide convenient sample preparations and electrical measurement capable of delineating angstrom scale electrical properties at ambient. Due to its small scale, however, minor disturbances to the SAMs system can result in major errors in measurement. Thus, a large number of current density measurements are needed to statistically account for stochastic perturbations. To examine such environmental effects, these measurements are collected over various seasons, times, and locations leading to complexity of the errors. After, various statistical plots are used to correlate performance and reliability with various metadata categories. Heatmaps of current density and voltage measurements tell of consistency and performance while scatterplots of circuit shorts tell of device reliability and error. With these plots, it was found that the season and time greatly affected the performance and reliability of alkanethiol SAMs irrespective of the substrate or molecular length. We hypothesize that humidity differences between seasons cause the above phenomena due to the interaction between water, electrodes, and SAMs. A regression-based Machine Learning algorithm was designed based on these trends to designate the most optimal season and time based on the SAM system chosen. In the future, the optimization of the algorithm along with experimental testing of the device against predictions will be performed. This research was supported by the National Science Foundation DMR-2150360 'REU Site: Materials Research with Data Science (MAT-DAT)'.

Feedback Controlled Self-Assembly of Elliptical Colloidal Particles: Navigating between Liquid, Nematic and Crystal States

Michelle Sandag, Lechuan Zhang, Alec Pellicciotti, Michael Bevan

Controlled self-assembly of ellipse-shaped colloidal particles holds promise for a range of applications, including stimuli-responsive coatings, advanced liquid-crystal displays, and new-generation electromagnetic devices (i.e. invisibility cloaks). While research into efficient control strategies for hard ellipse systems has accelerated in recent years, achieving an optimal controller remains challenging due to system heterogeneities, susceptibility to defects, and slow diffusion-limited relaxation. Here a precise control over the assembly and disassembly kinetics of ellipses is demonstrated across fluid, nematic, and crystal phase equilibria using closed-loop feedback-controlled actuation of an AC-electric field. By calculating order parameters from real-time particle arrangements and altering the applied field via feedback control, our controller enables the active reconfiguration of colloidal ellipses from any starting ensemble to a target microstructure in second-time scales. Further, an osmotic force balance model utilizing the local density profile is employed to accommodate varying numbers of particles, electrode gap widths, and particle dimensions, thereby expanding the controller application to a broad range of experimental conditions and potential technological implementations. Drawing from a systematic analysis of phase transitions and free energy landscapes, our method can serve as a basis for manipulating particle ensembles for use in various metamaterial applications.

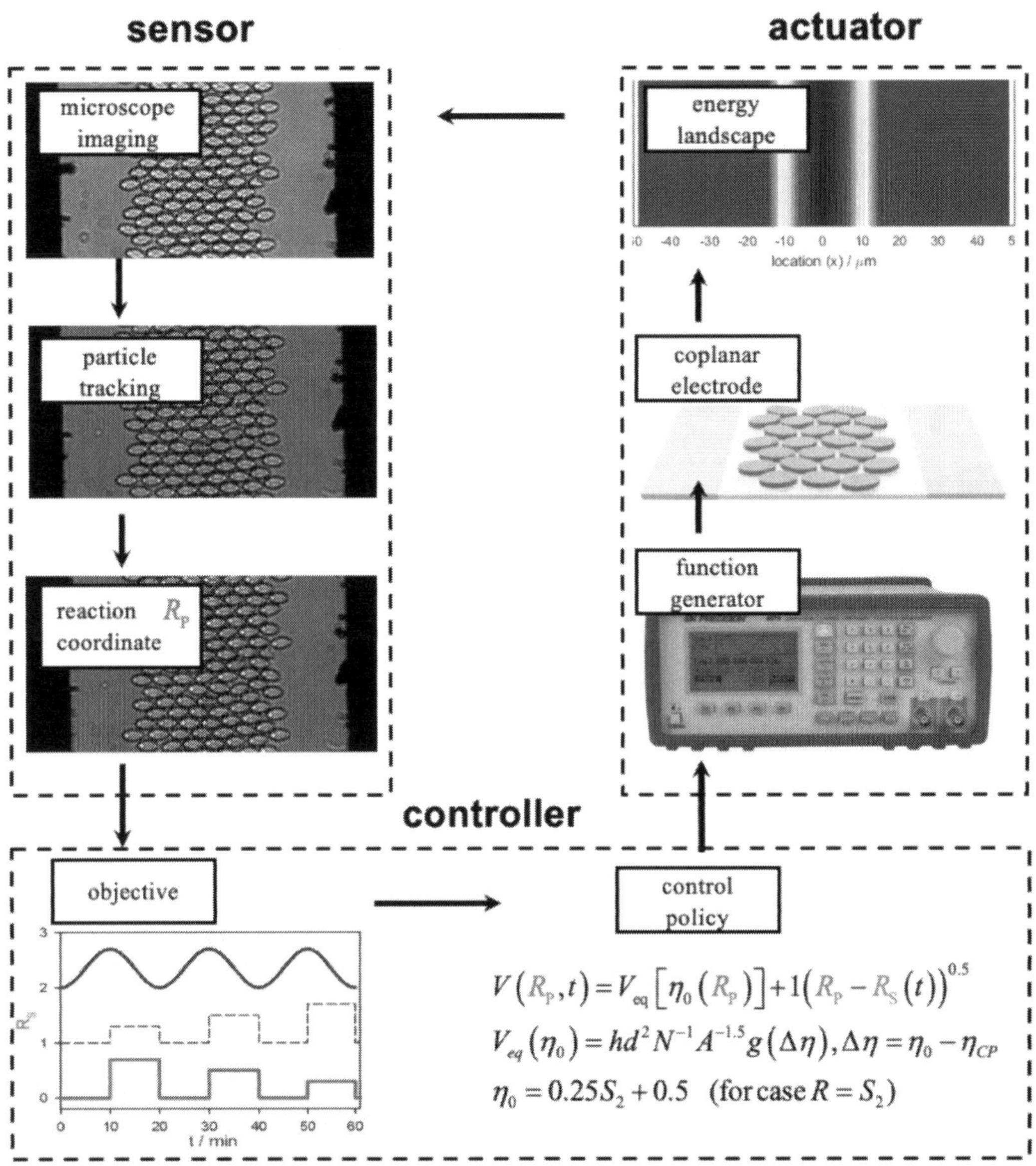
sensor
actuator
microscope imaging
particle tracking
reaction coordinate R_P
energy landscape
location (x) / μm
coplanar electrode
function generator
controller
objective
control policy
t / min
$V(R_P,t) = V_{eq}[\eta_0(R_P)] + 1(R_P - R_S(t))^{0.5}$
$V_{eq}(\eta_0) = hd^2N^{-1}A^{-1.5}g(\Delta\eta), \Delta\eta = \eta_0 - \eta_{CP}$
$\eta_0 = 0.25S_2 + 0.5$ (for case $R = S_2$)

Heparin Functionalization of Anodized Aluminum Oxide Membranes for Affinity-Based Nanoparticle Capture

Logan Martin, Kayla Daugherty, Barbara Knutson, Stephen Rankin

Nanoparticles are of growing interest as therapeutic devices, ranging from synthetic nanoparticle drug delivery systems to genetically engineered viral vectors such as adeno-associated viruses (AAVs). While these nanoparticle-based therapeutics have a high degree if efficacy for treating diseases, their separation from synthesis media is often a bottleneck in developing and scaling up processes for their manufacture. Centrifugation is frequently used at lab scale to separate nanoparticles, but this technology is not able to provide selective separations based on surface functionality and is difficult to scale. Adsorptive separations are more tunable based on charge and surface functional groups, but can be difficult to effectively design for the capture of 20-100 nm particles.

Here, anodized aluminum oxide (AAO) membranes are developed as a platform for affinity-based nanoparticle capture. These membranes provide a high density of straight, uniform pores through a chemically stable aluminum oxide framework that can be tuned by synthesis conditions. In the current study, 200 nm pores are used as a baseline. As a model affinity ligand, heparin is conjugated to the surface of the membranes. In the first step, membranes are functionalized with aminopropyl groups using aminopropyltriethoxysilane (APTES) to react with hydroxyl groups created on the AAO surface through a hydroxylation procedure. The density of amines in the resulting membranes is quantified using an acid orange II binding / release assay at pH values compatible with the chemistry of AAO. Heparin is subsequently attached via its carboxylic acid groups by using EDC-NHS coupling to form imide bonds. The density of attached heparin is measured by using binding of methylene blue (a cationic probe) followed by its release by charge screening.

To test the performance of the heparin-functionalized AAO membranes, studies are performed of the capture of virus-like nanoparticles.

The ability to capture nanoparticles is tested using silica nanoparticles (SNPs) with a size similar to that of AAV vectors (18 ± 4 nm), functionalized with amines to provide charged sites for interaction with heparin. The virus-like particles are fluorescently labeled to allow their capture and release to be readily quantified. Pressure-driven flux measurements with dead end filtration show the effects of functionalization on the flux of media with and without virus-like SNPs, and to demonstrate the capture effectiveness and capacity of the membranes towards virus-like SNPs.

Developing Deep Eutectic Solvent-Resistant Membranes for Efficient Lignin Recovery: A Comparative Study

Oscar Heft

The paper industry produces ~150 million metric tons of CO2 annually in the US alone. A new manufacturing process has been proposed that would reduce CO2 emissions due to the paper production process by ~65% by utilizing membrane separations rather than lignin recovery boilers. To accomplish this, ultrafiltration membranes must be developed that are 1) >95% effective at lignin filtration and 2) resistant to a novel hydrophobic Deep Eutectic Solvent (DES) used in the new process. Two potential polymers have been identified due to their hydrophobicity and resistance to DES: Polyvinylidene Fluoride (PVDF) and Polybenzimidazole (PBI). The objective of this work was to fabricate PVDF and PBI membranes to compare their permeability, rejection, and long-term DES resistance to make a recommendation on which polymer to further develop and functionalize.

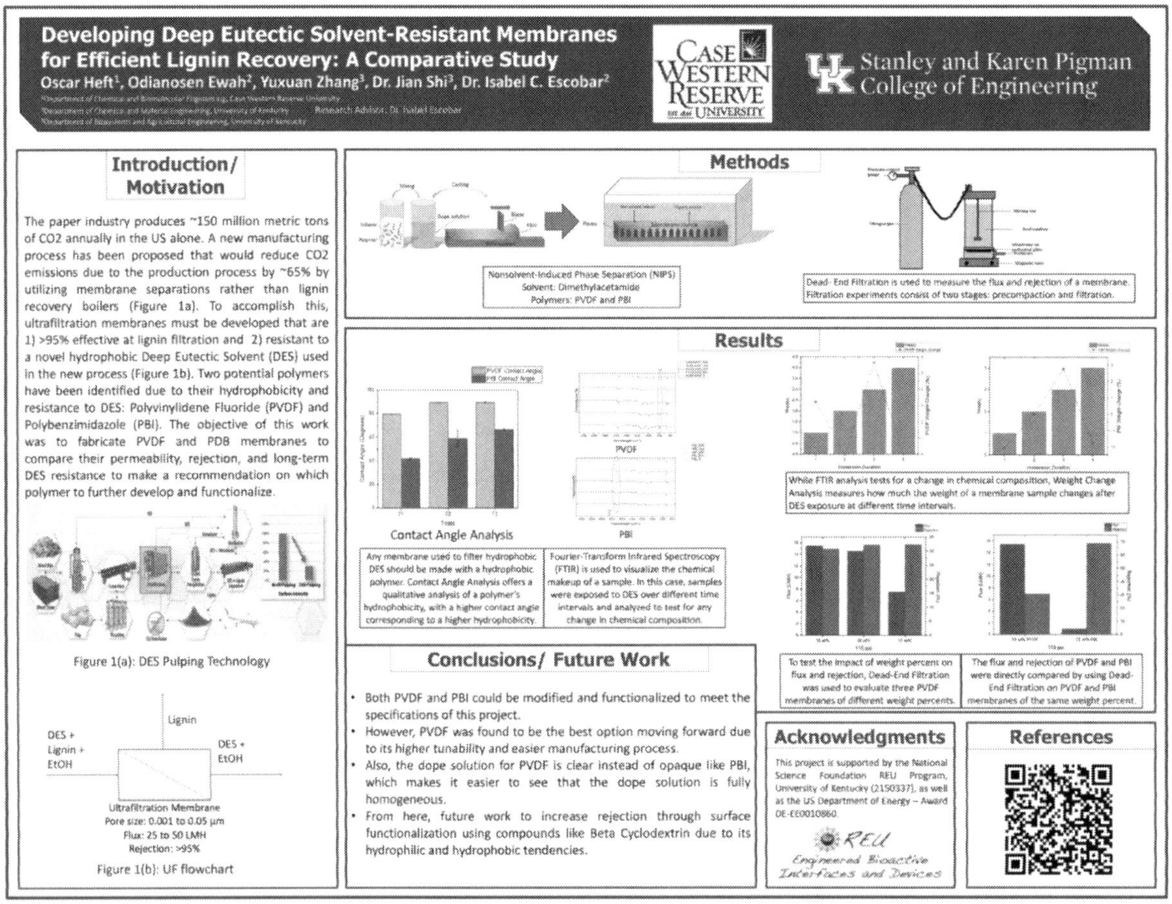

Functionalized Membranes for Organic Ion Capture to Nanoparticle Recovery

Jennifer Bukowski, Sam Thompson, Mara Leach, Dibakar Bhattacharyya

Separation of negatively charged nanoscale particles, viruses, and macromolecules have a variety of applications that affect human health. Negatively charged per- and polyfluoroalkyl substances, or PFAS, are of great concern in water systems. These "forever chemicals" are resistant to natural degradation, posing a threat to human health due to their ability to bioaccumulate. Recent EPA regulations regarding these potentially toxic substances have caused a push towards better understanding of these compounds and remediation techniques. While current water treatment methods such as coagulation, sedimentation, and disinfection are inefficient at removing PFAS, recent research in anion exchange has taken advantage of the electronegativity of these fluorinated compounds to show promising results in improved PFAS remediation. In other applications, one of the most common methods of gene delivery involves the use of viral vectors, or viruses that have been modified to carry the healthy copy of a gene into the body for therapeutic effect. Adeno-associated virus (AAV) vectors are a widely preferred platform for viral vector gene therapy. However, the challenge of separating and purifying full AAV nano sized (20 – 25 nm) capsids from empty capsids remains a major challenge that is often solved today with costly methods that prevent AAV viral vector therapy from being accessible to much of the population. Traditionally, full and empty AAV capsids are separated based on a slight difference in their isoelectric points—full capsids have an isoelectric point of ~5.9 and empty capsids have an isoelectric point of ~6.3. In these two and various other applications involving charged-based separations, resins and packed columns are the standard. However, this technology presents challenges due to limited diffusivity and extended absorption times. Membranes that have been functionalized to include active positively charged (similar to PFAS capture aspects) sites within their porous structure present an alternative platform for ion exchange. The addition of convective flow in membrane applications has the potential to increase the efficiency of the separation and decrease run times, saving on operations costs. We found that membranes functionalized with primary and quaternary amine-containing polymers were able to capture negatively charged dye and carboxylate-modified nanoparticles, which were used to model substances like PFAS and full viral vector capsids, better than pristine alternatives. Additionally, we explored different release mechanisms (i.e. high pH and high salinity solutions) for recovering the captured organics. These findings provide insight into a possible direction to continue research to improve the capture and recovery of negatively charged substances toward environmental and biotherapeutic applications. This research was supported by NSF EPSCOR Track 2 Program, NIEHS Superfund Research Program, and the National Science Foundation REU Program, University of Kentucky.

Portable Iron Filtration Made Possible: Use of High, Flux Low Pressure Ion-Exchange Membranes in Series

Courtney Wilmoth, Thomas McKean, Ranil Wickramasinghe, Jamie Hestekin

Iron and other heavy metals are harmful contaminants that can be found in many sources of drinking water, especially well water which is used by around 43 million Americans as a source of potable drinking water (EPA, 2024). As the commercial desire for portable water treatment systems increases, the scientific advancements for these systems must also increase. Microfiltration membranes allow for portable filtration due to their low-pressure requirements, but typically cannot remove metal ions via size exclusion. This work examines adding ion exchange functionality to microfiltration membranes to allow for the capture of metal ions without sacrificing low operating pressures that are necessary for portable use. The ion-exchange membranes are synthesized via ultraviolet grafting of glycidyl methacrylate intermediate onto microfiltration membranes made of polyethersulfone (PES). These intermediates are modified using sodium sulfite to establish positive fixed charges on the membrane surface. Permeability testing of a variety of ion-exchange microfiltration PES membranes with variable grafted weights of the intermediate, depict that increased grafted weight on the membrane surface leads to increased operating pressure and decreased flux. When stacked in series, these low-grafted-weight filters maintain a practical operating pressure. Membranes with grafted weight as low as 0.4 mg performed up to 69% static removal whereas membranes with 3.7 mg of grafted weight show up to 71% static removal. Operating with these membranes in series increases removal beyond additive functions of the grafted weight of individual membranes, shown by a comparison of a single membrane versus a membrane stack at similar total grafted weights. When a 1.7 mg and 1.2 mg grafted weight membrane are stacked in series the capacity increases to 19.3 mg/g compared to the capacity of a single membrane of 3.8 mg of grafted weight which has a capacity of 8.25 mg/g. The stacked series operates below 2 psi whereas the single membrane operates at 4 psi. Use of these ion-exchange microfiltration membranes with low grafted weight in portable water systems can allow for increased removal capacity of iron water filtration and removal of additional multivalent ionic species from water in a portable system.

Cyclone Separation of Rare Earth Elements from Coal Fly Ash after Supercritical CO_2 extraction

Keegan Privett, Eric Palmer, Sergio Martinez-Monteagudo, Isaiah Morones

Rare earth elements (REE), a group of elements comprising scandium, yttrium, and the Lanthanides, are used in a wide range of applications from electronics and magnets, to the handling of nuclear materials. The focus of this project is to produce gadolinium for the management of nuclear waste. gadolinium (Gd) is widely used in nuclear applications as a neutron poison, meaning that gadolinium readily absorbs neutrons which could cause fissions in other isotopes. Gd is found in small concentrations within the Earth's crust and is often found alongside other ores. Global production of pure Gd is estimated to be four hundred tonnes per year and is mostly produced in China.

The United States relies heavily on a steady supply of REE and, therefore, has a vested interest in increasing domestic production of REE. One promising domestic source of REE is coal fly ash. The U.S. currently produces millions of tons of coal fly ash per year. Preliminary results from batch extractions using organic acids in supercritical CO_2 have shown this method to be promising for yield and REE selectivity. This project aims to develop the means for continuous extraction of REE from coal fly ash. The fine particle size of coal fly ash makes many typical forms of filtration ineffective. However, there are many components of coal fly ash, such as silicon, aluminum, iron, and calcium, which could be separated from supercritical CO_2 by density.

Cyclones are regularly used in industry for the physical separation of mixtures through centrifugal force and differences in density. Less dense particles remain near the top of the vessel while denser particles fractionate to the bottom. There is very little existing data on the use of supercritical fluids in cyclone separation. One key issue in the design of a cyclone for this application is that the CO_2 must remain in the supercritical state to avoid the components of the coal fly ash reattaching to each other. To produce supercritical CO_2, a currently existing supercritical fluid extractor will be modified to pump supercritical CO_2 into a reactor vessel connected to the cyclone. Outlet pressure of the system will be carefully regulated to maintain the supercritical state and the appropriate flow through the system to achieve velocities necessary for the cyclone.

Microstructural Analysis of Metal-Embedded Carbon Molecular Sieve (CMS) Membranes for Sustainable Ammonia Synthesis

Alyssa Mize, Nhan Khuu, Oishi Sanyal

Ammonia (NH_3) is a key raw material for fertilizer production, traditionally produced through the energy-intensive Haber-Bosch process, which alone contributes about 1.8% of global CO_2 emissions. In contrast, catalytic membranes have the potential to reduce energy consumption in NH_3 synthesis while maintaining high NH_3 yields. Carbon Molecular Sieve (CMS) membranes, synthesized through high-temperature pyrolysis of fluorinated polyimide precursors, have demonstrated high performance in gas separation due to their bimodal pore size distribution, which offers excellent selectivity for NH_3/H_2 and NH_3/N_2. In this work, their application will be extended to reactive separation by incorporating metals into the membrane structure. The embedded metals (Fe, Ru) will serve as the catalytic sites, while the CMS membrane will facilitate the separation of NH_3 (2.6Å) from similarly sized gases H_2 (2.9Å) and N_2 (3.6Å). A variety of metal incorporation techniques and synthesis conditions will be evaluated, and the resulting materials will be characterized for both their separation and catalytic properties. A novel tool, viz. pressure-decay sorption, is being used to analyze the microstructure of these membranes, focusing on their sorption domains. The tool also probes changes in the microstructure due to high-temperature exposure and environments with 75-100% pure H_2. Additionally, characterization techniques such as X-ray Photoelectron Spectroscopy (XPS), Raman Spectroscopy, X-ray Diffraction (XRD), and BET CO_2 and N_2 physisorption will also be of focus.

Membrane Crossover Modeling for Electrochemical CO_2 Separations

Sarah Hernandez, Katelyn Ripley, Fikile Brushett

Adequate reduction of atmospheric carbon dioxide (CO_2) necessitates efficient and low cost carbon capture processes. Electrochemical CO_2 separation systems may provide a lower energy alternative to traditional thermochemical capture processes, but ongoing techno-economic evaluations are required to clarify further research directions and technology readiness. Key operating envelopes and cell designs relevant to these systems have yet to be studied in depth, creating the opportunity for preliminary modeling of these systems to define design specifications. While prior work in the Brushett group has investigated the performance and costs associated with a four-stage capture system, these models have yet to account for losses associated with crossover of redox-active species through the separator/membrane.[1] Electrochemical CO_2 capture systems may be able to withstand higher rates of crossover than other electrochemical systems (i.e., redox flow batteries) because the same redox-active species exist on either side of the separating membrane, resulting in no unrecoverable losses in the redox-active species itself. To verify this, cell modeling allows for understanding of how performance is impacted by crossover in the cell but also clarifying how lower costs can be achieved as these systems may tolerate more conductive but less selective separators/membranes.

The existing four-stage capture model includes three process units: an electrochemical cell for redox-active species activation and deactivation, an absorber for CO_2 binding, and a flash tank for release of concentrated CO_2. Using a CSTRs-in-series or "cascade" approach may allow for simplification of mathematical solving, perhaps reducing 2D concentration gradients into segmented 1D concentration gradients.[2] While we expect system efficiencies to decrease when accounting for crossover, the active species itself does not decay and thus will likely not contribute to shorter operating lifetimes. Ultimately, this electrochemical cell model can integrate into a previously-developed economic model to elucidate the cost trade-offs between conductivity and selectivity in separator/membrane selection in order to improve technology readiness.

(1) Clarke, L. E.; Leonard, M. E.; Hatton, T. A.; Brushett, F. R. Thermodynamic Modeling of CO2 Separation Systems with Soluble, Redox-Active Capture Species. Ind. Eng. Chem. Res. 2022, 61 (29), 10531–10546. https://doi.org/10.1021/acs.iecr.1c04185.

(2) Walsh, F.; Trinidad, P.; Gilroy, D. Conversion Expressions for Electrochemical Reactors Which Operate under Mass Transport Controlled Reaction Conditions - Part II: Batch Recycle, Cascade and Recycle Loop Reactors. IJEE 2005, 21 (5), 981–992.

Tunable Phase Separation of Detergents for the Crystallization of Membrane Proteins and Block Copolymers

Dominic Bujanos, Ronald J. Vogler, Katherine Kimball, Manish Kumar, Benny D. Freeman

Membrane proteins and the biomimetic membranes they produce are promising materials for membrane separations. The high protein packing in self-assembled 2D sheet structures is ideal for desirable characteristics such as high selectivity and permeability. One potential method for synthesizing these 2D sheets involves heating a mixture of proteins, amphiphilic block copolymers, and detergents. In this method, a critical component of 2D sheet formation is the phase separation of detergent at elevated temperatures, resulting in the aggregation of detergent micelles containing block copolymers and membrane proteins. However, some membrane proteins may be unstable and unfit for 2D sheet formation at extreme temperatures. To address this issue, we investigated using additives in detergent mixtures to reduce the temperature required for phase separation. The phase separation temperatures were determined using cloud point measurements of n-octyl-oligo-oxyethylene (Octyl-POE) detergent solutions containing various amounts of glucose. In these solutions, the phase separation temperature was lowered from 56°C to 29°C by increasing the glucose concentration from 0 to 2.2 M. [add data on lowest concentration where phase separation was observed and how it does not change much when Octyl-POE is present]. Adding 1.4 M of glucose to a solution of Outer Membrane Protein F (OmpF), poly(butadiene-b-ethylene oxide) (PB-PEO) and Octyl-POE enabled the formation of 2D sheets containing OmpF and PB-PEO at 45 °C. This temperature was far less than the necessary 60 °C required in the absence of glucose. These results showed the promising use of additives to decrease the required temperature for phase separation and the subsequent 2D sheet formation, enabling this method to be more broadly applied to other membrane protein-derived nanosheets.

Particle Molecular Layer Deposition Enhances Adsorption Capacity for CO2 Capture Materials through Amine Film Deposition

Michael Zaza, Hailey Loehde-Woolard, Bergen Evans, Katarina Odak, Kent Warren, Alan W. Weimer

Amine adsorbent materials retain traditional acid-base chemistry, operate under ambient conditions, and benefit from the presence of humidity. Current liquid-based functionalization techniques face limitations: grafted materials tend to have lower adsorption capacities, while impregnated materials suffer from poor regeneration due to the lack of covalent tethering. Vapor-phase grafting (VPG) of aminosilanes demonstrate reliable monolayer and sub-monolayer film formation, offering enhanced surface control, but this has not been extensively studied for CO2 capture applications. We propose that VPG, which exposes the substrates surface to a single reactant, can be extended to deposit larger quantities of amines by regenerating active groups on the particles surface through particle molecular layer deposition (MLD). Particle MLD utilizes a series of alternating gas-solid phase reactions to deposit a thin uniform coating. In our research, we focus on applying particle MLD to generate aminosilane films onto various powder substrates for the evaluation of CO2 capture materials.

The reactions occur in a custom-built vibrating fluidized bed particle MLD reactor system, which was validated through fluidization experiments and benchmark Al2O3 atomic layer deposition. We investigated the MLD chemistry of (3-aminopropyl)triethoxysilane and water, which deposited amine functional groups anchored by Si-O linkages. A design of experiments was implemented to optimize the number of cycles and deposition temperatures for generating sorbent materials. For comparison, we also generated materials using liquid phase solvent grafting and VPG. We employed BET surface area analysis, LECO elemental analysis, X-ray photoelectron spectroscopy (XPS), and thermogravimetric analysis (TGA) for film characterization. Research has demonstrated, in VPG processes, lower deposition temperature results in a higher amine content. However, for particle MLD materials, higher deposition temperature leads to increased adsorption capacity. Our results indicate that both the number of cycles and reaction temperature significantly affect the amine film's adsorption capacity. This observation provides a clear distinction from the traditional VPG processes, where the amine content decreases at higher deposition temperatures. Across the different supports used, particle MLD led to consistent densities of coating and consistent adsorption capacities on a surface-area basis. Our work demonstrates that particle MLD is a promising route to prepare solid adsorbent materials for CO2 capture applications.

Unlocking Potential for Hydrofloat™ Technology: Collector Screening for Enhancing Coarse Particle Recovery

Steven Lycans II

Unlocking Potential for HydroFloat™ Technology: Collector Screening for Enhancing Coarse Particle Recovery

SJ Lycans[a], Yucel Ozsoy[b], and Rick Honaker[b]

[a]Department of Chemical and Materials Engineering, University of Kentucky, Lexington, Kentucky 40506
[b]Department of Mining Engineering, University of Kentucky, Lexington, Kentucky 40506

Abstract:

The advancement of HydroFloatTM technology represents a major breakthrough in mineral processing industry, especially in coarse particle flotation, pushing the boundaries of the 'Elephant Curve' limitations. This cutting-edge technology has brought significant improvements to mineral separation by enhancing the recovery of valuable minerals from larger ore particles. Despite these advancements, its full potential has yet to be fully harnessed. A key to its success is rooted in the chemical principles of the flotation process, particularly the types of collectors used and their adsorption mechanisms. Enhancing these chemical interactions has the potential to greatly improve the performance of HydroFloatTM technology. This research highlights the impact of dithiocarbonate (DTC) and dithiophosphate (DTP) collectors, which are commonly employed to boost the flotation efficiency of sulfide minerals.

To explore their potential for enhancing flotation performance, eight distinct DTC and DTP collectors, developed and produced by Syensqo, were selected, each featuring unique chemical structures and varying carbon chain lengths. The methodology consisted of controlled flotation tests in which both the type and dosage of collectors were adjusted to evaluate their effects. The primary goal is to boost recovery rates and selectivity in coarse particle flotation by refining the chemical interactions between the mineral surfaces and these collectors.

The findings from this research will provide insights into how variations in collector chemistry could influence flotation outcomes. This exploration into the chemical nuances of flotation collectors is poised to contribute valuable data to the field, possibly leading to more effective collector designs tailored for specific mineral types and conditions, thereby enhancing the capabilities of HydroFloatTM technology.

Unlocking Potential for HydroFloat™ Technology: Collector Screening for Enhancing Coarse Particle Recovery

SJ Lycans[a], Yucel Ozsoy[b], and Rick Honaker[b]

a. Department of Chemical and Materials Engineering, University of Kentucky, Lexington, Kentucky 40506
b. Department of Mining Engineering, University of Kentucky, Lexington, Kentucky 40506

Abstract:

The advancement of HydroFloat™ technology represents a major breakthrough in mineral processing industry, especially in coarse particle flotation, pushing the boundaries of the "Elephant Curve" limitations. This cutting-edge technology has brought significant improvements to mineral separation by enhancing the recovery of valuable minerals from larger ore particles. Despite these advancements, its full potential has yet to be fully harnessed. A key to its success is rooted in the chemical principles of the flotation process, particularly the types of collectors used and their adsorption mechanisms. Enhancing these chemical interactions has the potential to greatly improve the performance of HydroFloat™ technology. This research highlights the impact of dithiocarbonate (DTC) and dithiophosphate (DTP) collectors, which are commonly employed to boost the flotation efficiency of sulfide minerals.

To explore their potential for enhancing flotation performance, eight distinct DTC and DTP collectors, developed and produced by Syensqo, were selected, each featuring unique chemical structures and varying carbon chain lengths. The methodology consisted of controlled flotation tests in which both the type and dosage of collectors were adjusted to evaluate their effects. The primary goal is to boost recovery rates and selectivity in coarse particle flotation by refining the chemical interactions between the mineral surfaces and these collectors.

The findings from this research will provide insights into how variations in collector chemistry could influence flotation outcomes. This exploration into the chemical nuances of flotation collectors is poised to contribute valuable data to the field, possibly leading to more effective collector designs tailored for specific mineral types and conditions, thereby enhancing the capabilities of HydroFloat™ technology.

Adsorptive Microparticles Formed By a Non-Solvent Induced Phase Separation Method for Metal Ion Removal

Natalie Crowner, Shukun Zhong, William Phillip

Growing populations, environmental conservation, and industrial prosperity all require the maintenance of a clean water supply. Adsorptive separation processes offer a solution for the efficient removal of dilute contaminants, such as heavy metals and PFAS. However, the inherently unsteady nature of these approaches increases cost and complexity. Therefore, to enable continuous steady operation, the idea of contacting a contaminated water stream and a slurry of adsorptive microparticles countercurrently was proposed. Here, we show that a polymer membrane casting solution containing 4, 6, or 8wt% polysulfone (PSF) and 2wt% poly(styrene)-block-poly(acrylic acid) (PS-PAA) dissolved in 2-pyrrolidone can be formed into spherical adsorptive microparticles using a Non-Solvent Induced Phase Separation (NIPS) technique without sacrificing functionality. To form the microparticles, casting solution microdroplets were pipetted directly into a dilute non-solvent bath. Using a static formation method, microparticles of average diameter of ~1000 μm and maximum binding capacity ~0.60 mmoles of copper bound per gram of adsorptive PS-PAA could be generated. A dynamic formation method results in particles with average diameter of ~650 μm, similar binding capacity, and a higher control of sphericity. These results demonstrate how microparticle size and shape can be manipulated through formulation method without altering the functionality of their adsorptive membrane composition. The results of this study will prompt further investigation into the factors that affect microparticle size and shape in an efficient NIPS formation process. Optimization of the adsorptive microparticle could lead to the creation of adaptable continuous adsorption systems for contaminant removal or resource recovery.

Enhancing Coarse Particle Flotation Efficiency on Metal Sulfides: A Comprehensive Adsorption Study on Novel Collectors

Dalton Hale

Enhancing Coarse Particle Flotation Efficiency on Metal Sulfides: A Comprehensive Adsorption Study on Novel Collectors

Dalton Hale[a], Yucel Ozsoy[b], and Rick Honaker[b]

1. Department of Chemical and Materials Engineering, University of Kentucky, Lexington, Kentucky 40506
2. Department of Mining Engineering, University of Kentucky, Lexington, Kentucky 40506

Abstract:

Coarse particle flotation has emerged as a breakthrough innovation in the mineral processing industry. Coarse particle flotation saves energy by reducing the need for extensive grinding, allowing for the recovery of valuable minerals at a larger particle size, which lowers overall energy consumption in the mineral processing cycle. A key factor in realizing its full potential lies in optimizing the chemical interactions involved in the flotation process, particularly the fundamental principles of adsorption, which directly influence recovery rates.

As part of a comprehensive project aimed at refining the performance of coarse particle flotation, this study investigates the adsorption mechanisms and underlying principles for both novel and commercially available collectors, with a focus on dithiocarbonate (DTC) and dithiophosphate (DTP). These collectors are commonly used in sulfide mineral flotation, but their behavior in the context of coarse particle recovery remains an area of active research. Understanding their adsorption dynamics is essential for improving flotation efficiency.

The experimental work was carried out on a range of particle size fractions with low surface area exposure of metal sulfides (less than 20%), including +850, -850+600, -600+300, -300+212, and -212+150 microns, with varying collector dosages to investigate the adsorption behavior of the collectors. These measurements provided critical insights into how the collectors interact with coarse particles, helping to identify the optimal conditions for maximizing recovery and selectivity.

The results of this research are expected to contribute significantly to the understanding of collector adsorption in coarse particle flotation, offering a scientific basis for the development of more effective collector designs. By tailoring collector chemistry to specific particle sizes and conditions, this study aims to further enhance the capabilities of coarse particle flotation and push the limits of mineral recovery.

Enhancing Coarse Particle Flotation Efficiency on Metal Sulfides: A Comprehensive Adsorption Study on Novel Collectors

Dalton Hale[a], Yucel Ozsoy[b], and Rick Honaker[b]

a. Department of Chemical and Materials Engineering, University of Kentucky, Lexington, Kentucky 40506

b. Department of Mining Engineering, University of Kentucky, Lexington, Kentucky 40506

Abstract:

Coarse particle flotation has emerged as a breakthrough innovation in the mineral processing industry. Coarse particle flotation saves energy by reducing the need for extensive grinding, allowing for the recovery of valuable minerals at a larger particle size, which lowers overall energy consumption in the mineral processing cycle. A key factor in realizing its full potential lies in optimizing the chemical interactions involved in the flotation process, particularly the fundamental principles of adsorption, which directly influence recovery rates.

As part of a comprehensive project aimed at refining the performance of coarse particle flotation, this study investigates the adsorption mechanisms and underlying principles for both novel and commercially available collectors, with a focus on dithiocarbonate (DTC) and dithiophosphate (DTP). These collectors are commonly used in sulfide mineral flotation, but their behavior in the context of coarse particle recovery remains an area of active research. Understanding their adsorption dynamics is essential for improving flotation efficiency.

The experimental work was carried out on a range of particle size fractions with low surface area exposure of metal sulfides (less than 20%), including +850, -850+600, -600+300, -300+212, and -212+150 microns, with varying collector dosages to investigate the adsorption behavior of the collectors. These measurements provided critical insights into how the collectors interact with coarse particles, helping to identify the optimal conditions for maximizing recovery and selectivity.

The results of this research are expected to contribute significantly to the understanding of collector adsorption in coarse particle flotation, offering a scientific basis for the development of more effective collector designs. By tailoring collector chemistry to specific particle sizes and conditions, this study aims to further enhance the capabilities of coarse particle flotation and push the limits of mineral recovery.

Influence of Hydrogen Bonding on CO_2 Capture in Choline and Ethylene Glycol Mixtures

Amelia Tomak, Ruth Dikki, Burcu Gurkan

There is a surge in the interest in carbon capture because of increasing global temperatures, which is a consequence of rising CO_2 emissions. One of the most commonly used absorbers for carbon capture is aqueous amine solution because of its high affinity and selectivity for CO_2 even under low concentrations. However, despite this high CO_2 affinity, aqueous amine requires high temperatures for regeneration, while also being highly volatile and corrosive. Therefore, alternative sorbents with low volatility and tunable binding enthalpy, such as eutectic solvents, are being studied for carbon capture. Eutectic solvents relevant to CO_2 capture are often composed of hydrogen bond acceptors (HBAs) and hydrogen bond donors (HBDs) with the ability to chemisorb CO_2. Given that the HBA and HBD compositions affect key solvent properties, such as volatility and CO_2 capacity, this study focuses on creating a better understanding of the impact of the HBD concentration on the CO_2 separation process, in particular the capacity and regeneration. To achieve this objective, mixtures of ethylene glycol (EG) HBD and choline 2-cyanopyrrolide ($[Ch]^+[CNpyr]^-$) HBA were examined as a function of HBA:HBD molar composition (1:3, 1:4, and 1:5). The samples were evaluated in terms of their CO_2 capacity and regenerability by performing absorption experiments at 25 °C and 1 bar of CO_2 followed by desorption at 50 °C. Products and the various CO_2 binding sites contributing to the overall capacity were identified and quantified by ^{13}C nuclear magnetic resonance (NMR) spectroscopy. A general decrease of the gravimetric capacity was observed with increasing HBD concentration through NMR analysis. Overall, while the concentration of HBDs did not directly impact the regenerability of the solvent, the increased HBD-HBD and HBA-HBD interactions at higher HBD concentrations resulted in a decrease in both the HBD-CO_2 and HBA-CO_2 adduct formation. This study showcases the design features in eutectic solvents, and primarily the compositional effects, in controlling the CO_2 absorption-desorption capability as relevant to CO_2 capture for emission mitigation and direct air capture efforts.

Thermodynamic and Kinetic Analysis of Impurity Retention in Dyed Pharmaceutical Crystals

Ricardo Otake, Gerard Capellades, Mitchell Paolello

Impurities in the crystal structure of an active pharmaceutical ingredient (API) can influence its properties, such as stability and solubility. Conventional methods to measure composition anisotropy or impurity concentration across different crystal faces are often destructive or require intricate calibration. By introducing colored impurities, such as dyes, visibility into impurity incorporation can be achieved, offering a non-destructive alternative based on model compounds.

The use of this technique is illustrated through thermal cycling of a saturated mother liquor with dye impurities, showing how composition anisotropy of single crystal can be analyzed over each cycle of dissolution and regrowth. Two host/guest systems were studied: acetaminophen with curcumin and potassium sulfate with acid fuchsin, under thermal cycling conditions between 25°C and 35°C to examine impurity distribution and dissolution-regrowth behavior. By using red-to-green pixel intensity ratios, from images taken under controlled lighting conditions, it can be used to assess impurity distribution across the crystal lattice without the destruction of the crystal samples.

When dye impurities are introduced, the thermodynamic and kinetic properties of crystallization become apparent during thermal cycling of each host/guest system. Thermal cycling alternates between dissolution and regrowth, revealing differences in impurity retention due to thermodynamic equilibrium and growth kinetics. In the acetaminophen/curcumin system, impurity-rich regions dissolve preferentially due to thermodynamic instability, leaving behind cleaner faces that serve as favorable sites for regrowth. This results in mass loss and reduced impurity retention, driven by the system's kinetics favoring cleaner surfaces during recrystallization. In contrast, the potassium sulfate/acid fuchsin system shows slower dissolution kinetics, maintaining most of its mass during cycling. However, regrowth leads to increased impurity incorporation in the cleaner crystal faces, as thermodynamic conditions promote impurity uptake in these regions. This system's slower impurity incorporation kinetics highlights a more gradual impurity retention process compared to acetaminophen.

The use of colored impurities enables non-destructive analysis of composition anisotropy, providing valuable insights into crystallization behaviors. These findings demonstrate the distinct mechanisms of impurity retention in different systems, emphasizing the importance of understanding the interplay between thermodynamics and kinetics to optimize crystallization processes in pharmaceutical applications.

Metal-Embedded Carbon Molecular Sieve Membranes for Sustainable Ammonia Production

Joseph Harrah, Nhan Khuu, Brhanu Kelali Desta, Jianli Hu, Oishi Sanyal

The research pertaining to this poster presentation is focused on developing a metal-embedded carbon molecular sieve (CMS) based catalytic membrane reactor for sustainable ammonia (NH_3) synthesis. In industry, it is most commonly produced via the Haber-Bosch process, a highly energy intensive process and by itself accounts for ~1.8% to global CO_2 emissions. Catalytic membranes allow the potential to reduce energy consumption and ensure high yields of NH_3. CMS membranes, synthesized via polyimide precursors subject to high temperature pyrolysis, have been studied as high-performance gas separation membranes. In this work, the range of application will be extended beyond to reactive separation by linking fluorinated-copolyimides and iron in membrane structure. The embedded metal acts as a catalytic site while the CMS membrane itself enables their special structure for separation of NH_3 (2.6 A°) from similar-sized gas H_2 (2.9 A°) and N_2 (3.6 A°) molecules. To synthesize these membranes, we analyze various metal incorporation techniques and conditions, as well as characterize these materials based on sorption and catalyst performance. The separation properties will be reported in sorption coefficients for NH_3, N_2 and H_2 by using different methods of making metal-embedded CMS membranes. These results will be used to elucidate the morphological structure of CMS as there are currently no other characterization methods that can analyze these structures.

Recovery of Critical Minerals from Spent Lithium-Ion Batteries Via Bioleaching Methods

Stina Daniel, Meltem Urgun-Demirtas, Haoran Wu

The sale of lithium-ion batteries (LiBs) has sky-rocketed in recent years. The International Energy Agency (IEA) estimates 145 million electric vehicles on the road by 2030[1]. Current recycling methods for spent LiBs, such as pyrometallurgy, hydrometallurgy, are costly, inefficient, and cause pollution. Elements such as lithium, cobalt, manganese, and nickel are critical minerals (CMs) needed to manufacture LiBs. However, the domestic supply of CMs is limited, raising concerns about their shortages and price volatility. Biohydrometallurgy offers an environmentally sustainable approach to recycling CMs[2], extracting them for use in the production of new LiBs through the bioleaching process. In this study, five different organic acids (i.e., lactic, succinic, citric, acetic, and gluconic acid) were examined in abiotic conditions to explore their abilities as biolixiviants on extracting critical minerals from spent LiBs. These acids were chosen because they can potentially be produced from organic waste streams via bioprocesses, which could significantly decrease the leaching cost and carbon footprint. The biolixiviants were mixed with black mass (ground up anodes and cathodes from spent LiBs) and agitated at 150 rpm. Periodical sampling was conducted to understand the kinetics of bioleaching at 40 ^{o}C. The samples were analyzed using inductively coupled plasma (ICP) to determine their elemental composition and assess the recovery rate of critical minerals from black mass. After the procedure was set with the biolixiviants, a reductant (i.e., ferrous sulfate) was added to aid in the efficiency and abundance of the recovery rate in the leachate. The results showed citric acid having a significantly higher recovery rate than any other biolixiviants in a reductant-free environment, and gluconic acid had the lowest recovery in this scenario. When a reducing agent is applied, the results show citric acid still achieved the highest recovery, while gluconic acid followed shortly behind, and the recovery for both was higher than without a reducing agent. The acetic acid leachate had the lowest recovery rate for the second experiment, with ferrous sulfate, and second to lowest in the first experiment, without ferrous sulfate. These results show that citric acid is the best leaching agent with greater than 80% recovery percent, but gluconic is viable as well when reducing agent is supplied. To continue further develop this bioleaching process, the promising biolixiviants identified in this study will be produced using specialized microbial communities from organic waste. To ensure the best results, future experiments could involve multiple microbes (producing citric and gluconic acid) and optimization of experimental conditions to maximize the recovery of critical materials from black mass. Our findings demonstrate the viability of biohydrometallurgy as a promising solution to critical mineral shortages, offering a more efficient, environmentally friendly, and cost-effective recovery method.

Reference:

[1] IEA (2021), Global EV Outlook 2021, IEA, Paris https://www.iea.org/reports/global-ev-outlook-2021, Licence: CC BY 4.0 (accessed 2024-09-17).

[2] Alipanah, M.; Reed, D.; Thompson, V.; Fujita, Y.; Jin, H. Sustainable Bioleaching of Lithium-Ion Batteries for Critical Materials Recovery. Journal of Cleaner Production 2023, 382, 135274. https://doi.org/10.1016/j.jclepro.2022.135274.

Maximizing Nutrient Recovery from Anaerobic Digest Using Membranes

Grace Neller, Paola Perez, Km Prottoy Shariar Piash

This research revolves around nutrient recovery from wastewater, specifically anaerobic digest (AD). AD was chosen as the primary feed sample because, ideally, this research could be tested and scaled up for real wastewater treatment plants where AD is prominent. AD collected from run-off and sewage systems contains potassium, nitrogen, and phosphorus, critical fertilizers' nutrients. However, the recovery of these nutrients are impeded due to the presence of other organic compounds. Compound separations boil down to charge and sized-based properties. Polyelectrolytes appeal to both charge and sized based differences in ions because of their pore sizes and alternating bilayer charges. This research focuses on designing polyelectrolyte-based nanofiltration membranes by analyzing the co-transport mechanism of organics and nutrient ($NH4+$, $K+$) ions through charged polyelectrolyte multilayer films for achieving simultaneous nutrient permeance and organic pollutant rejection. Current findings have indicated that surface charge predominantly facilitates nutrient permeance, while size-exclusion plays a vital role in retaining organic pollutants. Additionally, a unique ion desolvation based separation was observed where the membranes with covalent crosslinking enabled intra-nutrient ($NH4+/K+$) selectivity. This finding is currently being investigated further as both ions have the same hydrated radius, which with traditional size based separation, would not be achievable. This research aims to continue these investigations by comparing these findings when applied to real world feeds (AD).

Separating Stable Chloride Isotopes for Molten Salt Reactors

Alexandra Bradbury, Brandon Hunter, Richard Mayes, Laura H. Arias Chavez, Joanna McFarlane

37Cl- is the preferred isotope for thermal reactors, as chlorides are only candidates for secondary cooling loops. It is also required for molten salt reactors (MSRs), which generate electricity that can be used. Since 37Cl- only has a 24% abundancy for all Cl isotopes, the US supply is currently limited, and cannot be as widely utilized. Therefore, the goal of this project is to take a compound that is not highly limited, such as HCl, and extract the 37Cl- isotope from it. As of right now, the isolation of the 37Cl- isotope has been studied through anion exchange resins, but unfortunately there are handling issues, as well as the lack of cost and maintenance feasibility. Therefore, liquid-liquid extraction (LLE) is being studied as an alternative process. LLE is a process in which compounds are separated into an organic and aqueous phase, which can then be easily separated. When chloride ions interact with trioctylamine (TOA), the 37Cl- tends to be attracted to the TOA, whereas 35Cl- tends to be repulsed, remaining in its original solvent. The 37Cl- then interacts with NaOH, AgNo3, and 1-iodobutane through a series of separation and filtration processes in order to isolate the 37Cl- into 1-chlorobutane. The 1-chlorobutane is then analyzed using a GC/MS, as this provides the makeup of the compound, as well as the amount of each Cl isotope that was extracted. Once this process has been proven successful, the next goal is to optimize it through different solvent ratios, temperatures, and process times, to achieve the maximum amount of 37Cl- possible. This optimization is done through the program Aspen Plus V8.8, using the ELECNRTL methodology as it is best suited for aqueous and mixed solvent applications.

Dyeing Pharmaceuticals: Visualizing the Effect of Lattice Impurities on Crystal Solubility, Dissolution, and Fragility

Anne Nong, Claire Schleper, Abigail Martin, Mitchell Paolello, Fredrik L. Nordstrom, Gerard Capellades

Accidental formation of solid solutions between a crystallizing product and a process impurity is the primary mechanism by which unacceptable amounts of potentially toxic impurities incorporate into pharmaceutical crystals. These lattice impurities can drastically change the drug substance's solubility, dissolution behavior, and fragility. However, these effects are difficult to study when pure and impure crystals are visually indistinguishable from each other. This study uses dyes as model impurities, incorporated in pharmaceutical products to visualize how localized incorporation can lead to distinct localized behaviors within the crystal.

Acetaminophen solid solutions were grown in the presence of three colored impurities: acid fuchsin, curcumin, and sulforhodamine B. Grown products all presented different degrees of composition anisotropy, where specific faces of the crystal incorporated the impurity at higher levels. These crystals were analyzed for isothermal solubility, melting point depression, dissolution kinetics, and fragility upon compression.

Acetaminophen solid solutions presented a higher solubility and lower melting points than the pure material, in some cases reaching a 92% solubility increase with lattice impurity levels at the ppm range. Low levels of impurity (<200 ppm) were also found to play a significant role on crystal fragility, reducing the required force for breakage by an order of magnitude. Dissolution tests showed that impure regions of the anisotropic crystal always dissolve faster than pure regions, regardless of the impurity's solubility in the solvent and regardless of what faces incorporate the impurity. These observations were cross-validated to be generalizable with well-known dyed crystal system of K_2SO_4 with acid fuchsin.

Effects of process impurities on crystallization goes far beyond their potential role as toxic components in the drug substance. Drug substances retaining small amounts of lattice impurities can present vastly different solubilities, higher fragility, and anisotropic dissolution behaviors. This brings challenges for drug characterization, but also opportunities for solubility and dissolution control.

Optimized Solvent-Based Separation and Recrystallization of Co-Produced PHB Polymer and Indigo Pigments Synthesized Via E. coli Bioreactions

Zahin Kabir, Kobe Rogers

Abstract:

Ourobio's research and development efforts before Summer 2024 led to a breakthrough opportunity to co-produce indigo pigment alongside PHB (Poly-hydroxybutyrate) using fermentation. However, challenges in quantifying, extracting, and purifying the indigo pigment remained unsolved. The primary issue was the lack of an accurate quantification method, which was initially addressed by developing a UV/Vis spectroscopy method that proved unreliable. This was replaced by a more accurate LC-MS-based quantification approach. The LC-MS method involved constructing a standard curve based on peak areas of 95% pure synthetic indigo measured at 615 nm, using DMSO as the solvent. Additionally, the extraction process underwent major revisions, moving from an aqueous-based extraction method to a more efficient solvent-antisolvent method. The aqueous approach, which involved reduction, separation, and reoxidation, was replaced with a solvent-based approach using DMSO due to its high solubility for indigo and its ability to facilitate solvent removal in subsequent steps. After biomass was separated via centrifugation at 7000 rpm, indigo was precipitated using anti-solvents, namely ethanol and water, resulting in a final product that was approximately 80% pure.

LC-MS Method Development:

In the method development stage, a standard curve for indigo quantification was created. Triplicate stock solutions of pure indigo, sourced from Sigma-Aldrich, were solubilized in DMSO and diluted into five different concentrations, ranging from 0 to 100 micrograms of indigo per milliliter of DMSO. These solutions were evaluated using LC-MS, and the peak area of indigo at 615 nm was recorded. A linear correlation between the peak area and the indigo concentration was observed, with an R^2 value of 0.99, indicating high reliability and accuracy of the quantification method. This step established a reproducible method to quantify indigo concentrations in future samples with high precision.

Challenges with Aqueous-Based Extraction:

The initial extraction method used was an aqueous-based process, similar to the vat dyeing of textiles. Indigo was solubilized using a reducing agent and then reoxidized to a water-insoluble form after separation from the biomass. However, this process proved inconsistent. The aqueous extraction method presented challenges, including incomplete separation of biomass and difficulties with the reduction-oxidation cycle. Lower concentration samples (2.5 mg/mL to 7.5 mg/mL) often failed to reoxidize the leuco-indigo into usable indigo, making the samples essentially unusable. On the other hand, higher concentration samples had significant issues with biomass separation, resulting in poor indigo recovery. The aqueous-based method was ultimately deemed unreliable for consistent extraction of indigo pigment.

Transition to Solvent-Based Extraction:

The switch to a solvent-based extraction method marked a significant improvement. DMSO was chosen as the solvent due to its high solubility for indigo and its compatibility with subsequent solvent removal processes. In this improved method, DMSO was heated to remove any water that it had absorbed, ensuring maximum solubility for indigo. After heating, the indigo-laden solution was centrifuged at 7000 rpm for 30 minutes to separate biomass, followed by precipitation of the indigo pigment using cold ethanol and DI water as anti-solvents. This method not only enhanced biomass separation but also yielded purer indigo samples. A comparison between Dimethyl Carbonate (DMC) and DMSO as solvents showed that DMSO was significantly more effective in separating indigo from biomass, as evidenced by cleaner, off-white biomass pellets compared to the darker pellets in DMC-based extractions.

Figure Highlights:

- **Figure 1:** Standard curve for indigo quantification using LC-MS, showing a linear relationship between peak area and concentration.
- **Figure 2:** Results of aqueous-based extraction, highlighting the issues of incomplete oxidation in lower concentrations and poor biomass separation in higher concentrations.
- **Figure 3:** Comparison between Dimethyl Carbonate (DMC) and DMSO in solvent-based extraction. DMSO performed better in separating indigo from biomass.

Using LC-MS data, it was determined that indigo reached a maximum solubility of approximately 7 mg/mL in DMSO. Concentrations higher than this resulted in a sharp decline in accuracy, which was consistent with the findings in a 2015 study by Dr. Vizcaino Caston, who optimized similar extraction protocols. These findings further validated the choice of DMSO as the primary solvent for the extraction process.

Optimized Extraction Procedure:

The final extraction procedure was developed and refined based on the results of these trials. First, DMSO was heated to 100°C for 20-30 minutes to remove any absorbed water. Biomass was then dissolved in DMSO at a concentration of 7.6 mg/mL, heated, and stirred for 30 minutes to ensure complete dissolution of indigo without exceeding the boiling point of DMSO. After centrifugation at 7800 rpm for 10 minutes, the supernatant containing the indigo was decanted. If any remaining biomass was dark, a second extraction with DMSO was performed. Precipitation of indigo was achieved by adding ice-cold ethanol at three times the volume of the original solution, followed by DI water to aid in agglomeration. After allowing the solution to stir for 2 hours, the precipitated indigo was separated through centrifugation, and the pellet was washed multiple times with DI water to remove any traces of DMSO. Finally, the indigo was lyophilized overnight in a -80°C freezer to obtain a dry product.

Discussion and Results:

The initial challenges with aqueous-based extraction stemmed from difficulties in maintaining consistent redox reactions and biomass separation. The reducing solution, consisting of NaOH and fructose, was likely solubilizing some biomass along with the indigo, leading to inconsistent results. Precipitation using HCl as an oxidizing agent was inconsistent, sometimes yielding white biomass instead of indigo, further complicating the process. However, in later trials, it was observed that allowing samples to settle for several days resulted in some separation of biomass and indigo, prompting the introduction of centrifugation as a method to speed up this process.

The solvent-based extraction method, using DMSO, proved to be more reliable and efficient. Further experiments refined the extraction process, focusing on optimizing the concentration of biomass and adjusting solvent ratios. Higher biomass concentrations yielded more indigo, though a second extraction was sometimes necessary to dissolve all the indigo. Precipitation worked best with a combination of ethanol and DI water, as neither solvent alone was sufficient to precipitate indigo from the DMSO solution. Longer stirring times improved indigo precipitation, though stirring for over 2 hours led to product degradation. The drying process was also optimized through lyophilization, and multiple DI water washes effectively removed DMSO from the samples.

By the end of the summer, the final purity of extracted indigo reached approximately 70-80%, with PHB being the primary contaminant. This suggests that future work could focus on improving the purification process, potentially by applying PHB extraction protocols. Additional experiments could also explore the possibility of reusing DMSO to lower the costs associated with the extraction process. Moving forward, Ourobio plans to scale up production and deliver several grams of indigo to potential customers, continuing with the successful extraction methods developed during this research.

Figure 1. Indigo Quantification Standard Curve

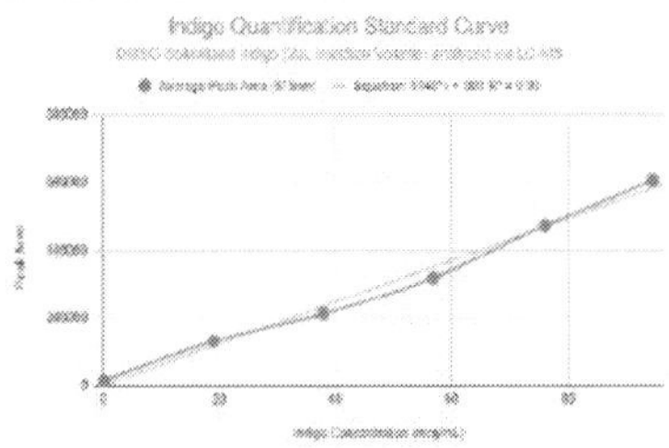

Figure 3. Dimethyl Carbonate Extraction Comparison

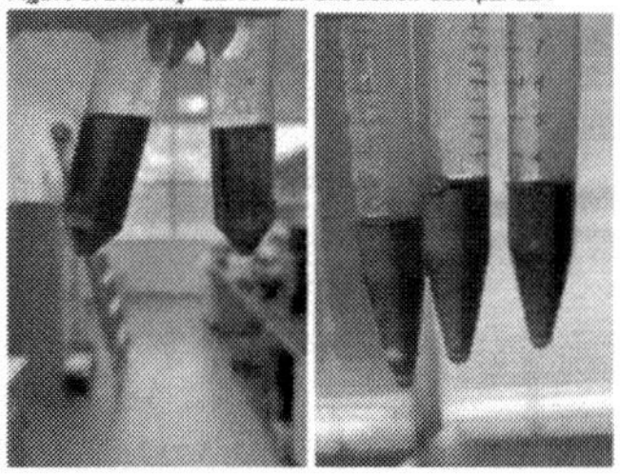

Figure 3. Example LC MS Data

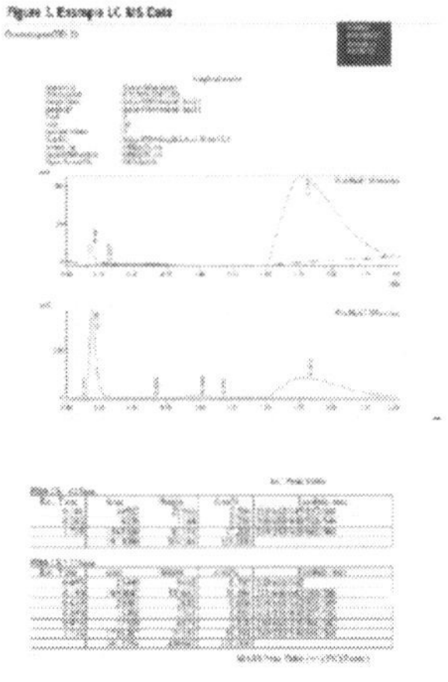

Figure 2. Aqueous Based Extraction Results

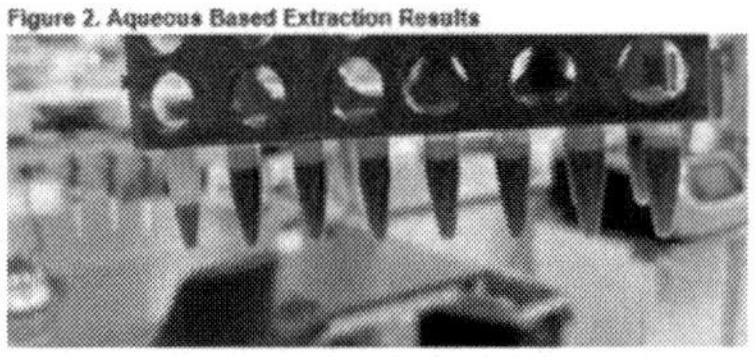

Figure 2. LC-MS Peak Areas Plotted with respect to absorbance

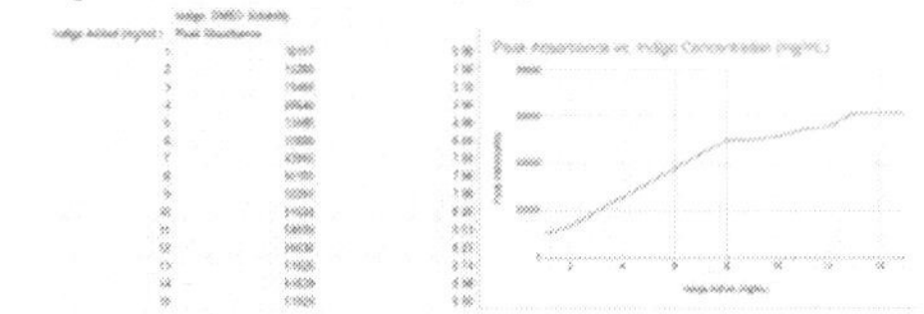

Figure 4. Final Indigo Recovery and Purity

Extraction Yields

	Initial Biomass Weight (mg)	Theoretical Indigo Content (mg)	Post Extraction Weight (mg)	Indigo Weight (mg) from Mass Spec	% Indigo Yield
1	760	38	41	30	[illegible]
2	770	38.5	44	[illegible]	[illegible]
3	[illegible]	[illegible]	38	[illegible]	[illegible]
4	760	38	[illegible]	27	[illegible]
5	760	38	41	30	[illegible]

A Model for Diffusiophoretic Particle Transport in Acid-Base Reaction Fronts

Ethan Coleman, Siamak Mirfendereski, Ankur Gupta

Diffusiophoresis is a fundamental transport mechanism, describing the tendency of micro-scale particles to move along the concentration gradients of chemical species in their environment. Diffusiophoresis plays a key role in many biological, reaction engineering and separations processes involving colloidal suspensions. These processes are abundant, and therefore, a fundamental understanding of diffusiophoresis is of great importance. It was recently shown in experiment, that diffusiophoretic particles are able to focus on the steady-state reaction front of a channel with acid and salt added on one side, and base added in the other. While this discovery provides an interesting insight into the diffusiophoretic process, there is currently little theoretical understanding of how this occurs. In order to better explore this phenomena, we have created a mathematical model describing the situation. This setup can be modeled by the steady Poisson-Nernst-Planck equation, solved for each individual species in the environment, in addition to special case boundary conditions regarding the location of the reaction front, and properties of the profiles at this point. The resulting concentration profiles are used to determine diffusiophoretic velocities that would be observed by particles added to the system, corroborating experimental results. In addition, a transient system can be considered in order to add depth to our understanding of the phenomena as a whole, beyond what was observed in experiment. It was shown that the location of focusing is highly dependent on added salt concentration, and so the transient system will provide insight into the propagation of the reaction front up until steady state is reached. This system can be solved numerically, and analytically in certain limits, in order to better explain this observation from first principles. Using this model, we can provide insight into the underlying principles governing this phenomena, and establish predictions about its applications in similar setups. The model is able to reasonably capture experimental results, and provide scalable insight into the process as whole.

Synthesizing Platinum Nanoparticles on Metal Organic Frameworks

Elizabeth Crawshaw, Nathan Thornburg, Alaba Ojo, Roozbeh Seifollahy Astaraee, John Regalbuto

Metal organic frameworks (MOF) are relatively new in the science of catalysis. With high internal surface area and structural stability, they are being explored as supports for supported metal catalysts. To date, the preparation method of Strong Electrostatic Adsorption (SEA), which has been widely applied to synthesize metal nanoparticles on numerous oxide and carbon supports, has not been applied to MOFs. SEA is a method that produces extremely small nanoparticles by exploiting the surface chemistry of the support to induce electrostatic attraction between charged surfaces and oppositely charged metal precursor complexes. This is done by controlling the pH of a metal precursor solution in relation to the support's point of zero charge (PZC).

We hypothesize that SEA can be used to synthesize ultrasmall platinum nanoparticles on carbon supports derived from MOF materials, when the support pore size becomes large enough to allow the internal diffusion of the hydrated metal precursors. To test this hypothesis, we are comparing metal uptake and reduced metal particle size over three different MOF-derived materials with increasing pore size distributions. These materials all have high PZCs, and the adsorption of anionic Pt chloride precursors has been measured in the low pH range. The metal nanoparticle size has been characterized by powder x-ray diffraction (XRD) and scanning transmission electron microscopy (STEM).

Preliminary results reveal metal adsorption trends among the three MOF materials consistent with significant diffusion limitations, but all materials given rise to very small Pt nanoparticles at relatively high Pt weight loadings.

Evaluating Novel, Amine-Free Switchable Hydrophilicity Solvents

Virginia Parparcen, John Oramas, Brian Morris, Lindsay Soh

Industries commonly use flammable and toxic volatile organic solvents for reactions, extractions, and separations. They require energy-intensive distillations to be separated, and represent health risks to workers due to their hazards. Switchable Hydrophilicity Solvents (SHSs) offer a solution, as they are able to switch between hydrophilic and hydrophobic behavior using an external trigger, allowing for a non-energy intensive separation. Anionic SHSs specifically use a carboxylic acid that switches between its neutral, hydrophobic state and its anionic, hydrophilic state. The switch is triggered by the addition of a base, which deprotonates the acid and forms a water-soluble salt. This reaction can be reversed by sparging with CO_2, which reprotonates the acid making it hydrophobic again. Previous work identified three main issues with current switchable solvents: the surfactant nature of long chain carboxylic acids, high water usage, and toxic counter-ions. Viable SHSs were identified, but the options are few and still limited by their water usage. The main goal of this project is to find new carboxylic acids with switchable behavior and to evaluate possible relationships between the structure and properties of the solvents. Since these novel solvents are not fully understood, studying their structure-property relationship can help develop a framework to identify new alternatives and their appropriate operating conditions for various applications.

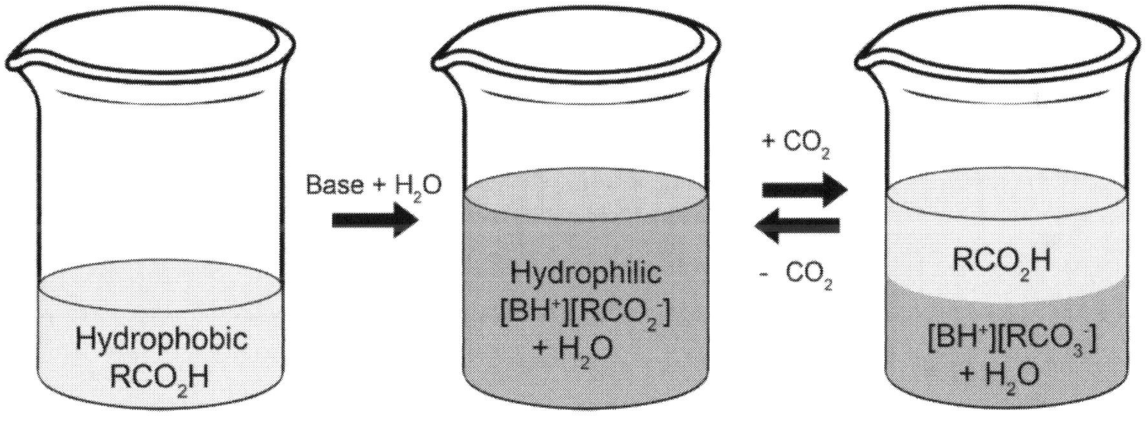

Extractive Distillation for Separating Refrigerants Using Ionic Liquids

Evanna Dominic, Julia Espinoza Mejia, Mark B. Shiflett, Aaron M. Scurto

Hydrofluorocarbons (HFC) refrigerants are widely used in refrigeration and air conditioning applications. Although HFCs have zero ozone-depleting potential compared to chlorofluorocarbons (CFCs), many HFCs have high global warming potential (GWP). The need to recycle and repurpose these refrigerants are critical for reducing the environmental impact. Traditional distillation methods are inadequate for separating azeotropic refrigerant mixtures, which requires alternative approaches such as extractive distillation using ionic liquids (EDIL). A pilot-scale EDIL for studying the separation of azeotropic refrigerant mixtures is located at the University of Kansas.

This research focuses on optimizing EDIL to separate azeotropic mixtures containing HFC-32 (difluoromethane, CH_2F_2), HFC-125 (pentafluoroethane, CHF_2CF_3), HFC-143a (1,1,1-trifluoroethane, CH_3CF_3), HFC-134a (1,1,1,2-tetrafluoroethane), and HFO-1234yf (2,3,3,3-tetrafluoropropene, $CH_2{=}CFCF_3$), with 99.5 wt% purity of each component. The separation of refrigerants such as R-410A (HFC-32/HFC-125 50/50 wt%), R-507 (HFC-143a/HFC-125 50/50 wt%), R-404A (HFC-125/HFC-143a/HFC-134a 44/52/4 wt%), R-407C (HFC-134a/HFC-125/HFC-32 52/25/23 wt%), and R-513A (HFC-134a/HFO-1234yf 56/44 wt%) are being studied using the EDIL. The purity of the distillate and bottom products are analyzed using a Buck Scientific, (model 910) gas chromatograph with a thermal conductivity detector, a HayeSep D packed column, and a Valco 0.5 ml sampling valve.

A process simulation using an ASPEN Plus rate-based model is in good agreement with the experimental results. Binary interaction parameters based on vapor-liquid equilibria for ten ionic liquids (ILs) with HFC-32 and HFC-125 were modeled using the Peng-Robinson equation of state with Boston-Mathias mixing rule (PR-BM). This allowed us to determine the best IL for separating R-410A based on a solubility-selectivity Pareto front. The VLE regression was in good agreement with the experimental data. The best IL for the separation of R-410A was 1-ethyl-3-methylimidazolium bis (trifluoromethylsulfonyl) imide $[C_2C_1im][Tf_2N]$. . A sensitivity analysis found the optimal operational parameters for achieving the highest distillate and bottom product purities was 20 feed stages, solvent-to-feed ratio of 6, and a reflux-ratio of 3.

Rate-based modeling requires measured transport properties such as density, viscosity, and surface tension for the HFC + IL mixtures. The surface tension for HFC-32 and HFC-125 in mixtures with the IL $[C_2C_1im][Tf_2N]$ under high-pressure and high-temperature conditions were measured using the pendant drop method over a temperature range from 293.15 to 373.15 K and a pressure range from 1 to 9 bar. The experimental data was successfully regressed using Li-Fu-Wang correlation.

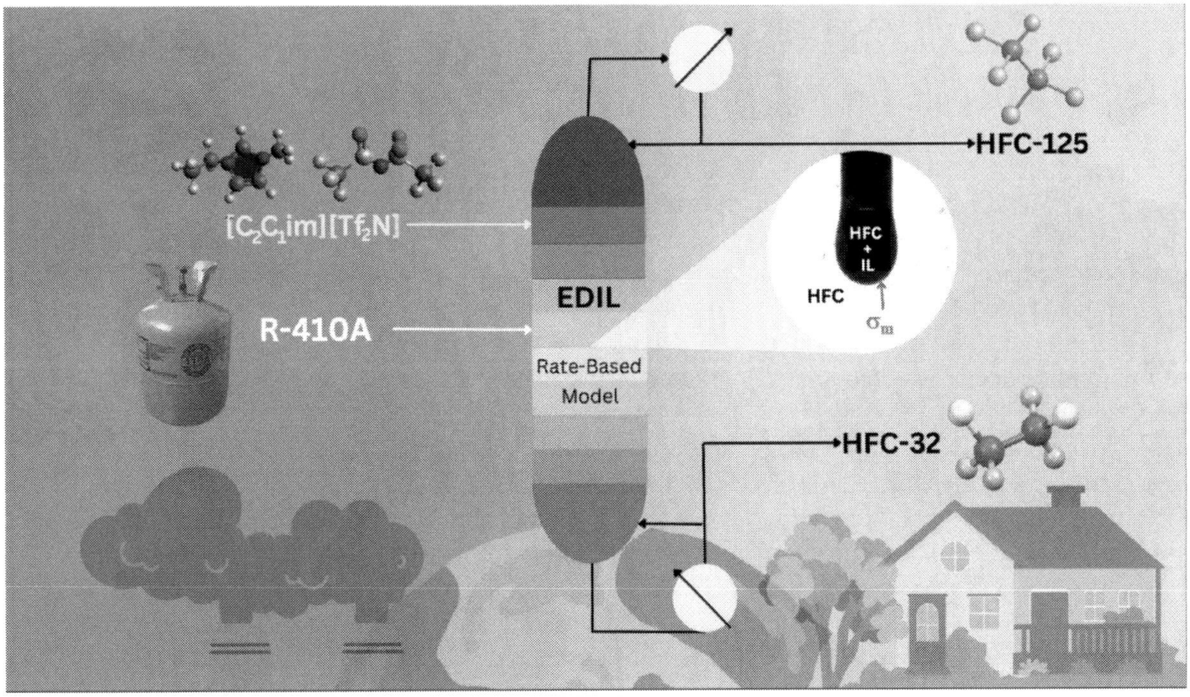
HFC-125
[C2C1im][Tf2N]
EDIL
HFC
+
IL
HFC
σm
R-410A
Rate-Based
Model
HFC-32

Optimizing Purity and Efficiency in Ethylene Oxide Production

Maijani Hall

This research explores the design and sensitivity analysis of an ethylene oxide production plant in Aspen Plus software, concentrating on the optimization of separation processes.The main objective is to find the optimal process conditions to maximize distillate purity and minimize reboiler duty. Separation processes are a fundamental part of transforming raw materials into value-added products. These processes are vital for product purification, driving economic impact in chemical operations and adapting to diverse industrial needs. Distillation is a purification process in which the components of a liquid mixture are separated based on their difference in boiling points or volatility. Properly designing a distillation column is crucial as it ensures separation efficiency, minimizes operational costs, enhances safety during operation and improves long term performance of the column. The experiments in this investigation were carried out using the sensitivity analysis tools of Aspen Plus v14. These computational tools help to analyze how changes in input variables affect output results. Based on this experiment, the optimal conditions of the column are a feed stage of 4 and a molar reflux ratio of 6. When compared with a base case of feed stage 7 and molar reflux ratio 10, the optimal design boasts a 7% decrease in reboiler duty with a 3% increase in distillate purity. These findings will have a positive impact in the field of engineering because they help production plants to save money by using the least amount of resources such as steam and cooling water to achieve the maximum product purity. Additionally, this investigation highlights how simply using advanced technology like simulation tools can help to considerably improve chemical processes on the industrial scale. The relevance of this project is substantiated by the fact that Mexico currently faces a notable shortage of ethylene oxide due to insufficient internal production and high dependency on imports. The results of this research demonstrate that ethylene oxide production can be more cost-effective, which can lead to an increase in local production and an economic boom in Mexico.

H2S Removal from Natural Gas Using Titania and Ceria Nanotubes/Zeolitic Imidazolate Framework-8 Nanocomposites Suspended in Water

Abdullah Alabbad, Mustapha Iddrisu, Sagheer Onaizi

Natural gas, a critical energy source, is consumed globally at a rapid rate, with a projected 220 trillion cubic feet (tcf) to be used by 2050 compared to 130 tcf in 2020. A challenge arises from hydrogen sulfide (H2S), a toxic gas present in sour natural gas reserves. H2S is both hazardous due to its toxicity and corrosiveness and explosive when mixed with air. Human exposure to H2S can lead to serious health issues, even at low concentrations. Additionally, it degrades equipment by causing hydrogen embrittlement and corrosion.

Amine absorption is the current method used for H2S removal, but it is energy intensive and corrosive. Therefore, alternative technologies are being explored, with zeolitic imidazolate frameworks (ZIFs) emerging as promising materials. ZIFs, which belong to the category of metal organic frameworks (MOFs), composed of zinc (Zn^{2+}) or cobalt (Co^{2+}) linked with imidazole ligands, are characterized by thermal stability, high surface area, and adaptability. Among them, ZIF-8 stands out for its ease of fabrication and high production yield, although it suffers from water instability. Recent studies suggest that modifying ZIFs can enhance their stability in water. This study presents the novel synthesis of TNTs/ZIF-8 (titania nanotubes) and CeNTs/ZIF-8 (ceria nanotubes) nanocomposites, which has not been previously reported in the literature to the best of our knowledge, highlighting the novelty of this study.

The TNTs@ZIF-8 nanocomposite was synthesized through a hydrothermal method, involving the dispersion of titania nanotubes in a zinc solution followed by the addition of 2-methyl imidazole. The CeNTs@ZIF-8 nanocomposite was prepared similarly with cerium nitrate and characterized through BET, XRD, FTIR, and SEM, confirming successful synthesis.

The performance of these nanocomposites in H2S capture was evaluated in a continuously stirred tank reactor (CSTR). Both materials were suspended (separately) in water while sour natural gas was bubbled through the suspension. The results showed that TNTs@ZIF-8 could completely eliminate H2S for up to 390 minutes (i.e., the breakthrough time) before reaching saturation at 475 minutes. CeNTs@ZIF-8 demonstrated breakthrough at 376 minutes and saturation at 675 minutes. These findings, illustrated in Figure 1, highlight the efficiency of these nanocomposites in desulfurization, making them promising for industrial applications.

This study provides a basis for organic-free, mild processes using submerged nanocomposites in CSTR systems, presenting a scalable and efficient alternative for natural gas desulfurization. However, further studies are required to explore the scalability and economic viability of these processes.

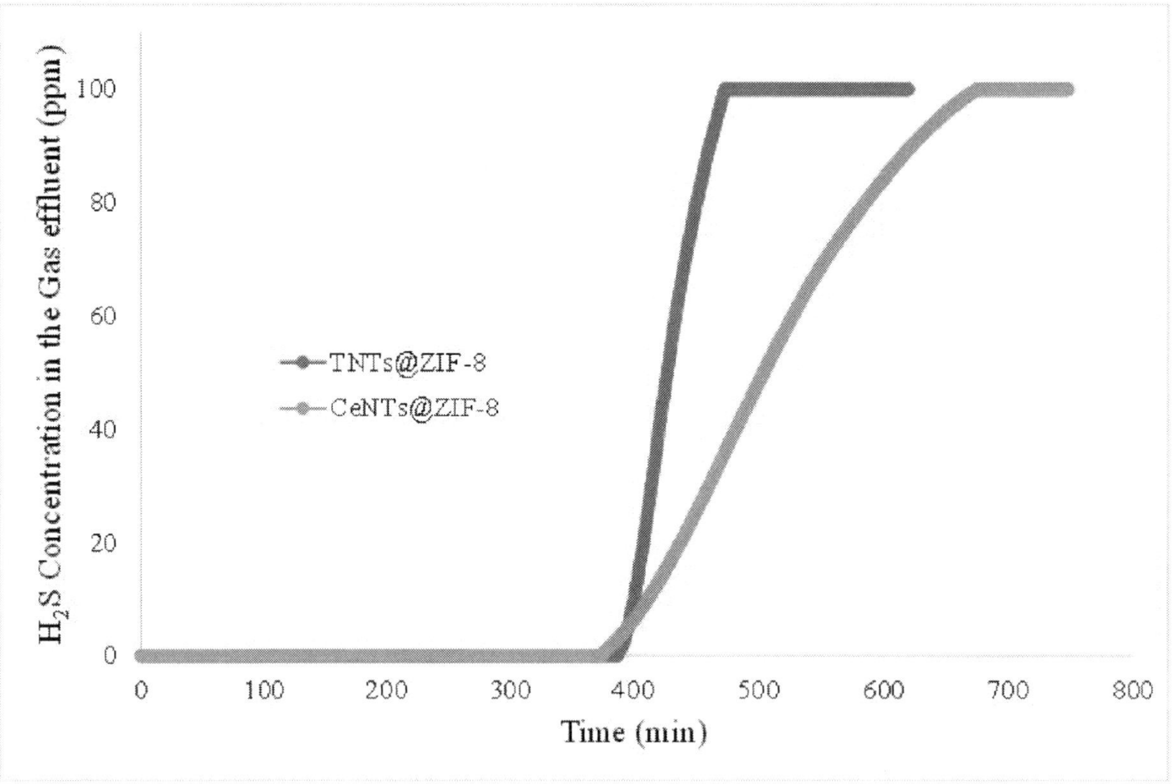

Figure 1: H_2S capture from natural gas using TNTs@ZIF-8 and CeNTs@ZIF-8 nanocomposites suspended in water. The inlet gas flow rate is 100 mL/min. The concentration of H_2S in the natural gas is 100 ppm The natural gas desulfurization process was conducted at ambient conditions using a continuously stirred tank reactor (CSTR).

Evaluation of Bed Height and the Use of Glass Beads in the Extraction of Microalgae Compounds By Supercritical Fluid

Leonardo Barbosa Martins, Eduardo Francisco, Gerti Weber Brun, Allan Morcelli, Laura Rocca Ribeiro

The environmental impact of industrial activity and the depletion of natural resources increase the need for changes in production chains. The exploration of microalgae is highlighted in this context, but its potential applications are limited by the high cost of cultivation and processing, requiring studies to optimize the technologies used. This study proposed to improve the production of valuable compounds for the nutritional and pharmaceutical industries from the microalgae Chlorella sorokiniana, such as carotenoids, chlorophylls and fatty acids, using supercritical extractions with mixtures of CO_2 and ethanol. The effect of height and filling of the extraction bed were investigated to understand the relationship between compound recovery, bed compaction in the extraction vessel and solvent-biomass contact under the tested conditions.

The Effect of Amine Location in Six-Carbon Amines on Desalination in Temperature Swing Solvent Extraction

Sadie Jarrell, Elizabeth Monti, Lauren Ward, Steven Weinman

Many industrial processes, as well as water treatment methods such as reverse osmosis and multi-stage flash, result in the creation of high salinity brines as byproducts, which can be up to ten times saltier than seawater. Current methods used to desalinate these solutions, such as thermal distillation, brine concentrators, and crystallization, either have high energy costs or a low through-put. As such, there is renewed interest in finding commercially viable methods to reduce the environmental impacts and costs of brine disposal. Temperature Swing Solvent Extraction (TSSE) is a desalination method that utilizes the temperature-dependence of certain solvents to selectively dissolve water out of brine solutions at lower temperatures and release it at higher ones. The focus of this project is to investigate the performance of six-carbon amine solvents in TSSE with the goal of understanding the structure-property-performance relationships between them. Six-carbon amine solvents have been tested for their desalination and water recovery capabilities using conductivity and mass measurements. Solvent behavior with various brine solutions has been characterized through change in volume, change in temperature, ion electrode, total organic carbon, and Karl-Fischer titration data. Results indicate that inclusion of a secondary amine rather than a primary or tertiary one is critical to the solvent's ability to selectively intake water over salt ions and release it upon the temperature change, with symmetrical molecules displaying the best desalination capabilities.

Building a Crystallization Kinetics Database through Automated Screenings

Lily Rhoads, Kennedy Tomlinson, Joshua M. Zaharof, Ibrahim Joel, Kevin P. Girard, Alpana Thorat, Gerard Capellades

Purpose:

Despite the large number of studies demonstrating the application of crystallization kinetic models for process development, reported kinetic parameters across literature often differ by several orders of magnitude, even for the same compound, temperature, and solvent composition. This is due to a large number of inconsistencies in common methodologies, including different simplifications to the supersaturation expression, different assumptions in mass balance and kinetic models, different analytical tools used to measure crystal growth, and the use of different seeding strategies and mixing approaches during kinetic screening.

Methods:

This work leverages our previously designed methods for the automated screening of nucleation and growth kinetics using a Technobis crystalline. Appr. 20 solute-solvent pairs have been screened for kinetics using consistent scales, mixing, crystal size assessment techniques, and models. The pairs include small molecule pharmaceuticals with different molecular sizes and complexities, as well as some inorganic salts that serve as a point of comparison.

Results:

The resulting database is presented as a kinetic log-log map of nucleation constants against crystal growth constants. This map is used to place new compounds in context with existing literature, and to study the relationship between solute structure, solvent composition, and crystallization kinetics.

Conclusions:

The developed kinetics database serves to standardize early crystallization screenings, presenting a point of comparison between new and existing solutes, filling a role similar to that of a Robeson plot for gas permeation. As the library keeps expanding, it will serve to facilitate early process development efforts as well as to answer open academic questions on what are the main factors driving a solute's tendency to nucleate and grow in a given solvent.

Keywords:

Crystallization, automation, kinetics, process design, modeling

Poly(1,3-dioxolane)-Based Terpolymer Membranes with 1-Hexyl-3-Methylimidazolium Bis(trifluoromethylsulfonyl)Imide for Carbon Capture and Light Paraffin Separation

Xavier Gatica, Justin J. Rosenthal, Isaac W. Tan, Adam N. Mann, Louise Marie C. Cañada, Joan F. Brennecke, Benny D. Freeman, Nathaniel Lynd

Rapid increases in anthropogenic carbon dioxide (CO_2) emissions have motivated research on carbon capture and sequestration (CCS), with CCS of post-combustion flue gas from fossil-fuel fired power plants being one of the most practical approaches. Compared to the already industrialized, yet energy-intensive, amine scrubbing technology, CO_2-selective membranes are expected to better reduce the cost of CCS operations in accordance with DOE targets. Here, we initially developed a dioxolane-based random terpolymer (PDXL) membrane that exhibited a CO_2/N_2 selectivity of 52 and a CO_2 permeability of 330 Barrer. However, for membrane-based CCS to be economically viable, the developed membranes must have as high of a CO_2 permeability as possible since membrane cost scales with membrane area. Therefore, an ionic liquid (IL) was also incorporated into the membrane which resulted in permeabilities of 530 Barrer with only slight losses in selectivity. Additionally, light paraffins, which have become more accessible in the U.S. due to an upsurge in fracking, currently require energy-intensive cryogenic distillation for separation and purification and could benefit from energy efficient advancements. The PDXL membranes also exhibited reverse-selectivity for C_3H_8/CH_4 and C_2H_6/CH_4 and were able to withstand industrial operating pressures (i.e., 7–30 bar), presenting a promising addition/alternative for bulk light paraffin separations. Overall, increasing the IL wt% in the membrane blends decreased the glass transition temperature, allowing for higher permeabilities of all gases and improvements in the light paraffin selectivity, but also resulting in depression of CO_2 selectivity and produced weaker and more stretchy membranes.

Exploring the Post-Separation Dynamic Behavior of Magnetic Nanoparticles through a Simple Image-Based Approach

Blair Dodge, Stefano Ciannella, Jenifer Gomez Pastora

Magnetic nanoparticles (MNPs), typically smaller than 100 nm, are usually synthesized from elements with magnetic properties like cobalt, nickel, iron, and their oxides. These particles may possess unique physical properties such as superparamagnetism, large surface area-to-volume ratios, and potential for surface functionalization, making them valuable for applications in targeted drug delivery, biosepararation, and environmental remediation. Magnetophoresis, which denotes the induced motion of MNPs through a viscous medium under a magnetic field, is one approach for manipulating and separating these particles.

Cooperative magnetophoresis can enhance the mobility of MNPs under certain conditions. Through magnetic dipole-dipole interactions, MNPs may aggregate into clusters that respond more effectively to external magnetic fields in terms of induced motion. This process is particularly useful for ultra-small particles (<50 nm), which typically exhibit low magnetophoretic velocities. Following, some applications require particle disaggregation accompanied by diffusion after magnetic separation to restore surface area and maintain their functional efficiency, especially in processes that involve pathogen capture and nanophotocatalysis, for instance. Despite its importance, this aspect of MNPs post-separation remains underexplored.

In this study, we examined the diffusion behavior of commercially available superparamagnetic iron oxide nanoparticles (SPIONs) of varying sizes (Ocean Nanotech, CA, USA) in the absence of a magnetic field (i.e. field free conditions). Pre-separation was conducted using a custom-built quadrupole magnetic sorter (QMS) that provides a constant magnetic gradient that induces MNPs magnetophoresis. To track SPIONs diffusion in an aqueous dispersion, we used a straightforward image-based approach capable of estimating mass concentration profiles based on image grayscale data, enabling real-time analysis of SPION dispersions at different initial concentrations.

This methodology can be applied as part of an analysis for recyclability of SPIONs in biocompatible fluids, with results relevant to biomedical and environmental applications. By fitting the collected concentration data into a diffusion model, we can estimate SPION diffusion coefficients based on particle size, according to the Stokes-Einstein equation. These findings contribute to a deeper understanding of MNP behavior in field-free conditions, aiding in the development of more efficient and sustainable magnetic nanoparticle technologies.

Applying Line-and-Groove Patterns to Nanofiltration Membranes

Catherine Martin, Lauren Ward, Steven Weinman

Nanofiltration (NF) membranes are a method of water/wastewater treatment that rejects divalent ions, suspended particles, viruses, and bacteria. NF membranes require a lower operating pressure than reverse osmosis, 3.5-16 bar, making them a less expensive option for the partial desalination of water and water softening. One issue with NF membranes is fouling, which is a solute build up that blocks the pores of the membrane. Membrane fouling decreases the permeance and rejection of the membrane over time. Membrane surface patterning is a method used to reduce membrane fouling. Our project goal was to understand the relationship between foulant size and pattern size and its influence on membrane fouling. This was accomplished by analyzing patterns that are smaller than (300-nm), equal to (10-μm), and larger than (50-μm) the foulant size (10-μm). The membranes were tested in a dead-end cell, and the feed solutions used to test the membranes were: deionized water, 2,000 ppm sodium sulfate, and a n-dodecane brine solution with 2,000 ppm n-dodecane and 35,000 ppm sodium chloride. The results from these experiments demonstrated that the flux recovery ratio was not significantly impacted by patterning, however the smaller than and larger than patterns decreased the rate of fouling by a small margin.

Chemical Modification of Polymer-Based Membranes for Carbon Dioxide Gas Separations

Zoe Reddecliff, Ryan Johnson, Joshua Moon

The effects of climate change are becoming more drastic as carbon dioxide continues to be released into the atmosphere. Reductions in industrial emissions along with the use of direct air capture technology are necessary to prevent further atmospheric warming. PEG-based polymer membranes present a potential solution for carbon dioxide separation as they are energy efficient, relatively inexpensive, and easy to synthesize and alter. Lewis bases are substituted onto ligands on the network's scaffolding through "click chemistry". These basic groups interact with carbon dioxide to promote facilitated transport, a process by which a gas is selectively passed through the membrane.

This study aims to push the boundaries of permeability and selectivity of CO_2 gas in polymer membranes by testing membranes with ligands of varying basicity, and thus varying strength of interactions with carbon dioxide. We hypothesize that stronger interactions with CO_2 would generally promote higher selectivity. However, if the interactions are binding, CO_2 may become trapped in the membrane and hinder selective diffusion. Future plans include characterizing the structure and transport properties of hydrated PEG-based membranes using sorption, dilation, and humidified gas permeation testing.

J. D. Moon et al. Macromolecules 2021, 54, 2, 866–873.

Tong, Zi Journal of Membrane Science 2017, 543, 202–11.

J. D. Moon et al. Journal of Membrane Science 2021, 639, 7197-7208

Additive Manufacturing Plant-Derived Char Meshes for Point-Source CO2 Capture

Nathan Smith

The need for complex activated carbon geometry, which can only be achieved with additive manufacturing, is necessary when looking to lower vehicle emissions. Adding carbon capture materials to tailpipes of cars will allow for point-source carbon capture, which is far more effective and efficient than trying to capture CO2 from the air after it has been released. To prevent the negative side effects of clogging a tailpipe, meshes are necessary. The variability of tailpipe size and shape would also be very expensive for existing manufacturing techniques, meaning additive manufacturing would be required. Additive manufacturing with varied polymeric carbon precursor ratio also allows pore microstructure to be controlled. Certain size pores have more selectivity for chemicals like arsenic or heavy metals, that are common in these processes.

Currently, 3D printed monoliths are made through the solgel method in which polymers are extruded, then activated. Microporosity is obtained strictly from this activation process. From this method surface areas can reach >1050 m^2/g and mean pore diameter at 1.3-1.4 nm. By using 100% plant derived materials, lignin and cellulose, and without using an activation step, the process is much safer and thus easier to manufactor. Although the solgel method attains higher surface area than biochars, biochars are seen to be much more efficient at removing impurities than the activated carbon chars. The reason for this is that biochar has the functional groups to adsorb anions far better.

As a result, 3D printing aqueous inks loaded with polymeric carbon precursors to make biochars offers a cost- and time-efficient route to mass producing carbon-based prints with complex geometries and specific pore size distribution. From prior research, pore size distribution can change based from the ratio of the polymeric precursors, solids loading, and printing conditions. When the ratio of cellulose is high (>80%), pores tend to be in the 10-20 micron range because of interparticle spacing. When phenolic resin is the most prevalent (>80%) large pores (100-500) are formed from trapped gasses. In an optimal ratio, small pores (1-5 micron) with a large amount of nanopores are created as the phenolic resin fills in the interparticle spacing like glue. By analyzing pore structure with Mercury Porosimetry and Physisorption I will be able to quantify these trends and make an ink prediction roadmap to guide ink composition and printing conditions for various filter applications.

AUTHOR INDEX